Encyclopedia of Cell-Cell Interactions

Encyclopedia of Cell-Cell Interactions

Edited by **Gloria Doran**

Callisto Reference

New York

Published by Callisto Reference,
106 Park Avenue, Suite 200,
New York, NY 10016, USA
www.callistoreference.com

Encyclopedia of Cell-Cell Interactions
Edited by Gloria Doran

International Standard Book Number: 978-1-63239-220-6 (Hardback)

Printed in the United States of America.

Contents

Preface

This book is a detailed and comprehensive medium helping students and researchers to understand cell-cell interactions. It describes various cell activities occurring within and outside the cells. Cell interactions are essential for the proper working of various organ systems. Cell adhesion, tissue regeneration, cellular interactions, metastatic cancer etc. are some of the activities and interactions occurring in human body. Some important features such as developmental, immune and neural cell interactions in both healthy as well as diseased states and state-of-the-art cell interactions assessment methods have also been discussed. This book will serve as a great resource to the scientists and professionals working in this domain.

All of the data presented henceforth, was collaborated in the wake of recent advancements in the field. The aim of this book is to present the diversified developments from across the globe in a comprehensible manner. The opinions expressed in each chapter belong solely to the contributing authors. Their interpretations of the topics are the integral part of this book, which I have carefully compiled for a better understanding of the readers.

At the end, I would like to thank all those who dedicated their time and efforts for the successful completion of this book. I also wish to convey my gratitude towards my friends and family who supported me at every step.

Editor

Section 1

Developmental Cell Interactions

Chapter 1

Transmitters in Blastomere Interactions

Yuri B. Shmukler and Denis A. Nikishin

Additional information is available at the end of the chapter

1. Introduction

Normal multicellular organism develops from fertilized egg cell, although the latter may give birth for twins too. Mechanisms that lead to the realization of these variants of the development may differ greatly, however, it naturally suggests itself that such deviation as twin formation is conditioned by this or that disturbance of cellular interactions. But the very fact of the existence of such interactions after classic experiments remained disputable until now. Naturally, far less is known on the suggested mechanisms of such interactions that provide the formation of integral organism or, correspondingly, their distortion that lead to the twins formation.

Neurotransmitters that are known as the mediators of cellular interactions in adult organisms are involved also in the regulation of various processes of embryonic development. This field remain little bit exotic for a majority of biologists although the researches in this field were started more than 50 years ago and brought a number of interesting facts and hypotheses. It was logically to suggest that neurotransmitters may take part in the embryonic cellular interactions together with the regulation of cleavage divisions and other processes of embryonic development.

2. Problem of the existence of early cellular interactions

Since the beginning the studies of cellular interactions at the early stages of embryonic development were hindered by the fact that very existence of those interactions was doubted. Really, August Weismann suggested his **germ plasm theory** in 1883 [1] that chromosome determinants are distributed nonuniformly and it in turn determines distinct prospective fates of embryonic cells. In particular, according to this concept already first division predetermines the fates of blastomeres as "left" and "right". Really, soon after Wilhelm Roux has carried out his pioneer experiment that became the starting point of experimental embryology [2]. Roux has denaturated one of two blastomeres of the frog *Rana* *fusca* by hot needle, as the result the intact blastomere formed half set of larval structures (Fig. 1). The conclusion was made that the development of embryonic cells is predetermined. Later such type of the development (mosaic) was found in a number of taxons where each blastomere forms the specific part of the definitive organism. Ideal example of such development is favorite subject of various researches the nematode *Caenorhabditis elegans*, where the fates of all more then 900 cells were traced [3]. Evidently that at least in such embryos substantial cellular interactions are absent at the early stages of the development.

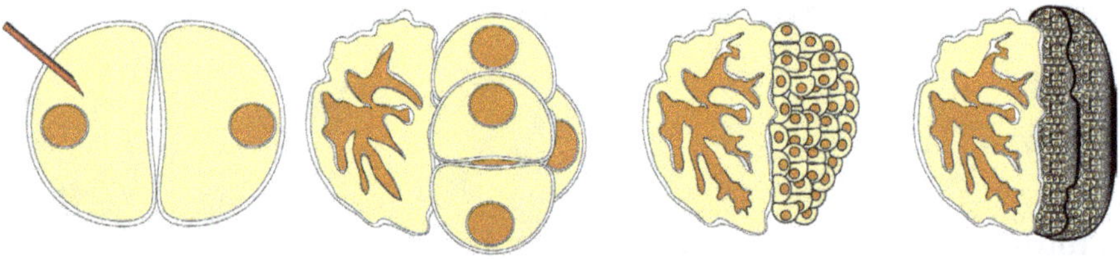

Figure 1. Roux's experiment. Denaturation of frog blastomere by hot needle. Left blastomere was killed by hot needle but right one prolong to cleave and form half set of structures

However, soon the death-blow to the universality of Weismann's theory was dealt by other Founding Father of experimental embryology Hans Driesch [4]. He demonstrated the ability of sea urchin isolated blastomeres to form quasi-normal half-size larvae (Fig. 2). Later McClendon carried out similar experiment in amphibian embryos with the same result [5]. On one hand, it supported Driesch's data, on the other it finally compromised Roux' result. Moreover the experiment similar to Roux' one was performed in sea urchin embryos at the end of XX century. In contempt to Roux' data it was shown that intact blastomere is able to form the diminished quasi-normal larval in spite of the death of sister blastomere [6]. So the data of Wilhelm Roux is believed artifact of defect experiment in all contemporary handbooks of embryology and developmental biology – *"Something in or on the dead blastomere still informed the live cells that it existed"* [1]. We shall return to the evaluation of this statement at the end of the present work.

Thus, amphibians and echinoderms as some other taxons, as distinct from mosaic ones, are able to form whole organisms from the blastomeres isolated at cleavage divisions. In particular, sea urchins preserve such ability, with some reserves, until 8 blastomere stage [7] whereas starfishes – until 32 cell stage [8]. Nevertheless prospective potencies of the blastomeres in the intact embryo are limited and there are no explanation for it except cellular interactions.

At the first glance classical Hans Driesch' experiment, that is the base of contemporary concept of the regulation of the development, proves this idea, however, it contain serious internal contradiction. On one hand, the result of Driesch' experiment evidences in favor of the existence of substantial cellular interactions, although, on the other, half-embryos at 4th cleavage division in his work showed the cleavage pattern of the half of whole embryo – 4 meso-, 2 macro-, and 2 micromeres (Fig. 2, [4]). As the matter of fact **such result principally**

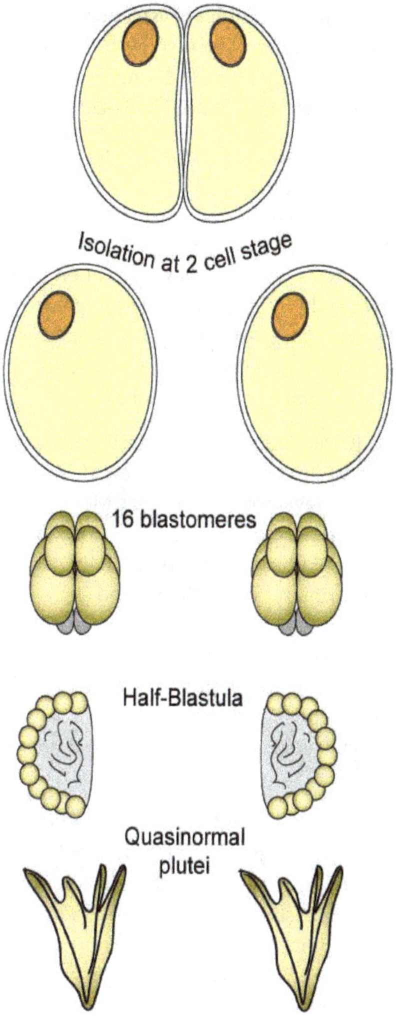

Figure 2. *Hans Driesch experiment* Isolated blastomere forms at 4th cleavage division half-set of cells – 4 meso-, 2 macro- and 2 micromere. Later they form half-size blastula with the blastocoel opened to outer medium. Then half-embryos undergoes "regulation" and form quasi-normal half-size plutei

do not differ from Roux' one except single difference – the ability of sea urchin embryos for the "regulation of the development" at the later stages. Really, such half-embryos first formed "opened half-blastulae" – hemisphere with the blastocoel opened to the outer medium that then closed and became indistinguishable from intact one. It was noted yet by Driesch [4]: "After the isolation one of two first *Echinus microtuberculatus* blastomeres it cleaved as half-embryo [Halbbildung], but form whole half-size organism. Recently it was confirmed by the experiments with blastomere isolation [9] and ricin microinjection [6]. After our own observations half-embryos may show first "partial" cleavage pattern, then form opened hemiblastulae, later (early blastula 2 – midblastula [10]) close the blastocoel in about 10 minutes and further develop into normal half-size embryos and plutei I [9].

Driesch' experiments in all possible variants were reproduces by a number of researchers. On the base of wide and scrupulous experiments on the isolation of the blastomeres and the fragmentation of fertilized eggs of sea urchin Swedish researcher Sven Hörstadius came to the conclusion on the independence of their cleavage pattern on any influence. He suggested the idea of the "micromere clock" – the micromere formation precisely at the moment of 4th cleavage division in the intact embryo independently on any experimental intervention [11, 12]. Properly speaking it means that every embryo is able "to count to four" that is quite strange. Nevertheless this concept persists in unaltered form until now in handbooks of embryology [13, 14] instead of a number of publications that deny it (see below).

In fact such result ruled out the role of direct blastomere interactions during cleavage divisions that would be able to limit the prospective potencies and "shift" the process of development regulation to later developmental stages. Mechanism of such "late regulation" is out of scope of present work, however, we need note that irreversible restriction of prospective potencies of sea urchin blastomeres occur at 4th cleavage division. In particular, micromeres further form the primary mesenchyme and then larval skeleton spicules [15]. Moreover, one of micromere quartets formed at 5th cleavage division stop the divisions at all and, probably, works as the pacemaker of further development [16]. Already at the next division (60 cell stage) the determination of prospective embryonic territories occur [17] and it is too short time for "regulation of the development". No publications of the possible mechanisms of such regulation was found in scientific literature.

In the meantime the facts that contradict the canonic concept were accumulated in parallel with the process of its consolidation. When Plough [18] reproduced Driesch' experiments he could obtain the regulation in the part of embryos only. The author explained this discrepancy with classic data by his imperfect technique as compare to Driesch' one. Many years later Marcus performed statistically assured study and confirmed Plough' data not Driesch' [19]. Harvey in her original work reported on the possibility of equal 4th cleavage division in sea urchin [20] but later, probably under the pressure of "public opinion", she specially stressed in her masterwork "American Arbacia" that half embryos **always** form 2 micromeres at 4th cleavage division [21]. So Driesch-Hörstadius concept preserved its dominating position at least until the end of sixties of XX century.

However the process of the revision of classic knowledge in this field was finally reinitiated. Katzuma Dan and his co-workers disproved the micromere clock concept when they have demonstrated that the suppression of any one cycle of changes of free sulfhydryl groups in sea urchin embryos leads to the delay of micromere formation to the next cell cycle [22-24]. Soon after the data were obtained on the possibility to evoke the functional isolation of blastomeres using short application of chemical substances during 1st or 2nd cleavage divisions (without mechanical isolation of the blastomeres) that lead to the formation of the specific aberrations of blastulae such as half-size blastulae, "Siamese twins" blastulae, 8-form embryos etc [25-27].

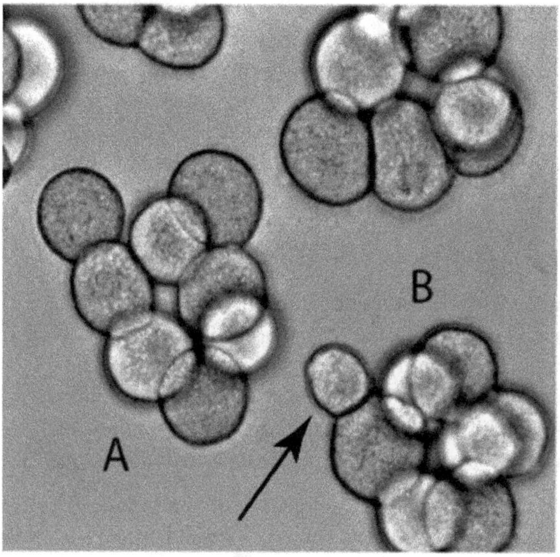

Figure 3. Half-embryos of P.lividus, isolated by glass needle
A – half-embryos, consisting of equal blastomeres at 4th cleavage division
B – unequal cleavage, the micromere is marked by the arrow

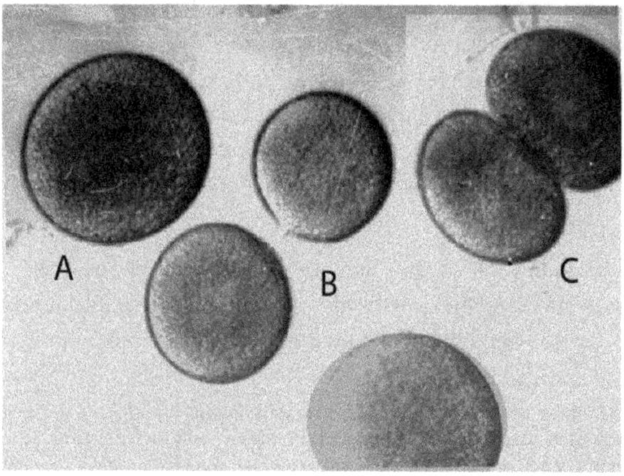

Figure 4. Figure 4 Adhesion of the *S.nudus* blastomeres
A – undivided egg cell; B - two blastomeres before adhesion; C – 2-cell embryo after adhesion

The problem of the existence of blastomere interactions was finally solved in the series of studies of blastomere isolation. More than 80 years after Driesch pioneer work it was carried out by the group of researchers who was not indoctrinated by old dogmas, probably, because there were no embryologists among them. In the very beginning of the work it was found that the isolation of sea urchin *Strongylocentrotus nudus* blastomeres (Japan sea) during 1st or 2nd cleavage division leads to the formation of half-embryos with two different cleavage patterns: partial, previously described by Driesch, and integral, when the embryo formed 8 equal blastomeres at the 4th cleavage division (Fig. 3), i.e. their cleavage pattern coincided with the whole embryo pattern but of a half size but not to half of intact embryo [9, 28]. Moreover in many cases when 4th cleavage division was unequal one micromere was formed only in contrast to classic results. The majority of half-embryos with equal 4th cleavage division could form the micromeres at the next stage.

During further studies in the embryos of the sea urchin *S.nudus* and sand dollar *Scaphechinus mirabilis* some logic was found in the formation of this or that cleavage pattern: equal 4th cleavage divisions predominantly occur in half-embryos isolated before "post-division blastomere adhesion" [25] (Fig.4), but unequal (with micormere formation) in half-embryos isolated some 10 – 15 minutes later when the adhesion was completed. The presence of this rule was then confirmed in a variety of sea urchin species (Tables 1 and 2)[1].

This rule was also confirmed by the experiments with multiple consecutive isolations of the blastomeres from the same *Sc.mirabilis* embryo. Twofold blastomere isolation lead to the increase of portion of half-embryos with equal 4th cleavage division by 13% as compare to single isolation and threefold – by 26% [29]. Analogous experiments in the embryos of *P.lividus* have shown that only 11,1% of twofold isolated blastomeres formed the micromeres simultaneously with intact ones, i.e. by 33,9% less then embryos isolated once before adhesion in 1st cleavage division. Such "accumulation" of the effect of the elimination of normal blastomere interactions, on one hand, evidences in favor of their importance in the determination of the pattern of early development and, on the other, on their **repeatability** during cleavage divisions. As concerns 3rd cleavage divisions there are some reservations because the isolation of the blastomere quartets at this moment is technically difficult because of spatial structure of the embryos. Rare cases of equal cleavage of such half-embryos (about 8%) at 4th division may be explained also by the specific peculiarity of this stage (see below).

Thus, the critical periods exist in the cleavage division cell cycles of sea urchin embryos when the processes limiting the prospective potencies of the blastomeres are realized. Evidently, these processes need to be mediated by the cellular interactions.

---

[1] It need be specified that three groups of sea urchin species could be distinguished by the relative contribution into cellular interactions of passive mechanical component (hyaline layer) and direct blastomere interaction [25]. Logic change of cleavage patterns of half-embryos was characteristic for the group of *S.nudus* also as sand-dollar *Scaphechinis mirabilis*. In the embryos of *Strongylocentrotus intermedius*, where the hyaline layer is far more important for the integrity of the embryo, half-embryos with equal 4th cleavage division were substantially less numerous. But equal 4th cleavage divisions were observed regularly even in these species.

Species	Moment of isolation	Number of embryos	Portion of half-embryos forming micromeres simultaneously with intact ones (%%)	Significance
S mirabilis	B_1	818	34,7±1,7	<0,001
	A_1	865	68,2±1,6	<0,001
	B_2	60	30,0±6,0	<0,001
	A_2	127	81,1±3,5	<0,001
S.nudus	B_1	56	23,2±5,6	<0,001
	A_1	26	92,3±5,2	<0,001
S.intermedius	B_1	62	79,4±5,1	<0,001
	A_1	48	83,3±5,4	<0,001
E.cordatum	B_1	22	27,3±9,5	<0,01
P.lividus	B_1	48	42,0±4,1	<0,01
	A_1	42	76,2± 5,3	<0,001
*P.lividus**	B_1	309	45,0±1,4	<0,001
	A_1	214	93,4±1,0	<0,001

Table 1. Influence of the moment of blastomere isolation on the cleavage pattern of half-embryos
* Adriatic population with absolute predominance of intact embryos prematurely forming micromeres (at 3rd cleavage division), therefore in this experiments the formation of the micromeres was controlled in 3rd but not 4th cleavage division
B – half-embryos, isolated before adhesion in 1st or 2nd cleavage division, A - half-embryos, isolated after adhesion in 1st or 2nd cleavage division

Species	Stages compared	Difference in portions of embryos with the same cleavage pattern (%%)	Significance
S. mirabilis	B_1 - A_1	33,5±2,3	<0,001
	A_1 – B_2	38,2±6,2	<0,001
	B_2 – A_2	51,1±6,9	<0,001
S. nudus	B_1 – A_1	69,1±7,6	<0,001
S. intermedius	B_1 – A_1	3,9±7,4	n.s.*
P.lividus	B_1 – A_1	24,2±6,2	<0,01
*P.lividus**	B_1 – A_1	38,4±1,7	<0,001

Table 2. Differences of cleavage patterns of sea urchin half embryos, isolated before or after blastomere adhesion in the 1st or 2nd cleavage divison
* not significant
Designation B and A as in Table 1

It can be added that real pattern of blastomere interactions are more sophisticated than simple consecutive signal exchange in the freshly formed contact zone. Time-lapse recordings of the development of sea urchin *S.nudus* embryos have shown that during 2nd cleavage division blastomeres first become rounded, zone of the tight adhesion in the furrow of 1st cleavage diminished, and then blastomere adheres along both 2nd and 1st cleavage furrows (Fig. 5), i.e. the interaction process repeats at the same place much times and the aggregated result is the consequence of the formation of three-dimensional structure of contacts.

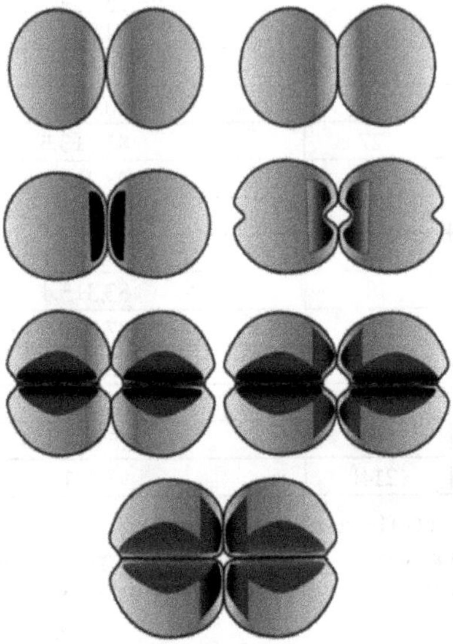

Figure 5. "Re-adhesion" of blastomeres during the 2nd cleavage division of *S. nudus* [29]. The dark areas are the adhesion zones.

Further studies have shown also that blastomeres isolated from the same embryo may have both the same or different cleavage patterns, and coinciding pattern may be both partial and integral. It follows thence, that no blastomere are "pacemaker" in the signal exchange and their interactions are **equal** but also **nonsynchronous**, at least their consequences.

A propos, on the base of above mentioned results it is possible to explain the reasons why Hans Driesch discovered only one variant of the development of half embryos – partial pattern. First, if the classic of experimental embryology has exerted scrupulousness and always performed isolation of blastomere after the full completion of their adhesion he **might** obtain exclusively partial pattern. Second, the results of Driesch' studies were influenced by more or less occasional choice of the species for the experiments (*Echinus microtuberculatus* and *Paracentrotus lividus*). These species have the specific feature: the blastomere isolation using Ca^{2+}-free sea water is impossible before the full completion of cleavage because it leads to the cell death and only after adhesion the isolation brings viable blastomeres. Discovery of the second cleavage pattern happened because experimenters did not wait for the completion of adhesion of blastomeres and that species used (S.nudus и Sc. mirabilis) allowed easy isolation both using Ca^{2+}-free sea water or simple mechanical isolation. Only recently the way to isolate blastomeres of *P.lividus* was found: the replacement of Ca^{2+}-free sea water for normal sea water just before isolation preserves the viability of the blastomeres, isolated before adhesion [30].

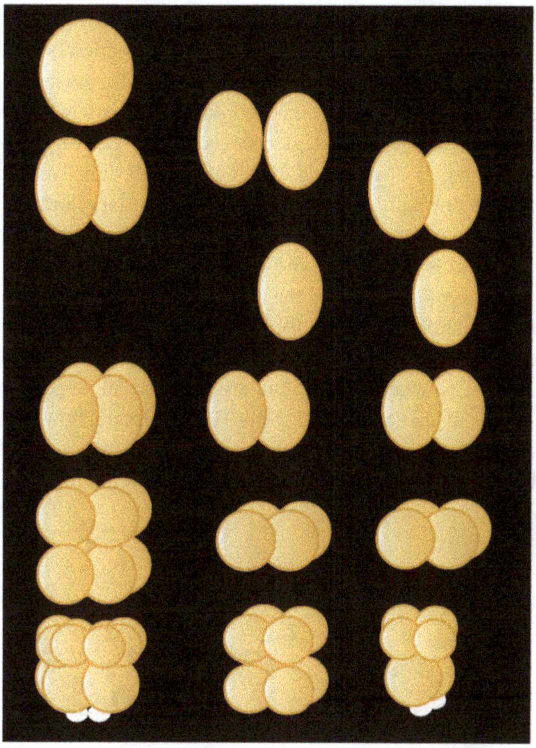

Figure 6. "Micromere model" of blastomere interactions
Left column – normal development of sea urchin, middle – cleavage pattern of blastomere, isolated before adhesion with equal 4th cleavage division, right - cleavage pattern of blastomere, isolated after adhesion with micromere formation simultaneously with intact embryos

Amusingly, the classical Driesch's statement on the ability of early sea urchin half-embryos to regulate their development was right but based on incomplete data concerning the cleavage pattern, disregard of internal contradiction, and specific phenomenon of late regulation of the blastula form (closing of opened half-blastula). So the correct conclusion was made on the basis of wrong premises. Now on the base of our own results we can suggest new and more complicated but more adequate "micromere model" that take into account the existence of substantial blastomere interactions instead of "micromere clock" (Fig. 6).

3. Possible mechanisms of blastomere interactions

Among the hyaline layer, providing the mechanical integrity of the embryo, the holistic development is grounded on the "direct blastomere signaling" [25]. Abstracting away from the nature of such signal for a time we can reconstruct the sequence of the events, leading to the formation of the micromeres as follows. The position of the furrow of the 1st cleavage division is predetermined by animal-vegetal axis, then the position of next cleavage furrow is determined by internal asymmetry of the blastomere (see for review [31]) formed under the influence of local intercellular signal (other word – Sax - Hertwig rule works). The above mentioned asymmetry is re-determined at the each next cleavage, including repeated signals from the contact zones of previous divisions (Fig. 7). Finally, after the 3rd cleavage division whose furrow is normally formed the at the equatorial plane the "critical mass" of vegetal cytoplasm evoke the asymmetric anchoring of the contractile ring of the 4th cleavage division. Even this process is situational and dynamic because the surprising observation which was made in *S.mirabilis* embryos: the furrow of the 4th cleavage division initially formed asymmetrically (as micromere will form there) but then the contractile ring migrated to about the middle of the blastomere and closed there [32]. Probably, there are some preferred sites of the furrow anchoring that are selected in dependence of cytocortex configuration. Let us recall that the processes of cellular interactions forming this or that cleavage pattern are multiple, non-synchronous and have own geometry, therefore the cleavage pattern of half-embryos is more or less stochastic but not unambiguous.

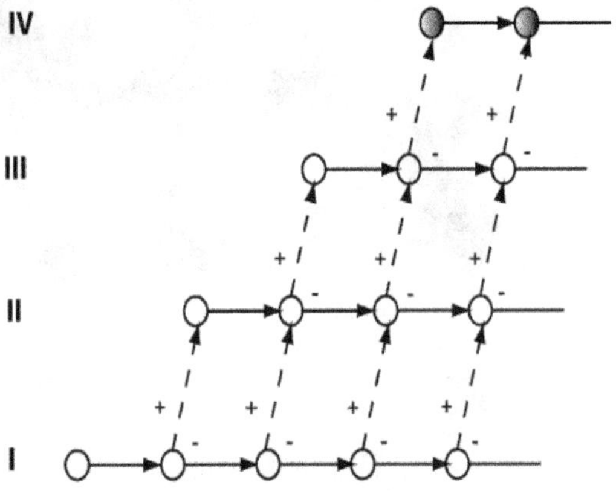

Figure 7. Scheme of the events, leading to the formation of the micromeres (from [9])
I – IV – numbers of cleavage divisions
"+" - realization of intercellular signal
"-" - absence of normal signal

Thus the sea urchin embryo really **is not able to "count to four"** but the micromere formation is determined by the realization of, at least, three consecutive intercellular signals (Fig. 7 [9, 29]).

One of the concepts, explaining the observed phenomena, was grounded on the geometric considerations, i.e. on the blastomere shape changes exclusively [33]. The cleavage patterns of half-embryos were scrupulously studied using the labeling of the cell surface with carbon particles and great diversity of the blastomere constellations was found. Nevertheless all this diversity resolves into three main variants: formation by half-embryos at the 4th cleavage division of one or two micromeres, or equal blastomeres only. However, this line of research got no further development, moreover no specialized structures were found in the contact zone using scanning electron microscopy [34] and no any new facts allow to discuss this approach.

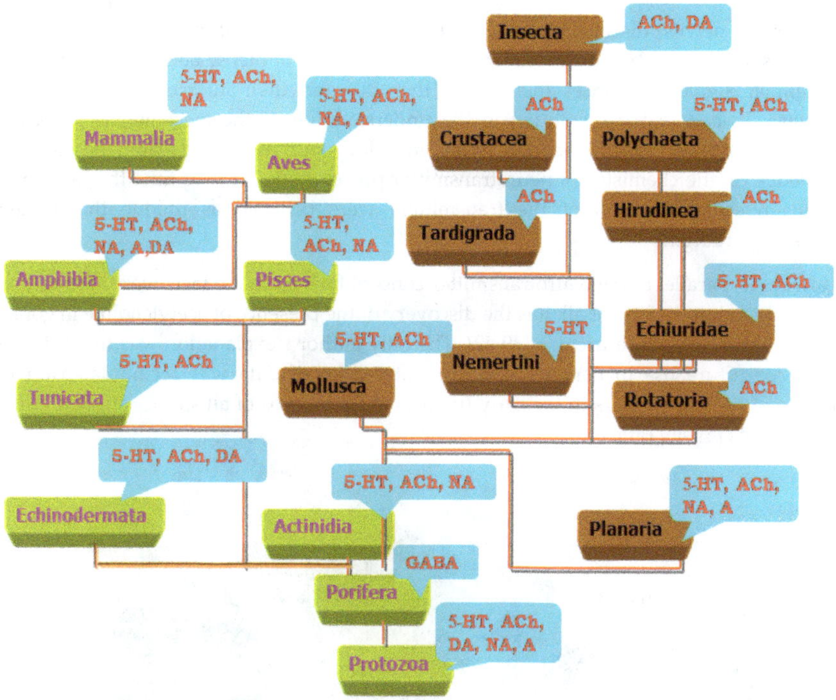

Figure 8. Transmitters in early embryos and Protozoa
5-HT – serotonin, DA – dopamine, NA – noradrenaline, A- adrenaline, ACh - acetylcholine

The idea on the possibility of interblastomere signaling via gap junction [35] failed because of that simple fact that such structures first occur in sea urchin embryos at 16-cell stage only [36, 37].

The investigations of the transfer of chemical signals between blastomeres occur more perspective and developed better. The possibility of the participation of prenervous transmitters in these processes became evident after the demonstration that transmitter antagonists are able to evoke the functional disturbances of cellular interactions [38]. Transient (10 – 15 minutes) action of serotonin antagonists before the end of post-division adhesion lead to the formation of Siamese Twins, half-embryos (including "opened half-blastulae") and 8-shaped embryos. This findings served as the base for studies of the effects of such substances in above mentioned "micromere model", although it was clear that at least in part their effects are due to the blockage of "post-division adhesion" [27].

4. Usual transmitters in unusual situation

Why the idea appeared on the participation of the transmitters such as serotonin, catecholamines and acetylcholine in embryonic cellular interactions far before the formation of nervous cells and even their precursors?

After Otto Loewi's discovery of neurotransmitter function of acetylcholine [38] the researches in this field became avalanche-type and brought immense new knowledge on the intercellular signal substances (now more than 40) and their intracellular transduction pathways. It lead to the revolution in the understanding of the mechanisms of various pathologies and, on the other hand, new pharmacology and therapy appeared, based on the knowledge on the chemistry of neurotransmitter processes. For a long time the nervous or nervous-muscular function of the transmitters were believed as unique that became sacrosanct paradigm.

As always in parallel to the neurotransmitter concept formation the facts were accumulated that not fitted into it. First of all, it is the discovery of the presence of acetylcholine in gonads and early embryos of sea urchins [39-42]. The only author's explanation was that it is "the supply for future use in nervous system". Later a lot of data was accumulated on the presence of the transmitters in the early (prenervous) embryos of all species studied also as in protozoans [43-46] (Fig. 8).

Figure 9. Principal scheme for first step of evolutionary transmitter origin (after [54])
1 – stream of substantial aminoacids into the cell; 2 – high threshold of key transmitter-synthesizing enzyme; 3 – transmitter-synthesizing enzymatic system; 4 – protein synthesis; 5 – newly synthesized transmitter; 6 – transmitter receptor, involved in the control of protein synthesis; 7 - flower

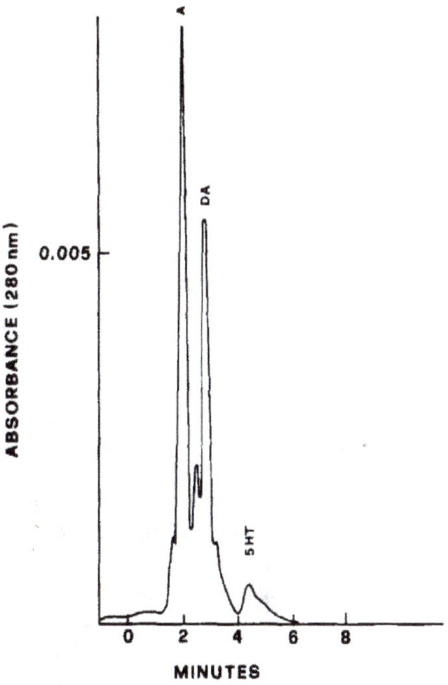

Figure 10. HPLC of the transmitters in unfertilized Paracentrotis lividus eggs
A – adrenaline, DA – dopamine, 5-HT – serotonin (from [56])

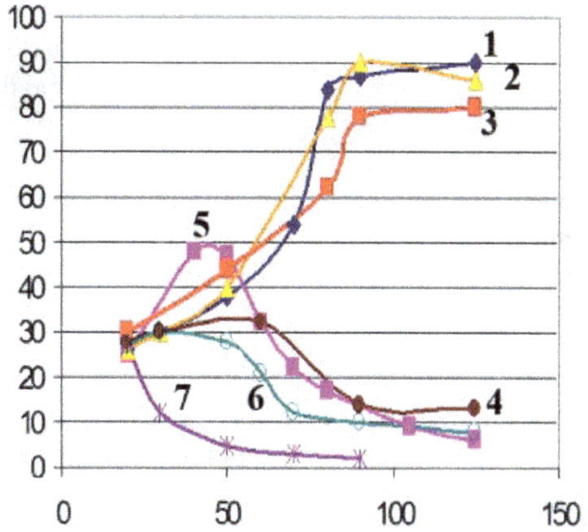

Figure 11. Effects of β-adrenergic (1-3) and serotonergic antagonists on the rigidity of sea urchin Paracentrotus lividus early enbryos (from [46]).
1 – alprenolol (400 μM), 2 – propranolol (200 μM, 3 – dichloroisoproterenol (500 μM), 4 – cyproheptadine (35 μM), 5 – inmecarb (20 μM), 6 - DPTC (75 μM), 7 – cytochalasine B (10 μM, for comparison). Abscissa: time from fertilization (min); ordinate: rigidity (din x cm^2/μm)

Components of transmitter system such as receptors and corresponding enzymes were also found everywhere in animal kingdom [44-49] even in Prokaryota [50, 51] although it is impossible to exclude secondary origin of such receptors as the result of the interactions with highly developed host organisms.

First attempt to elaborate the concept that could connect all the data on the transmitters was shot in the middle of XX century by outstanding comparative physiologist Khachatur Koshtoyantz [52]. He advanced the idea that neurotransmitter function is the result of evolution of original intracellular mechanisms of the metabolism regulation which can persist in any changed forms in the embryos of the contemporary species. Pioneer experiments of Buznikov, former student of Koshtoyantz, have shown the serotonin regulation of nudibranch velliger ciliary motility [53] and the ability of transmitter antagonists to block specifically sea urchin cleavage divisions [54] that confirmed the functionality of embryonic transmitters. Later this concept was developed on the base of Koshtoyantz original idea and the data accumulated [55], coming from the fact that some classic transmitters are the metabolites of the substantial aminoacids. Such aminoacids as phenylalanine and tryptophan cannot be synthesized by the animal cells and, thus, are the limiting point in the process of protein synthesis. High threshold enzyme that transforms aminoacid residues into the form that, on one hand, cannot be used in the protein synthesis and, on the other hand, could be recognized as the signal molecule together with the receptor molecule form the intracellular probe of the levels of the component, limiting the protein synthesis (Fig. 9). If both threshold concentrations triggering the enzyme, transforming the aminoacid, and the sensitivity of proteins (prospective receptor) to such transformed aminoacid (prospective transmitter) are sufficiently high, this offer the possibility of control over intracellular levels of substantial aminoacids in the cell. In other words, it is easier for cell to detect even a few of transmitter molecules (transformed aminoacid) then measure the absolute levels of regular aminoacids. An increase in the concentration of certain transmitter (transformed aminoacid) to the threshold levels would then indicate that total aminoacid concentration attained the level sufficient for successful protein synthesis. According to the number of substantial aminoacids there are corresponding number of their derivatives, performing the functions of the transmitters (phenylalanine – dopamine, catecholamines; tryptophan – tryptamine, serotonin, histidine – histamine etc). It is noteworthy that just in Protozoa and early embryos of multicellular organisms Dale principle: "one neuron – one transmitter" does not work. It was shown that protists and early embryonic cells may contain more than one, up to four, transmitter simultaneously (Fig. 10, see also for review [44]). Our recent study has shown the simultaneous presence of dopamine, noradrenaline and serotonin in zygotes and cleaving embryos of *Xenopus laevis* (Shmukler, Nikishin, unpublished data). Such "metabolic hypothesis" can also explain the multiplicity of neurotransmitters and finally solve the problem quoted by Kandel and a number of authors [57-60]: «Why do neurons have different transmitters when any one transmitter could in fact mediate all the required electrical signals?» All previous attempts to answer this question were limited to various features of the process of nervous signaling organization but ignored evolutionary aspect of the problem.

5. Embryonic transmitters – functions and specific features

Soon after discovery of the serotonin ability to regulate embryonic ciliary motility, the specific effects of transmitter antagonists onto cleavage divisions in various species, first of all echinoderms, were found then confirmed in all taxons studied [43, 44, 61, 62]. Antagonists of serotonin, catecholamines and acetylcholine added soon after the fertilization blocked the cleavage divisions and their effect could be prevented or weakened by the addition of specific transmitter. Most probably, transmitter antagonists have multiple effects onto cleaving embryos, in particular the triggering of cell cycle is influenced (44, 46, 55, 63] and the state of cytoskeleton [64], interestingly, in the latter serotonin and catecholamines worked as antagonists of each other (Fig. 11). Probably, serotonin also takes part in the control of closing of the contractile ring [65] of cleavage furrow and further adhesion of blastomeres [27]. At the later stages transmitters takes part in the control of left-right asymmetry formation, larval ciliary motility, gastrulation, cranio-facial and heart morphogenesis etc [46, 87, 88, 100]. These transmitter functions are realized simultaneously and/or consecutively all over ontogenesis (Fig. 12) [46]. Thus, neurotransmitter function itself is *ultimus inter pares* only.

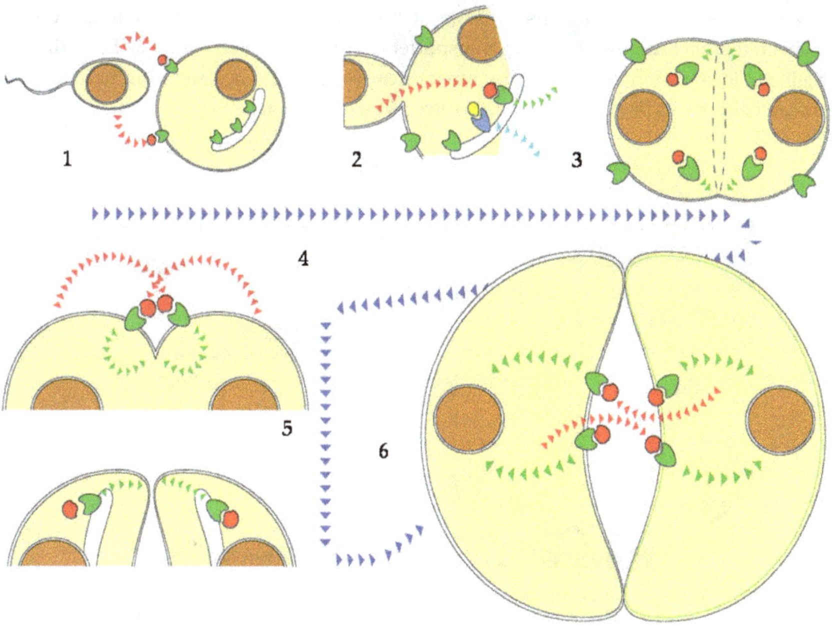

Figure 12. Transmitter control of the 1st cell cycle of sea urchin embryo.
1 – interaction with surface membrane receptors at fertilization; 2 - interaction with intracellular receptor at fertilization and triggering of cell cycle; 3 – control of the state of cytocortex via intracellular receptors; 4 – control of the completion of cleavage furrow via surface membrane receptors;
5 – control of post-division adhesion of blastomeres via intracellular receptors; 6 – direct exchange with interblastomere signals via surface membrane receptors.
Red and yellow signs – transmitters, green and blue signs - receptors

Many transmitter-related effects in embryogenesis are coupled to **intracellular** receptors [43, 44, 63, 66]. For the first time ever it was marked occasionally during the studies of the effects of transmitter antagonists, especially serotonin, it was found that the embryostatic activity depends on their ability to penetrate the cytoplasm from the medium [27, 44]. First the difference was noted between tertiary and quaternary serotonin analogues [43] but then the direct dependence was found between the embryostatic effect of indole derivatives and their lipophily [67]. On the base of these data Buznikov suggested non-trivial idea on intracellular localization of receptor link of embryonic transmitter process [43] that remain strange for physiologists until now despite the results of direct experiments with microinjection of transmitter receptor ligands into the cells of early *Xenopus* embryos [63, 66]. Microinjection of propranolol (antagonist of β-adrenoreceptors) and atropine (antagonist of m-cholinoreceptors) evoked transient block of cleavage divisions in *Xenopus* embryos that could be weakened by the addition of corresponding transmitters. Specific binding of radiolabeled ligands by microsomal fraction of *Xenopus* embryos were also demonstrated [68]. Data on the intracellular localization of transmitter process were obtain also in other subjects [69-71]. Recently the expression of the components of embryonic serotonergic system was shown that allow us to suggest the scheme of such intracellular receptor mechanism that includes receptor and transporters of the transmitter (Fig. 13). For a time the intracellular localization of embryonic transmitter mechanisms became new paradigm for the researchers in this field until new data forced to withdraw from it.

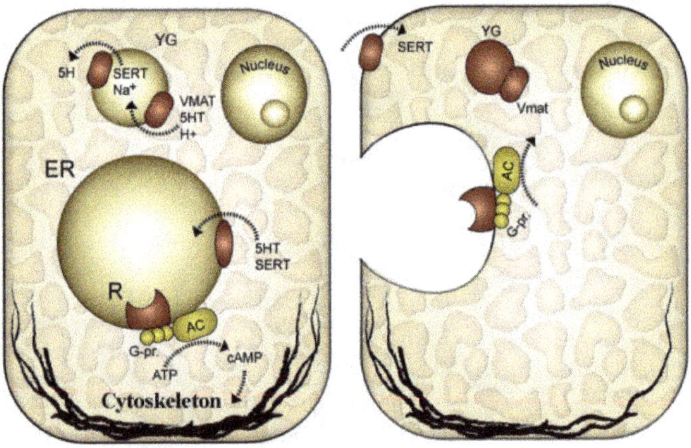

Figure 13. Hypothetic scheme of intracellular embryonic receptor mechanisms
Left – transmitter receptor is localized on the inner surface of endoplasmic reticulum (ER) being coupled to G-protein and adenylate cyclase (AC). Transport of the transmitter to ER and out there into cytoplasm is supplied by the transporters SERT and VMAT. Supposed place of transmitter synthesis (serotonin in present case) is yolk granule (YG). Right – imbedding of receptor structure from the left into the surface membrane gives usual membrane receptor complex

In spite of such non-trivial localization the embryonic transmitter mechanisms show the features similar to the classic ones. Effects of serotonin can be imitated by cyclic nucleotides, i.e. they weaken embryostatic effects of serotonin antagonists [72-74] and evoke the increase in cAMP levels in embryonic cells [75]. At the same time the ability of transmitter ligands to influence the activity of protein kinase C and intracellular free calcium ion levels was also shown in early sea embryos [65, 76, 77].

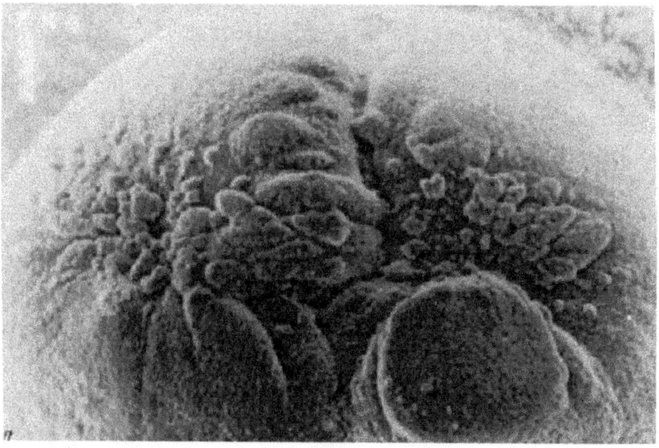

Figure 14. "Bubbling" of blastomere surface of X. laevis after cAMP microinjection (scanning electron microscopy) (from [80])

The presence and functional activity of both transmitters and second messengers inside the embryonic cells may seem excessive but only at a superficial glance. First, second messengers are effective at relatively short distances (in case of IP3, no more than 20 μm and about 3 μm for calcium ions) whereas the diameter of, for example, sea urchin egg is about 100 μm or even greater [78, 79]. Moreover microinjection of cAMP and calcium ions caused diffuse "surface bubbling" (Fig. 14, [80]) whereas microinjection of adrenaline into the blastomeres of Xenopus laevis merely accelerated cleavage furrow formation [63]. Thus transmitter, at least in this case, is more "targeted" messenger as compare to second ones that are able to activate a number of the effectors.

We should note that intracellular localization of embryonic transmitter receptors is in a good agreement with original Koshtoyantz' idea that "cell-keeping" function of the transmitters is evolutionary archetypal. Nevertheless, the fate of this paradigm is the same as the fates of other paradigms which seemed unbreakable. At the early nineties of XX century first facts were found that not fitted into previous paradigm. Some phenomena such as effects of neuropharmaca in "micromere model" (see below) and "phorbol syndrome" in sea urchin early embryos in contrast to previously studied effects on cleavage division were evoked by transmitter ligands poorly penetrating embryonic cells [81]. Correspondingly, the specific binding of radiolabeled ligand of serotonin receptor 8-OH-DPAT in the conditions that maximally restrict the penetration of ligand into the cytoplasm (0°C, short incubation) was shown [82, 83]. Later another effects of serotonergic ligands, poorly penetrating the cells of sea urchin embryos, in particular the influence of ligands on the levels of intracellular free

Ca^{2+} and inward currents [30, 65, 84]. Thus, most probably embryonic cells contain not only few transmitters but also multiple receptors that differ in their localization. In the next section the complexity of this system will be additionally sophisticated by the diversity and multiplicity of the receptor types.

6. Expression of the components of transmitter system in embryogenesis

For a long time the studies of embryonic transmitter systems developed using mainly physiological, biochemical and rarely cyto- or immunocytochemical approaches, whereas molecular biology data were sparse and rare. Only during last decade some studies appeared that confirmed by these methods the presence of the expression of the components of transmitter systems, first of all transmitter receptors in early embryos.

It was shown that already during cleavage divisions embryos of various species expressed mRNAs of transmitter receptors (Fig. 15, Table 3). Several types of the receptors to the same transmitter can be expressed simultaneously besides receptors to other transmitters. In particular, the expression of serotonin receptor type 4 also as several n-cholinoreceptor subunit was shown in early sea urchin embryo whereas in clawed frog embryo – serotonin receptors type 2 and 7 along with β-adrenoreceptor. Together with the data on the specific binding of transmitter ligands (44, 68, 79, 83) it suggests the presence of corresponding receptor proteins too. Identity of transmitter receptors' sequences in embryos and in adults is the indirect argument in favor of genetic unity of embryonic and definitive transmitter mechanisms, including cellular interactions. The expression of other components of serotonergic system such as transporters SERT and VMAT and enzymes of serotonin synthesis was found in clawed frog Xenopus and mammalian embryos [62, 85-87].

Species	Gene	Reference
Mouse Mus musculus	HTR1D	[90]
	HTR5	[91]
	HTR7	[92]
	β-AdR	[93]
Caenorhabditis elegans	HTR2C	[94]
Danio rerio	HTR1A	[95]
Sea urchin Paracentrotus lividus	HTR4	[97]
	nAChR α6	
	nAChR α10	
	nAChR α7	[96]
Clawed frog Xenopus laevis	HTR2C	[85]
	HTR7	
	β-AdR	[98]

Table 3. Accumulated molecular biology data on the expression of transmitter receptors during the early embryogensis of various species (from [85]).

Our study of other serotonin receptors in *Xenopus* embryos has shown that HTR1A is detectable beginning at late blastula stage only, HTR3A and HTR4 – at the beginning at the neurula stage, and HTR2B after hatching. The entire set of serotonin receptors is expressed at the tadpole stage only. Pharmacological and immunohistochemical data support the possible participation of 5-HT3 and 5-HT4 receptors in establishing the left-right asymmetry in *Xenopus* [88]. The 5-HT_3 receptor is a ligand-dependent ionic channel whose functional activity is associated with the presence of HTR_{3A}-subunits [89]. Therefore, the absence of HTR3A expression during early embryogenesis results in the absence of *de novo* formation of the functional 5-HT_3 receptors at this stage. The controversy between our data and those reported by Fukumoto et al. [88] can be explained by the difference in the sensitivity between mammalian and amphibian serotonin receptors to the same ligands. Fukumoto et al. [88] also reported the presence of HTR4 transcripts during the early stages of the development using *in situ* hybridization (ISH). Taking into account the higher specificity and sensitivity of RT-PCR [85] compared with ISH, a false-positive result of the ISH is more probable than a false-negative RT-PCR result.

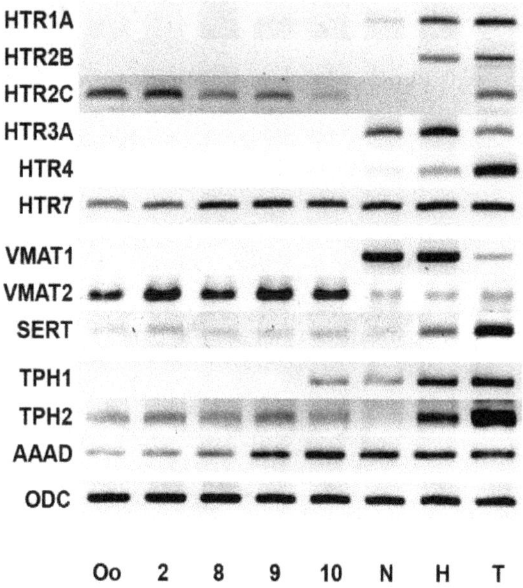

Figure 15. Temporal expression of serotonergic system components during Xenopus laevis development (from [82].
Oo – oocyte, 2 – 2-cell embryo (stage 2), 8 – midblastula (stage 8), 9 – late blastula (stage 9), 10 – early gastrula (stage 10), N – neurula (stage 15), H – hatched larvae (stage 33-35) and T – tadpoles (stage 57). ODC is the endogenous control. The serotonin receptors HTR2C and HTR7, vesicular transporter VMAT2, sodium-dependent transporter SERT, enzymes of synthesis tryptophan hydroxylase (TPH2) and aromatic aminoacid decarboxylase (AAAD) are expressed during the early stages of development (from [85])

It is intriguing that we found a wide diversity of serotonin receptor types that are expressed during early development (Table 1) when summarizing both our data and the literature data, with a few coincidences – the 5-HT2C in *Xenopus* and *C. elegans*, and the 5-HT7 in *Xenopus* and mouse. It is possible to reliably predict that serotonin receptors are also expressed in the embryos of species not yet studied in this regard, similarly to serotonin and other transmitters that were discovered in embryos of all species investigated. However, the types of serotonin receptor evidently are not strictly determined, although their mechanisms would be highly conserved.

Thus the transmitters and all main components of their systems which are characteristic for definitive organisms are present in early embryos and are functionally active all over early development (all over whole ontogenesis too).

7. Again to embryonic cellular interactions

Interestingly, prenervous transmitter mechanisms were not considered as the candidate to the role in embryonic cellular interactions at the start of these researches although it evidently offered. Probably, the paradigm of intracellular functions of embryonic transmitters influenced the approach of Founding Father of this scientific field Prof Buznikov who suggested that "*transmitters may take part in early cellular interactions by certain way but it is improbable that they are embryonic intercellular signal substances itself*" [99]. The ability of the serotonin antagonists to suppress blastomere adhesion (that lead to the functional isolation of the blatomeres) did not contradict this concept. Nevertheless it was highly inviting to check the ability of the prenervous transmitters to participate in the process named by Vacqueir and Mazia [25] "direct interblastomere signal exchange" that really influence the prospective fates of the cells of embryo.

8. The transmitter effects in micromere model

Addition of the serotonin to blastomeres, isolated before post-division adhesion, significantly increased the portion of half embryo with partial cleavage pattern, when unequal 4th cleavage occur (Table 4), meaning serotonin imitates the interblastomere signal in intact embryo. It is noteworthy that it was the first early embryonic model that allowed to obtain own effect of the transmitter but not its antagonists since original experiment in nudibranch veliger. The same concentrations of serotonin onto the blastomeres isolated after adhesion did not influence the cleavage pattern of half-embryos since, probably, it added nothing to natural signal already received by blastomere. In turn, serotonin antagonists had no effect in half-embryos, isolated before adhesion but after adhesion significantly increased the portion of half-embryos with integral cleavage pattern (with equal 4th cleavage division), i.e. imitated the avoid of interblastomere signal (Fig. 16). Along with serotonin antagonists the blocker of serotonin and catecholamine reuptake imipramine occur highly effective (Table 4). It is important that the delay of micromere formation under the action of neuropharmaca was found also in intact embryos in the special experiments [27].

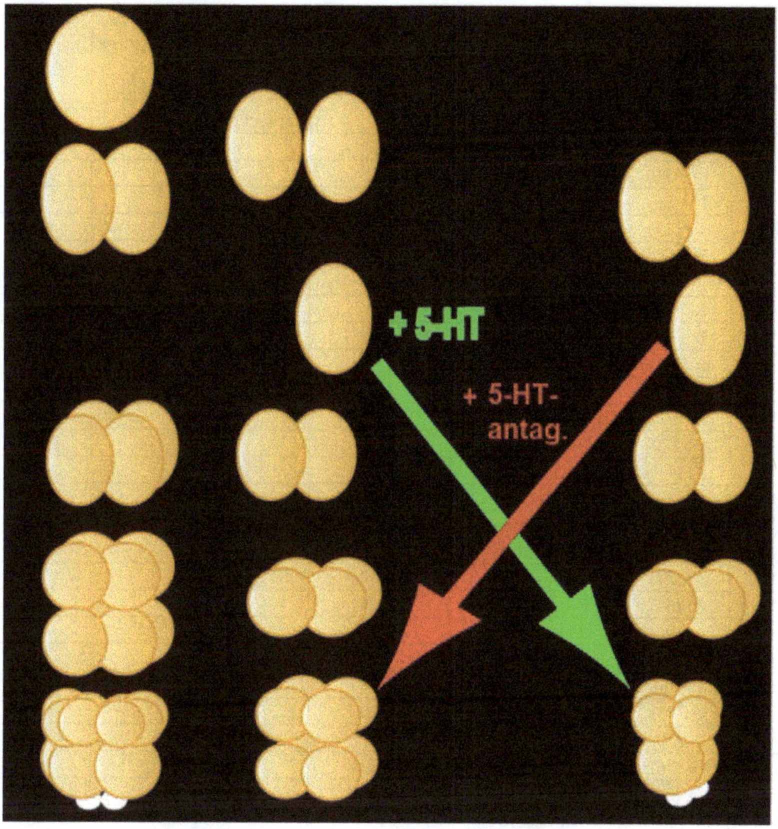

Figure 16. Effects of serotonin and its antagonists in "micromere model"
Left column – cleavage pattern of intact embryo; middle – prevailing pattern of half-embryos, isolated before adhesion, right – prevailing pattern of half-embryos, isolated after adhesion. Arrows show the possibility to influence the pattern type by serotonergics

The works with "micromere model" were started at the time when neither contemporary ligands nor transmitter classification existed [32], so recently we needed to repeat some experiments using new ligand (mainly belonging to agonists of 5-HT_3-receptors) in classic subject – sea urchin *Paracentrotus lividus* embryos [30]. These experiments confirmed that in this model ligands, poorly penetrating the cells, are equally or more effective than their lipophilic analogues [82, 83], i.e. the receptors involved are localized on the surface membrane in contrast to that regulating cleavage divisions and blastomere adhesion ones. This statement is supported also by the ability of surfactants to disturb the cellular interactions and blastomere formation in intact embryos of sea urchins [27, 32] and by the results of serotonin ligand binding assays in the conditions maximally restricting the penetration of ligand into the cell [82].

Species	Moment of isolation	Substance	Concentration (μM)	Change of portion of half-embryos, forming micromeres simultaneously with intact embryos (%%)	Significance
S. mirabilis	B_1 (317)		55	+14 ± 4	<0,001
	B_{12} (180)	Serotonin	55	+12 ± 6	<0,05
	B_{123} (86)		55	+14 ± 4	<0,01
	B_1 (115)	Tryptamine	250	+13 ± 6	<0,05
	B_1 (73)	Carbacholine	275	- 8 ± 8	n.s.*
	B_1 (53)	ATP	360	+ 3 ± 9	n.s.
	B_1 (51)	Dopamine	260	+ 6 ± 10	n.s.
	B_1 (84)	Papaverine	50	+34 ± 6	<0,001
	B_1 (101)	cAMP	270	+ 7 ± 5	n.s.
	B_1 (107)	cGMP	270	- 8 ± 6	n.s.
	B_1 (92)	dibutyryl-cAMP	210	+ 41 ± 6	<0,001
	A_1 (170)	Imipramine	5	-32 ± 5	<0,001
	A_1 (77)	Cyproheptadine	60	-21 ± 6	<0,05
	A_1 (85)	Inmecarb	25	-34 ± 8	<0,001
	A_1 (62)	Inmecarb methiodide	25	-26 ± 8	<0,001
	A_1 (71)	Aminazine	15	- 3 ± 8	n.s.
	A_1 (64)	Propranolol	135	+ 2 ± 7	n.s.
	A_1 (96)	Gangleron	32	-10 ± 7	n.s.
	A_1 (48)	Quatelerone	400	- 2 ± 9	n.s.
	B_1(86)	Valinomycine	5,4x10^{-3}	+22 ± 9	<0,05
	A_1(89)	Ouabaine	1000	-25 ± 9	<0,01
	A_1(70)	Triftazine	49	-28 ± 13	<0,05
S.nudus	B_1 (36)	Serotonin	112	+24 ± 12	<0,05
P.lividus	A_1 (53)	Inmecarb	50	- 1± 12	n.s.
	A_1 (203)	Inmecarb methiodide	40	-30 ± 10	<0,05
	A_1 (27)	KYuR-14	100	0	n.s.
	A_1 (43)	KYuR-14 methiodide	100	-17 ± 7	<0,05
	A_1 (137)	Imipramine	60	−34 ± 4	<0,001
	A_1 (97)	3-Tropanylindole carboxylate methiodide	100	−25 ± 7	<0.001
	A_1 (96)	3-Tropanylindole carboxylate hydrochloride	100	−12 ± 1	<0.001
	A_1 (82)	Quipazine	100	+24 ± 1	<0.001

Table 4. Effects of chemical substances on the cleavage patterns of sea urchin half-embryos
* not significant
Designation B and A as in Tables 1 and 2. B_{12} – embryos, isolated before adhesions in 1st and 2nd cleavage divisions. B123 - embryos, isolated before adhesions in 1st – 3rd cleavage divisions. Number of embryos in parenthesis Inmecarb and KYuR-14 also as their quaternary analogues are the originally synthesized indole derivatives

Ligands of acetylcholine- and adrenoreceptors had no significant effects in this model of blastomere interactions, although, as was mentioned above, we detected the expression of the subunits of nicotinic cholinoreceptor in the early sea urchin *P.lividus* embryos [97]. At the same time in the whole-cell patch-clump experiments in cleaving *P.lividus* embryos nicotine and n-AChR-agonists epibatidine and methylcarbamylcholine evoked small inward currents in a few cases only, whereas 5-HT_3-receptor agonists (5HTQ, SR 57277A, quipazine, methylquipazine) stably induced inward currents that were maximal during the formation of furrows of 1st and 2nd cleavage divisions both at microapplication or addition into experimental chamber [30]. Such discrepancy may be because of two circumstances: first, receptor ligand were elaborated and tested in mammalian models and their specificity in echinoderm embryos is insufficient, especially if take into account high homology between nicotinic acetylcholinoreceptor and serotonin receptor type 3. The similarity of 5-HT_3- and n-AChR is quite considerable, in particular, aminoacid sequence of rat 5-HT_{3A}-receptor (P35563) coincides to rat α_{10}-n-AChR subunit sequence (NP_072161) by 89%, and by 96% - to the sequence of n-AChR α_6-subunit precursor (NP_476532) [97]. Second, it is surprising but ligands of 5-HT_3-receptor are highly effective in 5-HT_4-receptor although former is ligand-dependent ionic channel whereas latter is metabotropic receptor [101, 102], whose expression in early sea urchin embryo was recently shown [97].

The suggestion that the effects of serotonin agonists are mediated by 5-HT_4-receptor is supported by the data as follows. 5-HT4-receptor is known to activate adenylate cyclase [101]. Papaverine (blocker of phosphodiesterases) and dibutyryl-cAMP have similar and even more pronounced effects as serotonin in "micromere model", i.e. it is possible that in the present case serotonin triggers the signal transduction pathway via adenylate cyclase. Moreover serotonin activates adenylate cyclase in sea urchin embryos [73] Furthermore, the adenylate cyclase activity in the early sea urchin embryos which first was localized at membrane of endoplasmic reticulum then transferred to the microvilli in the contact zone and after adhesion increased greatly at the places of the closest contact of the membranes of sister blastomere [103]. Similar data were obtained in mammalian embryos [104].

9. Concepts of chemical blastomere interactions

The first attempt to form the concept on the transmitter-based mechanism of embryonic cell-cell interactions was made yet in 1981 but ruling paradigm of intracellular localization of embryonic transmitters lead to the formation of pretentious construct that included isolation of the signal molecules from the receptors in the same cell [35] and involvement of gap junctions. Soon occur this concept is totally insufficient because it was found that gap junctions first appear in sea urchin embryo at 16-cell stage only [105]. It forced us to revise the concept especially taking into account the data on the existence of surface membrane transmitter receptors obtained to date.

Figure 17. Allegory of protosynapse
Tomcat – receptor; mouse – transmitter; green lawn – interblastomere cleft; bars – adhesion contact, isolating interblastomere cleft from outer medium
The probability of transmitter–receptor interaction is higher in the interblastomere cleft then at free blastomere surface.

10. Protosynapse

The idea on the exchange with chemical signals between blastomeres suggested itself because of accumulation of various data on unusual concentrations of transmitters and related substances as gangliosides [109, 110] and products of adenylate cyclase activity [103] in the contact area of blastomeres. Suggestion on the localization of transmitter receptors at the surface membrane of blastomeres [81-83] became impulse to elaborate the new concept.

Similar situation is suggested in case of cholinergic interaction of gametes that both contain acetylcholine and corresponding receptors, taking part in the fertilization [96, 107]. The second was the astonishing fact that the main way of transmitter inactivation in the embryonic cells are the transport of the transmitter molecules to outer medium because of low activity or absence of MAO (enzyme of serotonin and catcholamine degradation) [96, 108]. Recently the absence of the expression of MAO A at early stages of *Xenopus* development was shown [85].

Transmitter-driven blastomere adhesion shorten the distance between blastomeres and creates the interblastomere space which is the prerequisite for further intercellular signal exchange. The leakage of the transmitters from the interblastomere compartment is restricted by adhesive contacts and the concentration of transmitter in the interblastomere cleft remain increased as compare to free blastomere surface.

Coming from above mentioned and the fact that both blastomeres of regulative embryos are equal in properties and prospective potencies the suggestion was made on the existence of **double-side symmetric structure of signal exchange.** The presence of transmitter receptors in blastomere surface membrane, including interblastomere contact zone, make possible such structure, where both blastomeres are: i) the source of signal substance, ii) its target, and iii) the obstacle for leakage of the transmitter, named **"protosynapse"** [82, 83] (Fig. 17).

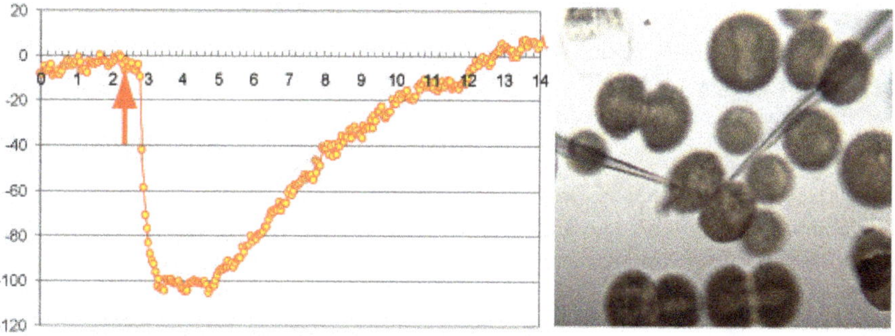

Figure 18. Effect of 5-HTQ microapplication into contact area of 2-cell *P.lividus* (from [30])
Left – current, evoked by single pulse of 5-HTQ (serotonin agonist), red arrow - the moment of application; abscissa - current (pA), ordinate – time (sec); right – experimental arrangement, left pipette is for whole cell patch, right pipette – for application

Coming from such point of view it is not significant whether transmitter receptors are equally distributed over the blastomere surface or they are concentrated at the contact area because the physiological response is due to the difference of transmitter concentrations in the contact zone and at the free blastomere surface. However, this problem was solved too using microapplication of serotonin agonists to the contact area and to the free surface of whole-cell patch-clumped sea urchin blastomeres [30]. It was shown that the application of the agonist into the interblasomere cleft before the end of adhesion evoked significantly more pronounced inward currents with substantially shorter latent period, then the application to the free surface of blastomere or after the end of adhesion (Fig. 18). Thus, localization of the transmitter receptor in the interblastomere cleft is more probable (Fig. 19), although such localization can be the result of secondary specialization of contact blastomere surface.

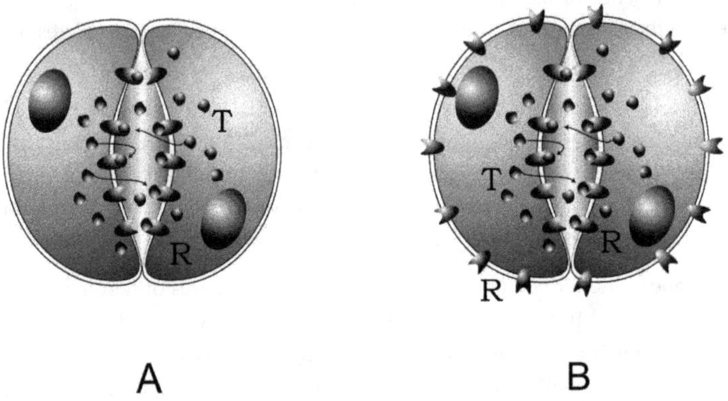

Figure 19. Serious protosynapse scheme (A – variant with transmitter receptors localized in interblastomere cleft, B – with equal receptors distribution over whole cell surface); R – receptor, T- transmitter (from [30])

Increased transmitter concentration in the interblastomere cleft and concentration of the receptors here can be the base for the formation of the primary cellular asymmetry which may thus determine the position of further cleavage plane orientation.

The existence of such structure together with the data on the genetic identity of embryonic transmitter receptors to those from adult organisms allows us to suggest that protosynapse is evolutionary predecessor of definitive synaptic structures. As the matter of fact, it is quite to remove the transmitter from one cell and the receptor – from another in protosynapse scheme to get classic synapse.

Protosynapse concept allows the analysis of all previous data on the blastomere interactions, including classic ones, and eliminate some historical injustice. As was already mentioned above, there are the complex of roots that Hans Driesch could find only one type of cleavage pattern after blastomere isolation. From the point of view of the concept under consideration late isolation, most probable in Driesch's experiments, means that blastomeres quite long remained in contact with each other, i.e. **had time to receive normal intercellular signal** and then developed with the partial cleavage pattern (Fig. 20A). Later development of such half embryos into opened half-blastula and then to closed one and quasinormal larva are the result of yet unstudied processes, not directly coupled to early blastomere interactions.

Really the same situation is reproduced in Roux experiment because although denaturated blastomere cannot be nether source, nor target of the transmitter it remains the obstacle for the leakage of transmitter from interblastomere cleft, thus the situation for intact blastomere remain unchaged as compare to intact embryo (Fig. 20B). Therefore, it is time to rehabilitate the experiment of one of founders of experimental embryology and exonerate Roux from guilt in artifact experiment.

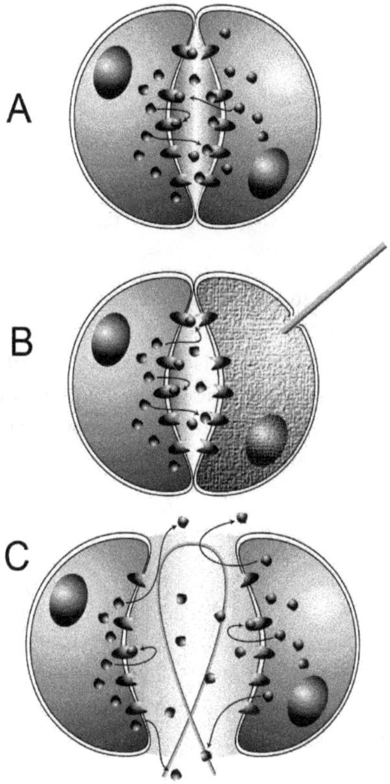

Figure 20. Analysis of various experiments on the blastomere isolation
A - Normal development and Driesch experiment; B – Roux' experiment;
C – early blastomere isolation (after [32])

In frame of the protosynapse concept it is possible to explain the fact earlier never considered. The formation of only one micromere was quite frequent. Taking into account non-synchrony of interblastomere signal realization and specificity of the geometry of interblastomere space and transmitter receptors' distribution there it is clear that adequate signal, changing the state of cytocortex, could be received by part of contact blastomere surface only that lead to the pattern formation with only one micromere.

Finally, the most easy explainable is the pattern of equal 4th cleavage division because in this case the blastomere is isolated before accumulation of enough transmitter in the interblastomere space and, correspondingly, the receiving of the adequate signal (Fig. 20C).

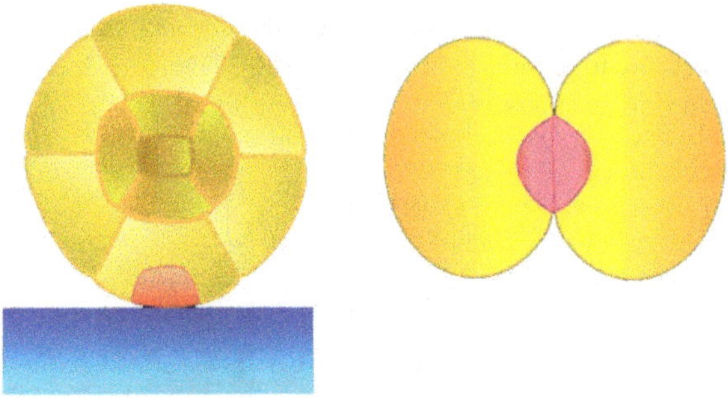

Figure 21. Origin of cell asymmetry
Left - Wolpert' concept of cell interaction with the substrate; right – protosynapse, forming internal asymmetry of the cell

The protosynapse concept also is in a good agreement with Wolpert's idea on the origin of multicellularity [111] and allow to simplify it. After Wolpert the base of origin of the cell asymmetry, needed for their specialization is the contact of part of cell with the substrate, where the zone is formed with the specific conditions that could differ from other cell surface, in particular, various regulatory substances may accumulate. The protosynapse concept allows to exclude the substrate because its role can be played by sister blastomere (Fig. 21).

11. Conclusion

So, we hope that problem of the existence of blastomere interactions is finally solved and all historic misunderstandings in this field are eliminated. Wilhelm Roux and Hans Driesch were both great researchers but they could not be irreproachable because they were first and they have created new science, invented new methods, and discovered amazing facts. It was far easier to correct their impreciseness some 90 years later.

The role of the transmitters in early development, especially in the embryonic cellular interactions, is proved too, although it attracted relatively low attention of biologists. Maybe it is because this field is "too physiological for embryologists and too embryological for physiologists". Anyway great insight of Koshtoyantz and immense Buznikov's work are still developing by their students and followers. At the same time readers can observe that great deal of the data in this field were obtained quite long ago and although they did not lose their value need in revision and renovation.

New pharmacology and molecular biology have brought new knowledge that sometimes contradicts original ideas of the researchers who carried out pioneer experiments in this field. Now studies of transmitters in the development looks like the battlefield after tank breakthrough, when fighting front got far ahead, new strategic targets appeared but a lot of enemies' firing point remained far in the rear that needs in wide and scrupulous further works.

Author details

Yuri B. Shmukler* and Denis A. Nikishin

N.K.Koltzov Institute of Developmental Biology, Russian Academy of Sciences, Moscow, Russia

Acknowledgement

Authors are grateful to Mr E.Yu.Shmukler for the help in the preparation of Figures. This work was supported by the Russ. Fund for Basic Research grant 11-04-01469-a for Y.S. and D.N.

12. References

[1] Gilbert SF (2006) Developmental Biology, 6th Edition, Sinauer Associates, Inc., Publishers, Sunderland, 751 p.

[2] Roux W (1888) Über die künstliche Hervorbringung halber Embryonen durch Zerstörung einer der beiden ersten Furchungszellen, sowie über die Nachtentwicklung (Postgeneration) der fehlden Köreperhälfte. *Virchows Arch. Path. Anat. u. Phys.*, 114: 419 - 521

[3] Wilkins AS (1986) Genetic Analysis of Animal Development. New York, Wiley

[4] Driesch H (1891) Entwicklungmechanische Studien. I. Der Werth der beiden ersten Furchungszellen in der Echinodermentwicklung. Experimentelle Erzeugung von Theil- und Doppelbildungen. *Z. wiss. Zool.* 53: 160 – 184

[5] McClendon JF (1910) The development of isolated blastomeres of frog's egg. *Am. J. Anat.* 10: 425 – 430

[6] Khaner D (1993) The potency of the first two cleavage cells in echinoderm development: the experiments of Driesch revisited *W.Roux's* Arch Dev Biol. 202:193-197

[7] Hörstadius S. 1973. Experimental embryology of echinoderms. Oxford, Claredon Press, 192 pp.

[8] Dan-Sohkawa M, Satoh N. (1978) Studies on dwarf larvae developed from isolated blastomeres of the starfish. Asterina pectinifera. J Embryol Exp Morphol. 46: 171-85.

[9] Shmukler YuB., Chaylakhian LM, Smolyaninov VV, Bliokh ZhL, Karpovich AL, Gusareva EV, Naidenko TKh, Khashaev ZH-M, Medvedeva TD (1981a) Cellular interactions in early sea urchin embryos. II.Dated mechanical separation of blastomeres. Ontogenez, 12: 398 – 403 (in Russian)

[10] Buznikov GA, Podmarev VI (1975) Sea Urchins (Strongylocentrotus dröbachiensis, S.nudus, S.intermedius). In: Subjects of developmental biology. Moscow, Nauka Publishers pp. 188-216 (in Russian)

[11] Hörstadius S (1937) Investigation as to the localization of the micromere-, the skeleton-, and the entoderm-forming material in the unfertilized egg of *Arbacia punctulata. Biol. Bull.*, 73: 295 – 316

* Corresponding Author

[12] Hörstadius S (1939) The mechanics of sea urchin development studied by operative methods. *Biol. Rev.* 14: 132 - 179

[13] Hörstadius S (1973) Experimental embryology of echinoderms. Oxford, Claredon Press, 192 pp.

[14] Czihak G (1973) The role of astral rays in early cleavage of sea urchin eggs. *Exptl Cell Res.* 83: 424 – 426

[15] Okazaki K (1975) Spicule formation by isolated micromeres of the sea urchin embryo. *Amer. Zool.*, 15: 567 – 581

[16] Parisi E, Filosa S, De Petrocellis B, Monroy A (1978) The pattern of cell division in the early development of the sea urchin. Paracentrotus lividus. Dev Biol. 65: 38-49

[17] Davidson EH, Cameron RA, Ransick A (1998) Specification of cell fate in the sea urchin embryo: summary and some proposed mechanisms. Development 125: 3269-3290

[18] Plough H (1927) Defective pluteus from isolated blastomeres of Arbacia and Echinarachnius. Biol. Bull., 52: 373-393

[19] Marcus NH (1979) Developmental aberrations associated with twinning in laboratory-reared sea urchins. *Dev. Biol.*, 70: 274 – 277

[20] Harvey EB (1940) A new method of producting twins, triplets and quadruplets in *Arbacia punctulata* and their deveopment. *Biol. Bull.*, 78, 2, 202 – 216

[21] Harvey EB (1956) The american Arbacia and other sea urchins. Princeton, New Jersey, Princeton Univ. Press

[22] Dan K. 1972. Modified cleavage pattern after suppression of one mitotic division. *Exptl Cell Res.*, 72, 1, 69 – 73

[23] Dan K., Ikeda M. 1971. On the system controlling the time of micromere formation in sea urchin embryos. *Develop., Growth & Differ.*, 13, 4, 285 – 301

[24] Sakai H. & Dan K. 1959. Studies on sulfhydryl groups during cell division of sea urchin egg. I. Glutathion. *Exptl Cell Res.*, 16, 1, 24 – 41

[25] Vacquier VD, Mazia D (1968a) Twinning of sand dollar embryos by means of dithiothreitol. The structural basis of blastomere interactions. *Exptl Cell Res.*, 52: 209 - 219

[26] Vacquier VD, Mazia D (1968b). Twinning of sand dollar embryos by means of dithiothreitol. Roles of cell surface interactions and of the hyaline layer. *Exptl Cell Res.* 52: 459 - 468

[27] Buznikov GA, Shmukler YuB (1978) The effect of neuropharmacological drugs on interactions between the cells in the early sea urchin embryos. Ontogenez 9: 173-178

[28] Shmukler YuB, Chailakhyan LM, Karpovich AL, Khariton VYu, Kvavilashvili ISh (1981) Cellular interactions in early sea urchin embryos. I. The existence of different cleavage patterns of sea urchin half-embryos. Ontogenez 12: 197-201

[29] Shmukler YuB (2010) A "Micromere Model" of Cell–Cell Interactions in Sea Urchin Early Embryos. Biophysics 55: 399–405

[30] Shmukler YuB, Silvestre F, Tosti E (2008) 5-HT-receptive structures are localized in the interblastomere cleft of Paracentrotus lividus early embryos. Zygote 16: 79–86

[31] Rappaport R (1986) Establishment of the mechanism of cytokinesis in animal cells. Int Rev Cytol. 105: 245-81
[32] Shmukler YuB (1981) Cellular interactions in early sea urchin embryos. III. Effects of neuropharmaca on the cleavage pattern of half-embryos of *Scaphechinus mirabilis*. Ontogenez 12: 404-409
[33] Bozhkova VP, Nikolaev PP, Petryaevskaya VB, Shmukler YuB (1982) Cellular interactions in early sea urchin embryos. IV. Spatial orientation of the planes of blastomere divisions. Ontogenez 13: 596-604
[34] Schroeder TE (1988) Contact independent polarization of the cell surface and cortex of free sea urchin blastomeres. Dev. Biol. 125: 255-264
[35] Buznikov GA, Shmukler YuB (1981) The possible role of "prenervous" neurotransmitters in cellular interactions of early embryogenesis: a hypothesis. Neurochem.Res., 6: 55-69
[36] Yazaki I, Dale B, Tosti E (1999) Functional gap junctions in the early sea urchin embryo are localized to the vegetal pole. Dev Biol 212): 503-510
[37] Andreuccetti P, Barone Lumaga MR, Cafiero G, Filosa S, Parisi E.Cell junctions during the early development of the sea urchin embryo (Paracentrotus lividus). Cell Differ. 1987 Mar;20(2-3):137-46
[38] Loewi O (1921). Über humorale übertragbarkeit der Herznervenwirkund. I: Mittellung. Pflügers Arch 189: 239-242
[39] Numanoi H (1953) Studies on the fertilization substances. IV. Presence of acetylcholine-like substance and cholinesterase in echinoderm-germ cells during fertilization. Scient. Papers Coll. Gen. Educ. Univ. Tokyo, 3: 193 – 200
[40] Numanoi H (1955) Studies on the fertilization substances. VI. Formation of acetylcholine-like substance in echinoderm eggs during fertilization. Scient. Papers Coll. Gen. Educ. Univ. Tokyo, 5: 43 – 54
[41] Numanoi H (1959a) Studies on the fertilization substances. IX. Effect of intermediates split from lecithin in sea urchin eggs during fertilization. Scient. Papers Coll. Gen. Educ. Univ. Tokyo, 9: 297 – 301
[42] Numanoi H (1959b) Studies on the fertilization substances. VII. Effect of acetylcholine esterases on development of sea urchin eggs. Scient. Papers Coll. Gen. Educ. Univ. Tokyo, 9: 279 – 283
[43] Buznikov GA. (1967). Low-molecular regulators of embryonic development. Science, Moscow. (in Russian)
[44] Buznikov GA. (1990) Neurotransmitters in Embryogenesis. Harwood Academic Publ., Chur
[45] Buznikov G.A. (2007). Preneural transmitters as regulators of embryogenesis. Current state of the problem. Russ. J. Dev. Biol.. 38: 213-220
[46] Buznikov G.A., Shmukler Yu.B., Lauder J.M. 1996. From oocyte to neuron: do neurotransmitters function in the same way throughout development? Cell. Molec. Neurobiol 16: 532-559

[47] Delmonte Corrado MU, Ognibene M, Trielli F, Politi H, Passalacqua M, Falugi C (2002) Detection of molecules related to the GABAergic system in a single-cell eukaryote, Paramecium primaurelia. Neurosci Lett 329: 65-8

[48] Drews U (1975) Cholinesterase in embryonic development. Prog Histochem Cytochem 7: 1-52

[49] Falugi C, Amaroli A, Evangelisti V, Viarengo A, Corrado MU (2002) Cholinesterase activity and effects of its inhibition by neurotoxic drugs in Dictyostelium discoideum. Chemosphere 48: 407-14

[50] Stacy AR, Diggle SP, Whiteley M. (2012) Rules of engagement: defining bacterial communication Curr Opin Microbiol 15:155-61

[51] M.B. Clarke, D.T. Hughes, C. Zhu, E.C. Boedeker, V. Sperandio The QseC sensor kinase: a bacterial adrenergic receptor Proc Natl Acad Sci USA, 103 (2006), pp. 10420–10425

[52] Koshtoyantz KhS (1963) Problems of enzyme chemistry of excitation and inhibition and the evolution of functions of nervous system. AN SSSR Publ, Moscow (in Russian)

[53] Buznikov GA, Manukhin BN (1960) Influence of serotonin on the embryonic motility of nudibranch. Zh. obsh. biol. 21: 347 – 352 (in Russian)

[54] Buznikov GA (1963) Use of tryptamine derivatives for the study of the role of 5-hydroxytryptamine (serptonin) in the embryonic development of invertebrates. DAN SSSR 152: 1270 – 1272

[55] Shmukler YuB, Buznikov GA (1998) Functional coupling of neurotransmitters with second messengers during cleavage divisions: facts and hypotheses. Perspect. Dev. Neurobiol. 5: 469-480

[56] Renaud F, Parisi E, Capasso A, and De Prisco EP (1983) On the role of serotonin and 5-methoxytryptamine in the regulation of cell division in sea urchin eggs. Dev. Biol 98: 37 - 47

[57] Kandel ER (1979) Cellular insights into behavior and learning. The Harvey lectures. Ser 73. N.Y. pp. 19-92

[58] Sakharov DA (1990) Neurotransmitter diversity: functional significance. Zh.evol.biokhom.fiziol 26: 734 - 741

[59] Van Valen LM (1982) Why is there more than one neurotransmitter. Behav. Brain Sci 5: 294-295

[60] Bloom FE (1984) The functional significance of neurotransmitter diversity. Am. J. Physiol 246: C184-C194

[61] Capasso A, Parisi E, De Prisco P, De Petrocellis B (1987) Catecholamine secretion and adenylate cyclase activation in sea urchin eggs. Cell Biol.Int. Rep. 11: 457-463

[62] Basu B, Desai R, Balaji J, Chaerkady R, Sriram V, Maiti S, Panicker MM (2008) Serotonin in pre-implantation mouse embryos is localized to the mitochondria and can modulate mitochondrial potential. Reproduction. 135: 657-669

[63] Shmukler IuB, Grigor'ev NG, Buznikov GA, Turpaev TM (1984) Specific inhibition of cleavage divisions in Xenopus laevis in propranolol microinjection]. Dokl Akad Nauk SSSR 274: 994-997

[64] Grigor'iev NG (1988) Cortical layer of cytoplasm – possible place of the action of prenervous transmitters. Zh. Evol. Biokhim. Fiziol 24: 625 – 629

[65] Shmukler YuB, Buznikov GA, Whitaker MJ (1999) Action of serotonin antagonists on cytoplasmic calcium level in early embryos of sea urchin Lytechinus pictus. Int.J.Dev.Biol. 42: 179-182

[66] Shmukler YuB, Grigoriev NG, Buznikov GA, Turpaev TM (1986) Regulation of cleavage divisions: participation of "prenervous" neurotransmitters coupled with second messengers. Comp. Biochem. Physiol 83C: 423-427

[67] Landau MA, Buznikov GA, Kabankin AS, Kolbanov VM, Suvorov NN, Teplitz NA (1977) Embryotoxic activity of indole derivatives. Khim-farm. Zh 11: 57 – 60

[68] Shmukler YuB, Grigoriev NG, Moskovkin GN (1988) Adrenoreceptive structures in the early clawed frog (Xenopus laevis) embryos. Zh. Evol. Biokhm.fiziol 24: 621-624

[69] Brandes LJ, LaBella FS, Glavin GB, Paraskevas F, Saxena SP, Nicol A, and Gerrard JM (1990) Histamine as an intracellular messenger. Biochem. Pharmacol. 40: 1677-1681

[70] Brandes LJ, Davie JP, Paraskevas F, Sukhu F, Bogdanovic RP, and LaBella FS (1991) The antiproliferative potency of histamine antagonists correlates with inhibition of binding of [H3]-histamine to novel intracellular receptors (HIC) in microsomal and nuclear fractions of rat liver. Agents & Actions. - Suppl. 33: 325-342.

[71] Brandes LJ, Bogdanovic RP, Tong J, Davie JR, and LaBella FS (1992) Intracellular histamine and liver regeneration: high affinity binding of histamine to chromatine, low affinity binding to matrix, and depletion of a nuclear storage pool following partial hepatectomy. Biochem. & Biophys. Res. Comm. 184: 840-847

[72] Shmukler YuB, Buznikov GA, Grigoriev NG, Mal'chenko LA (1984) Influence of cyclic nucleotides onto the sensitivity of sea urchin early embryos to cytotoxic neuropharmaca. Bull. Exp. Boil. med 97: 354-355

[73] Capasso A, Creti P, De Petrocellis B, De Prisco P, Parisi E (1988) Role of dopamine and indolamine derivatives in the regulation of sea urchin adenylate cyclase. Biochem. Biophys. Res. Comm. 154: 758 –764

[74] Carginale V, Capasso A, Madonna L, Borelli L, Parisi E (1992) Adenylate cyclase from sea urchin eggs is positively and negatively regulated by D-1 and D-2 dopamine receptors. Exptl Cell Res 203: 491-494

[75] Sadokova IE (1982) Dynamics of cyclic nucleotide content in the developing embryos of sand dollar Scaphechinus mirabilis. Ontogenez 13: 435 – 440

[76] Buznikov GA, Marshak TL, Malchenko LA, Nikitina LA, Shmukler YuB, Buznikov AG, Rakic Lj, Whitaker MJ (1998) Serotonin and acetylcholine modulate the sensitivity of early sea urchin embryos to protein kinase C activators. Comp. Biochem. Physiol. 120A: 457-462

[77] Harrison PK, Falugi C, Angelini C, Whitaker MJ (2002) Muscarinic signalling affects intracellular calcium concentration during the first cell cycle of sea urchin embryos. Cell Calcium 31:289-97.

[78] Allbritton NL, Meyer T, Stryer L (1992) Range of messenger action of calcium ion and inositol 1,4,5 - triphosphate. Science 258: 1812 – 1815

[79] Rasmussen E, Barrett P (1984) Calcium messenger system: an integrated view. Physiol. Rev 61: 938-984

[80] Shmukler YuB, Grigoriev NG, Martynova LE (1987) Changes in cell surface of Xenopus laevis blastomeres at microinjection of cAMP and calcium ions.DAN AN SSSR 294: 507-510

[81] Buznikov GA, Koikov LN, Shmukler YuB, Whitaker MJ (1997) Nicotine antagonists (piperidines and quinuclidines) reduce the susceptibility of early sea urchin embryos to agents evoking calcium shock. Gen.Pharmacol 29: 49 – 53

[82] Shmukler YuB (1992) Specific binding of [H^3]8-OH-DPAT by early embryos of sea urchin Strongylocentrotus intermedius. Biol. Membr 9: 1167-1169

[83] Shmukler YuB (1993) On the possibility of membrane reception of neurotransmitter in sea urchin early embryos. Comp. Biochem. Physiol 106C: 269-273

[84] Shmukler YuB, Tosti E (2002) Serotonergic-induced ion currents in cleaving sea urchin embryo. Invertebr. Reprod. Dev., 42, 1, 43–49

[85] Nikishin DA, Kremnyov SV, Konduktorova VV and Shmukler YuB (2012) Expression of serotonergic system components during early Xenopus embryogenesis. Int. J.Dev.Biol 56: 000-000

[86] Amireault TP, Dubé F (2005a) Serotonin and its antidepressant-sensitive transport in mouse cumulus-oocyte complexes and early embryos. Biol Reprod. 73(2): 358-365

[87] Fukumoto T, Blakely R, Levin M (2005a) Serotonin transporter function is an early step in left-right patterning in chick and frog embryos. Dev Neurosci. 27: 349-363

[88] Fukumoto T, Kema IP, Levin M (2005b) Serotonin signaling is a very early step in patterning of the left-right axis in chick and frog embryos. Curr Biol 15: 794-803

[89] Boyd GW, Low P, Dunlop JI, Robertson LA, Vardy A, Lambert JJ, Peters JA, Connoly CN (2002) Assembly and cell surface expression of homomeric and heteromeric 5-HT3 receptors: the role of oligomerization and chaperone proteins. Mol Cell Neurosci. 21: 38-50

[90] Veselá J, Rehák P, MihalikK J, Czikková S, Pokorný J, Koppel J. (2003) Expression of serotonin receptors in mouse oocytes and preimplantation embryos. Physiol Res 52: 223-228.

[91] Hinckley M, Vaccari S, Horner K, Chen R, Conti M. (2005) The G-protein-coupled receptors GPR3 and GPR12 are involved in cAMP signaling and maintenance of meiotic arrest in rodent oocytes. Dev Biol. 287: 249-261

[92] Amireault P, Dubé F (2005b). Intracellular cAMP and calcium signaling by serotonin in mouse cumulus-oocyte complexes. Mol Pharmacol. 68: 1678-1687

[93] Čikoš Š, Veselá J, Il'kova G, Rehák P, Czikková S, Koppel J. (2005) Expression of beta adrenergic receptors in mouse oocytes and preimplantation embryos. Mol Reprod Dev. 71(2): 145-153.

[94] Hamdan FF, Ungrin MD, Abramovitz M, Ribeiro P (1999) Characterization of a novel serotonin receptor from Caenorhabditis elegans: cloning and expression of two splice variants. J Neurochem. 72: 1372-1383

[95] Nikishin DA, Ivashkin EG, Mikaelyan AS, Shmukler YB (2009) Expression of serotonin receptors during early embryogenesis. Simpler Nervous Systems, IX East European Conference of the International Society for Invertebrate Neurobiology. p. 70 (Abstr)

[96] Falugi C, Diaspro A, Ramoino P, Russo P, Aluigi MG (2012) The sea urchin, Paracentrotus lividus, as a model to investigate the onset of molecules immunologically related to the α-7 subnit of nicotinic receptors during embryonic and larval development. Current Drug Targets, in press

[97] Nikishin DA, Semenova MN, Shmukler YB (2012) Expression of transmitters receptors during early development of sea urchin Paracentrotus lividus. Rus. J. Dev. Biol., 43(3), 000-000

[98] Devic E, Paquereau L, Steinberg R, Caput D, Audigier Y (1997) Early expression of a beta1-adrenergic receptor and catecholamines in Xenopus oocytes and embryos. FEBS Lett. 417: 184-190

[99] Buznikov GA (1979) Biogenic monoamines if prenervous period of phylogenesis and ontogenesis In: Catecholaminergic neurons (TV Turpaev and AYu Budantzev, eds) Nauka Publ. pp 5 – 16

[100] Beyer T, Danilchik M, Thumberger T, Vick P, Tisler M, Schneider I, Bogusch S, Andre P, Ulmer B, Walentek P, Niesler B, Blum M, Schweickert A (2012) Serotonin Signaling Is Required for Wnt-Dependent GRP Specification and Leftward Flow in Xenopus. Current Biology 22: 33–39

[101] Peroutka SJ (1997) 5-Hydroxytryptamine receptor subtypes. In: Serotonin Receptors and their Ligands (Eds. B. Olivier, I. van Wijngaarden and W. Soudijn) Elsevier Science B.V., pp. 3-13

[102] Glennon R.A., Dukat M., Westkaemper R.B. Serotonin receptor subtypes and ligands // Psychopharmacology: the Fourth Generation of Progress. Lippincott Williams & Wilkins, 1995

[103] Rostomyan MA, Abramyan KS, Buznikov GA, Gusareva EV (1985) Ultracytochemical Электронно-цитохимическое revelation of adenylare cyclase in early sea urchin embryos. Tzitologia 27: 877-881

[104] Vorbrodt A, Konwinski M, Solter D, Koprowski H (1977) Ultrastructural cytochemistry of membrane-bound phosphatases in preimplantation mouse embryos. Dev.Biol., 55, 117-134

[105] Yazaki I, Dale B, Tosti E (1999) Functional gap junctions in the early sea urchin embryo are localized to the vegetal pole. Dev Biol. 212: 503-10.

[106] Baker PC, Quay WB (1969) 5-hydroxytryptamine metabolism in early embryogenesis, and the development of brain and retinal tissues. Brain Res. 12: 272-295.

[107] Falugi C. and Prestipino G. 1989. Localization of putative nicotinic cholinoreceptors in the early development of Paracentrotus lividus. Cell. Molec. Biol. 35, 147 –161

[108] Markova LN, Buznikov GA, Kovacević N, Rakić L, Salimova NB, Volina EV (1985) Histochemical study of biogenic monoamines in early (prenervous) and late embryos of sea urchins. Int. J. Dev. Neurosci. 3: 493 – 500

[109] Zvezdina ND, Sadykova KA, Martynova LE, Prokazova NV, Mikhailov AT, Buznikov GA, Bergelson LD. (1989) Gangliosides of sea urchin embryos. Their localization and participation in early development. Eur J Biochem186: 189-94

[110] Mikhailov AT, Prokazova NV, Zvezdina ND, Kocharov SL, Malchenko LA, Buznikov GA, Bergelson LD (1981) Immunochemical study of gangliosides at the cell surface of sea urchin embryos. Differentiation 18: 43-50

[111] Wolpert L (1994) The evolutionary origin of development: cycles, patterning, privilege and continuity. Development Supplement 79-84

Chapter 2

Cell Interaction During Larval-To-Adult Muscle Remodeling in the Frog, *Xenopus laevis*

Akio Nishikawa

Additional information is available at the end of the chapter

1. Introduction

Amphibian metamorphosis provides an excellent model to study remodeling of the body. This phenomenon is characterized by overall remodeling of the body plan (i.e. larval body) which was once established in early embryogenesis. This metamorphic organ remodeling is induced by thyroid hormone (Gudernatsch, 1912; Kaltenbach, 1953) and a larval body is thus converted into an adult one (Ishizuya-Oka and Shi,Y.B, 2007; Miller, 1996). During this period, most of preexisted larval body organs, i.e., 'larval-specific organs' such as tail and gills degenerated (Nishikawa and Hayashi, 1995) and new 'adult-specific organs' such as fore- and hindlimbs formed (Brown et al., 2005). This cell replacement is thought to be essential for amphibian metamorphosis and deeply involved in various fundamental biological processes such as cell growth, programmed cell death, differentiation, morphogenesis and cell-cell and cell-environment interactions (Nishikawa, 1997; Shibota et.al., 2000; Shimizu-Nishikawa et al., 2002; Yamane et al., 2011). How are these remodeling-events regulated by metamorphic hormones, triiodothyronine (T_3) and thyroxine (T_4)? It will provide a great value for developmental and endocrinological research to understand the regulatory mechanism. This is because, thyroid hormone work not only for inducing amphibian metamorphosis but also for triggering metamorphosis in the fishes , such as flounder (Miwa and Inui, 1987) and conger eel (Kitajima et al., 1967), and also in sea urchin (Chino et al., 1994). The process of metamorphosis in the anuran is usually coupled with biphasic development (ancestral life-history). However, in species of direct developer such as the Puerto Rican tree frog, *Eleutherodactylus coqui*, the larval period is eliminated and the metamorphic period is juxtaposed with the embryonic period (Callery et al., 2001). In this case (direct development), interestingly, the metamorphic process is regulated by thyroid hormone (Callery and Elinson, 2000).

The reason for adopting thyroid hormone as a metamorphosis-inducing trigger by many organisms is involved in the fact that thyroid hormone is a ligand for nuclear receptors (transcription factors) and can directly cause tissue-specific gene expression changes in the same way as steroid hormone work in insect metamorphosis. Thus, in many metamorphosing organisms, their rebuilding from larval to adult body were achieved through switching of gene expression from larval to adult program by regulation of hormonal concentration and spatiotemporal expression of the receptors. For example, in developing limb buds, the expression of thyroid hormone receptor (TR)-β increases at early metamorphic period but down-regulates at metamorphic climax stage. While, in the tail, the TR-β upregulation occurs at metamorphic climax when tail shortening occurs (Yaoita and Brown, 1990).

A key feature of thyroid hormone action during metamorphosis is the multifaceted nature. The hormone promotes some tissue cells to grow and differentiate but induces the other ones to stop proliferation and degenerate (cell death). Although it is obvious that each program (growth or death) is triggered by thyroid hormone action, it still unknown how it causes the different reactions, death or live, only to a particular (larval or adult-type) cells among many TR-β-expressing cells. This is one of the most important issues in developmental biology field. In other word, it is important to understand not only the thyroid hormone actions but also the mechanisms by which some cells are programmed to commit to a specific cell fate (larval or adult type cells). It would be important to focus on cell-to-cell interactions or cell-to-cellular environments for the analysis of regulatory mechanism of cell fates. The environmental factors include nutrients, growth factors, hormones, cytokines, morphogens and extracellular matrices. There are basically two types of the target cells for such factors, i.e. adult and larval type cells. The adult type cells (or adult precursor cells) start their adult gene program by the actions from such factors while the larval type cells respond to the same factors to stop larval program and activate cell death program. In order to clarify this completely different mode of hormonal actions in organ remodeling during metamorphosis, it would be necessary to fully analyze not only hormonal response of target cells but also the interaction between target cells and surrounding cells.

Muscle remodeling also occurs during anuran metamorphosis (Nishikawa and Hayashi, 1994). It is of great interest to study muscle remodeling from larval to adult type during anuran metamorphosis from the aspect of molecular and cellular interactions. This is because, during anuran metamorphosis there are three different muscle changes, 1) degeneration of larval muscle in the tail (Kerr et al., 1974), 2) formation of new muscle (secondary myogenesis) in developing limbs, and 3) conversion of muscles from larval-type to adult-type in the trunk region (Ryke, 1953), which provide a useful system for analyzing programmed muscle cell death and initiation of adult program of myogenesis. In addition to this, it is also important from the viewpoint of evolutionary adaptation of myogenesis in the transition from fish to tetrapod trunk (Glimaldi et al., 2004). Thus, this chapter concentrates on *Xenopus laevis* muscle remodeling during metamorphosis, and in particular, the following 5 points are described in detailed. 1) Characterization of muscle contractile proteins in larval- and adult-type muscles and temporal and special progress of cell replacement by adult muscle formation and larval muscle death (Nishikawa and Hayashi, 1994, 1995; Kawakami et al., 2009). 2) Characterization of programmed muscle cell death in the trunk and the tail of *Xenopus laevis* tadpoles (Nishikawa and Hayashi, 1995; Nishikawa et al., 1998). 3) Characterization of the differences between larval- and adult-type myoblasts and their responses to hormonal signals in vitro (Shibota et al., 2000). 4) Regulatory mechanism for muscle cell fates, death or adult differentiation, by interaction between two types of myoblasts (larval and adult types) (Shimizu-Nishikawa et al., 2002). 5) Regulatory mechanism of adult type myogenesis by interaction between myoblasts and notochord (notochord-suppression) and spinal cord cells (spinal cord-promotion) (Yamane et al., 2011). The rationale for studying a cell-to-cell interaction during larval-to-adult muscle conversion in the amphibian is that such an analysis will provide important insights into the processes occurring during the differentiation (and/or adult organogenesis) of other stem cells such as mammalian embryonic and tissue (adult) stem cells.

2. Muscle remodeling during metamorphosis of *Xenopus laevis*

In evolutionary history, amphibians are in the process of evolving from aquatic fishes into terrestrial vertebrates. There are many changes in body organs between aquatic and terrestrial species. The most obvious example of such change would be seen in the epidermal changes in the skin. The stratified and keratinized (or cornified) epidermis in whole body skin is one of the phenotypes macroevolutionaly-acquired in amphibians but not fishes and became a well-established characteristic feature of tetrapoda, i.e. terrestrial vertebrates. Since amphibians have an aquatic larval period, the skin of the larva is a fish type of non-cornified epidermis. During metamorphosis, differentiation of a terrestrial type epidermis (i.e. stratified and cornified epidermis) occurs and complete transformation of whole body skin with terrestrial-type cornified (keratinized) epidermis is achieved after metamorphosis. Thus, in the amphibian, it is a unique feature that macroevolutional (phylogenic) changes from fish to terrestrial type are replicated during metamorphic (ontogenic) changes. Other than skin, is there any organs that are converted from larval (or aquatic) to adult (or terrestrial) type during metamorphosis? Just as the skins had evolved so as to protect whole body from drying in a terrestrial environment, it would be needed for the muscles to evolve from aquatic to terrestrial type providing with increased muscle strength so as to overcome the intense gravitational force in the terrestrial environment.

From this point of view, Nishikawa and Hayashi (1994) analyzed electrophoretically the difference in profiles of muscle contractile protein between larval and adult dorsal muscles in the frog *Xenopus laevis* (Fig.1) and found that there are significant differences in isoform expression of tropomyosins (TM) and myosin heavy chains (MyHC). The isoforms of MyHC switched from larval (higher mobility on SDS-electrophoresis) to adult type (lower mobility) and adult-specific β-TM appeared during metamorphosis in addition to preexisting α-TM, resulting in the 1:1 ratio of α:β. These changes mean that adult-specific muscles are newly formed during metamorphosis. These isoforms-transitions starts quickly from stage 54 (early metamorphosis) in the hindlimb and slowly from stage 57 (mid metamorphosis) in the body (dorsal muscles). On the other hand, the regression of adult phenotypes occurs in the tail during metamorphosis. Further immunohistochemical examination (Fig.3 A and Fig.2 B) revealed that (1)before metamorphosis, only a small number of muscle fibers at dorsomedial (DM) part of dorsal muscle expressed "adult type" muscle isoforms; and (2)during metamorphosis, the area expressing adult type isoform gradually expanded from anterior to posterior direction (axial gradient) with increase in adult type muscle fibers. Thus, we can clearly understand at a cellular level the three distinct muscular changes during *Xenopus* metamorphosis: (1) rapid adult muscle differentiation (limbs); (2) gradual muscle replacement from larval to adult type (dorsal muscles); and (3) disappearance of preexisting larval-specific muscles (tail).

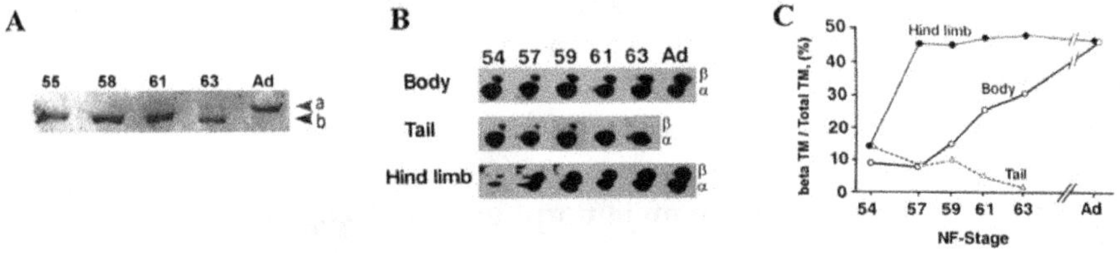

A: Developmental changes in MyHC (myosin heavy chain) isoform expression. The trunk dorsal muscles from tadpoles at various stages (55, 58, 61 and 63) and an adult frog (Ad) were applied to SDS-polyacrylamide gel electrophoresis (PAGE). The arrowheads (a and b) represent the positions of adult- and larval-type MyHC isoforms, respectively. B: Developmental changes of α- and β-tropomyosin (TM) expression. Muscle each from body, tail, and hindlimb were dissected from stages 54-63 tadpoles and a young frog (Ad) and applied to IEF (isoelectric focusing)/SDS-PAGE. The positions of α-TM and β-TM isoforms are indicated by α and β, respectively. C: Changes in ratios of TM isoforms during metamorphosis. The ratio (%) of β-TM to total (α and β) TM content was calculated from densitometric analysis of B and is shown on the ordinate. The abscissa shows developmental stages. The figures were modified from original paper (Nishikawa and Hayashi, 1994).

Figure 1. Expression patterns of adult type muscle contractile proteins during *Xenopus laevis* metamorphosis.

There are two possibilities for the dorsal muscle isoform conversion during metamorphosis (Fig.2A). In the hypothesis 1 (H1), the isoform transition occurs by cell replacement with anterior-posterior proliferation of adult-type myoblasts and death of preexisting larval–type fibers. On the other hands, in hypothesis 2 (H2), the switch in gene expression from larval to adult program occurs within the same cells without larval cell death. If the former (H1) is the case, myoblasts proliferation should occur with anterior-posterior gradient just before the isoform transition in dorsal muscles. Examination of this point by the assay of DNA synthetic activity (Nishikawa and Hayashi, 1994) and PCNA (proliferating cell nuclear antigen) expression (Kawakami et al., 2009) revealed that the cell proliferation activity is higher in the anterior than the posterior dorsal muscles during early metamorphosis just prior to the isoform changes.

These proliferation activities are well-matched the observations that small portions expressing β-TM (adult muscle area) appeared first at dorsomedial (DM) parts of dorsal muscles and the area gradually expanded to overall dorsal muscles (Fig.3 A, B). The DM parts correspond to the "cord" of the tadpole axial muscles which is reported by Elinson et al. (1999). The "cords" in tail portion also express adult muscle isoforms but this (β-TM^{+}) regions never increased during metamorphosis. On the other hands, in β-TM (-) regions, i.e.

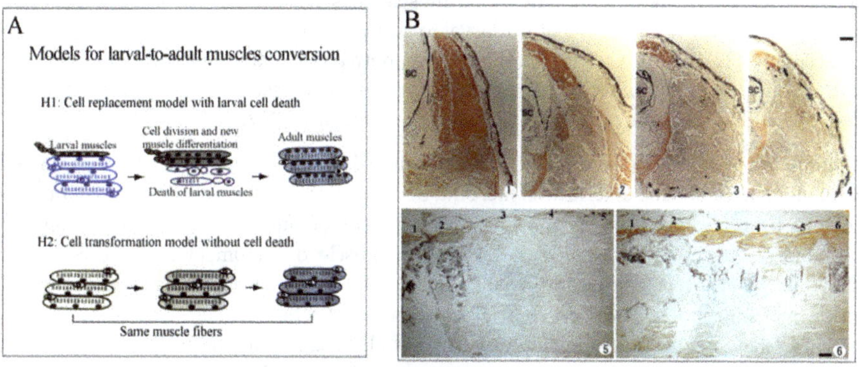

A: Models for larval-to-adult muscle conversion. In hypothesis 1 (H1), adult-type myoblast proliferation newly occur within a small portion of the muscles during early metamorphosis with a consequent increase in the number of adult muscle fibers, and also the death of preexisting larval muscle fibers occur during metamorphic climax. In hypothesis 2 (H2), all muscle cells change their gene expression from the larval program to the adult one without any cell death. B: Photomicrographs of the immunostained sections with β-TM-specific antibody. 1-4: Cross sections (from anterior to posterior) of dorsal muscles of stage 63 (climax stage) tadpole. Bar=100 μm. SC: spinal cord. 5-6: Sagittal sections of dorsal muscles at prometamorphosis; stage 59 and 61, respectively. Bar=200 μm. The number shows the arrangement of muscle segment from anterior to posterior in this order. The figures were modified from original paper (Nishikawa and Hayashi, 1994).

Figure 2. Muscle conversion models (A) and the β-TM patterns in dorsal muscle during *X. laevis* metamorphosis (B).

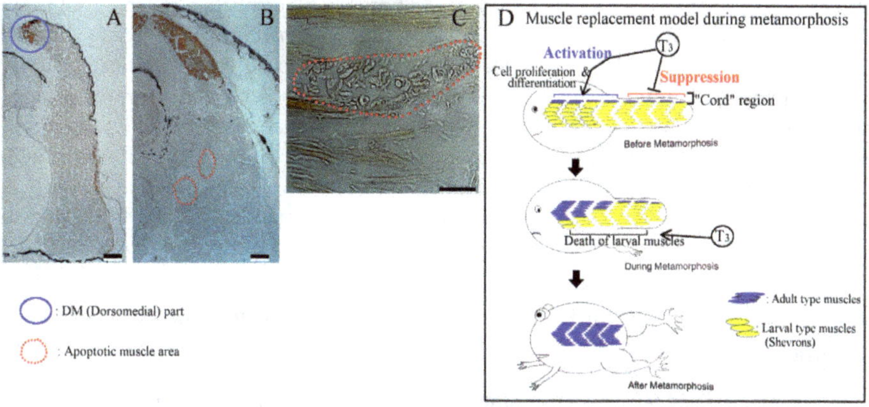

A and B: Cross sections of dorsal muscles from pre-metamorphic (A; stage 53) and metamorphic climax (B; stage 63) tadpoles. The Bars=100 μm. The brown colored areas were β-TM-positive adult muscle areas. C: Frontal section of tail muscle from tadpoles at stage 63. The apoptotic muscle fragments (sarcolytes) were β-TM-negative. The bar=50 μm. D: A schematic model of the mechanism for the muscle replacement during *Xenopus laevis* metamorphosis. Figures were reformed from the original papers (Nishikawa and Hayashi, 1994, 1995).

Figure 3. Immunohistochemical analysis of metamorphosing *X. laevis* tadpoles with β-TM (A-C) and a possible model for muscle replacement during metamorphosis (D).

larval muscle areas, many apoptotic dying muscle cells were observed (Fig.3 B, C). From these results, it has been found that the larval to adult muscle remodeling is achieved by cell replacement (H1: "cell replacement model") with new proliferation of adult myoblasts and death of preexisting larval cells, not by changing gene expression program from larval to adult one within the same cells (H2) (Nishikawa and Hayashi, 1994).

In summary, we can understand the muscle remodeling in *Xenopus laevis* as "evolutional adaptation" from aquatic to terrestrial life. And the evolutional processes of terrestrial muscles seem to be replicated by the ontogenic development of anuran amphibians. In anuran amphibians, one of the most unique features in muscle development is the presence of "chevrons", larval-type muscles in axial muscles, with a fate of apoptosis triggered by T_3 during metamorphosis. The tadpole axial muscles are attached on both side of notochord, showing "chevron" structures in side view (Das et al., 2002). At the most dorsal and ventral parts in each chevron, there are "cord" structures that consist of two dorsal and two ventral parallel rows of slow muscle bundles connected by collagen fibers that run the length of the tail (or trunk–to-tail)(Elinson et al., 1999). These parts ("cord" muscles) persist until the very end of tail resorption and are known not to be a direct target for T_3-action of cell death-induction (Das et al., 2002). While the "cord" consists of slow muscle fibers, the "chevron" is composed of fast muscles. Therefore, these two muscles (chevron and cord) can be distinguished by the fast (or slow) fiber-specific antibodies such as F59 for the fast isoforms of myosin heavy chains and S46 (or S58) for the slow isoforms. The "chevron" muscles, i.e. the larval-type muscles, exist not only in the tail but also in the dorsal muscles of the tadpole trunk region and all these muscles are to die in response to T_3-upregulation. Interestingly, "cords" muscles, which can be considered as "adult-type" muscles by judging from their T_3-response, also exist both in trunk and tail regions. The adult muscle differentiation in the dorsal trunk "cords"(i.e. trunk DM parts) is to be promoted by T_3 while the tail "cord" region, which is thought to contribute to the tension for tail resorption, is to be destroyed finally during metamorphosis (Elinson et al., 1999; Das et al., 2002). Thus, the difference of developmental fates between tail and trunk muscles are caused by differential regulation of DM portions, activation or suppression (Fig.3 D: cell replacement model on adult muscle differentiation). Therefore, it would be important to analyze the interaction between myogenic stem cells (i. e. larval and adult myoblasts) and non-myogenic cells (or extracellular matrices) in order to understand the mechanism of the activation (or suppression) of trunk (or tail) "cord" muscles.

3. Mechanism of programmed muscle cell death and macrophage phagocytosis

Death of regenerating tails in anuran amphibians has attracted interests of many researchers and Kerr et al. are not exceptional, who define the specific term "apoptosis" for the processes of non-accidental and active cell death (Kerr et al., 1974). They observed the tail cell death in *Xenopus laevis* by electron microscopy 38 years ago and proved morphologically that the tail cell death during anuran metamorphosis is achieved by the apoptotic processes including modified formation of apoptotic bodies (i.e. sarcolytes). The sarcolytes are formed by the internal fragmentation of muscle fibers not by the usual surface budding (Fig.4 A).

Nishikawa and Hayashi (1995) examined whether the death in the trunk dorsal muscles occurs with the same processes as seen in the tail cell death and found the processes are the same between the two. The apoptotic dying muscles first appeared near the base of the tail in early climax stage of metamorphosis (stage 59) when T_3 level is quite high, and thereafter expanded in an anterior direction in dorsal body and posteriorly in the tail. The direction of area-expansion of larval cell death was thus opposite to that of adult muscle differentiation. This relationship between cell death and differentiation activities is very interesting in that cell death occurs at first in the place where the adult differentiation occurs at last. As to the signals which regulate the axial gradient in adult muscle differentiation, the involvement of the factors from the nerve (spinal cord) and the notochord cells is suggested as described below (section 6).

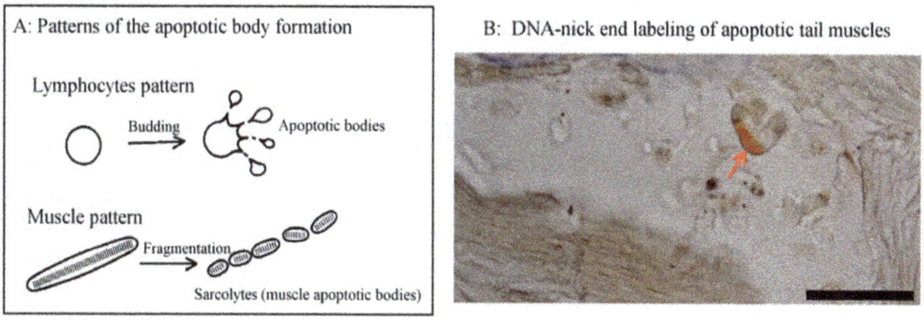

A: In case of non-muscle cells such as lymphocytes, the apoptotic bodies are formed by cytoplasmic buddings. On the other hands, in case of muscle cells, the apoptotic bodies (sarcolytes) are formed by fragmentation of muscle fibers. B: Frontal section of tail muscle from tadpoles at stage 63 was analyzed by TUNEL method (DNA-nick end labeling). There is an apoptotic muscle fragment with a TUNEL$^+$ nucleus (a red arrow). The bar shows 50 μm. The figure was reformed from the original paper (Nishikawa and Hayashi, 1995).

Figure 4. Patterns of two different apoptotic bodies (A) and a TUNEL (DNA-nick end labeling) detection of apoptotic tail muscle cells during *Xenopus laevis* metamorphosis (B).

Oligonucleosomal DNA fragmentation is thought to be a good biochemical evidence of the apoptotic cells (Kaufmann et al., 2000). Nishikawa and Hayashi (1995) detected electrophoretically the nucleosomal DNA fragmentation in *Xenopus* tail and trunk dorsal muscles at stage 63 when the muscle degradation occurs most frequently. The apoptotic enzymes which direct the DNA fragmentation is reported as the "endonucleases" such as DNase I (Rauch et al., 1997), DNase X (Los et al., 2000), DNase II (Krieser and Eastman, 1998), CAD/DFF40 (Enari et al., 1998; Liu et al., 1997), DNase γ (Shiokawa et al., 1994; Shiokawa and Tanuma, 2001) and endonuclease G (Parrish et al., 2001). Among these endonucleases, DNase γ was originally found by Shiokawa et al. (1994) as the enzyme which is activated during apoptosis of rat thymocytes and the cDNA encoding human DNase γ was cloned (Shiokawa et al., 1998). The homolog of this gene was also cloned from *Xenopus laevis* (*xDNase γ*) and the expression was found to be upregulated in the tail during metamorphosis (Shiokawa et al., 2006). In studies with mouse C2C12 cells, it was found that DNase γ is the endonuclease responsible for DNA fragmentation in apoptosis associated with myogenic differentiation (Shiokawa et al., 2002). Furthermore, DNase γ was found to be expressed in rat pheochromocytoma PC12 cells during naturally occurring apoptosis which associated with neuronal differentiation (Shiokawa & Tanuma, 2004). These finding suggest that DNase γ has an important function for the selective deletion of unnecessary cells that could not successfully differentiate. If we apply this idea to the metamorphic apoptosis in *X. laevis*, DNase γ might contribute to the selective deletion of unnecessary larval type cells during metamorphosis. Although the upregulation of DNase γ mRNA level occurs in metamorphosing tadpole tail of *X. laevis*, it is not determined which tissues in the tail are responsible for the upregulation of DNase γ gene. As to this point, the TUNEL analysis (Fig.4 B) suggested a possibility that apoptotic dying muscles express DNase γ. On the other hand, the possibility cannot be excluded that DNase γ is expressed in macrophages (and/or neutrophils) which phagocytose and digest the apoptotic muscle fragments. Because, it was shown that DNase γ–like activity dramatically increased in a leukocyte fraction of the liver (a major hematopoietic organ) during metamorphic climax stage in *X. laevis* (Nishikawa et al., 1997).

These are four cellular components, i.e. two types of myogenic stem cells (larval and adult myoblasts) and two types of muscle fibers (adult and larval muscles), and these are responsible primarily for the muscle conversion during *Xenopus* metamorphosis. Among these, the cells responsible to the oligonucleasomal DNA laddering are the larval muscle fibers because they shows some apoptotic features, i.e., sarcolytes formation and DNA breakage ($TUNEL^+$ reactions)(Fig.4 B). Such muscle cell death with apoptotic DNA laddering was found to be induced dose-dependently by thyroid hormone (T_3) in tadpole tail organ cultures (Nishikawa and Hayashi, 1995). Not only death of matured larval muscles but also that of undifferentiated larval stem cells (myoblasts) is needed for the coordinated muscle conversion during metamorphosis. From this point of view, Shibota et al. (2000) examined whether the programmed cell death occurred at stem cell (i. e. myoblast) level with cell culture technique and found that larval myoblast-specific cell death was induced by thyroid hormone in a dose-dependent manner (Shibota et al., 2000: see Section 4).

Another feature of the apoptotic processes is a phagocytosis by macrophages of the dying apoptotic bodies. In regenerating tail of *X. laevis*, phagocytosis of sarcolytes (muscle apoptotic fragments) was frequently observed (Kerr et al., 1974). So, do macrophages operate for the deletion of dying muscles during metamorphosis in the trunk region as well as in the tail? From this point of view, Nishikawa et al. (1998) analyzed the macrophage distribution in tadpole axial muscles (both in trunk and tail regions) using with macrophage-specific monoclonal antibodies (Ham56). The results showed that $Ham56^+$ cells (macrophages) appeared not only in the tail but also in the trunk dorsal muscles of metamorphosing tadpoles. Electron microscopic observation revealed that the macrophages ingested the fragment of dying trunk muscle fibers in large quantities. Interestingly, macrophage number markedly increased at late metamorphic climax stage when muscle cell death most frequently occurred and decreased at the completion of metamorphosis. In other words, the distribution and change in the number of macrophages were the same as those of muscle apoptotic bodies (sarcolytes) during metamorphosis. Analysis with Western blotting suggested that Ham56 recognizes *Xenopus laevis* homologues of mouse attachmin (Tomita and Ishikawa, 1992) which is non-specific adhesion proteins in macrophages. The expression of Ham56 antigens was found to increase with macrophage phagocytosis at the late climax stage, thus, Ham56 antigens would be essential for macrophage-dying cell interaction. Furthermore, cell culture studies with isolated *Xenopus* tail cells suggested that macrophage differentiation and its phagocytic activity were regulated by thyroid hormone (Ochi and Nishikawa, unpublished data). It still however remains unsolved how macrophages recognize the muscle cells that have a death fate (i.e. larval type muscles) or whether some signals from macrophages are needed for the induction of the first step of muscle apoptosis. As for macrophage actions for the muscle cell systems, Sonnet et al. (2006) reported that human macrophages rescue myoblasts and myotubes from apoptosis through a set of adhesion molecular systems. In this case, the signal from macrophages is a survival signal but not a death-signal. Thus, the spectrum of macrophage actions seems fairly-broad and these actions should be totally clarified in the future.

On the other hands, researchers in French (Demeneix' group) have been greatly contributed to the finding of cell death regulators for the muscle cell death during metamorphosis. The method was developed by de Luze et al. (1993) to introduce genes directly into tail muscles and using this method Sachs et al. (1997) found that somatic gene transfer with a mouse *bax* into *Xenopus* tadpole tail muscles induced apoptosis and T_3 treatment significantly increased *Bax* transcription. After that Sachs et al. (2004) further cloned a cDNA encoding *Xenopus laevis bax* (*xlbax*) and found that the gene expression increased during metamorphosis and was experimentally up-regulated by T_3-treatment. Also, overexpression and antisense experiments showed that *xlbax* is a regulator of muscle fiber death in the tail during *Xenopus* metamorphosis. In addition to the importance of Bax-regulation, Rowe (2005) clarified that the activation of caspase 9, one of the components in mitochondria-dependent cell death pathway, was pivotally involved in tail muscle apoptosis during *Xenopus* metamorphosis by using overexpression system of dominant-negative *caspase 9*. Furthermore, Du Pasquier et al. (2006) showed that developmental cell death during *Xenopus* metamorphosis involves Bid cleavage and Caspase 2 and 8 activations.

We can see from the above that cell-death signaling in the tail muscle of metamorphosing *Xenopus* tadpoles is nearly identical to that in mammalian mitochondria-dependent cell death system. However, cell death in *Xenopus* tail has distinguishing features in that it is under the control of thyroid hormone. Therefore, it would be an important subject to examine whether thyroid hormone regulates metamorphic muscle cell death not only at a muscle fiber level but also at a stem cell (myoblast) level. For such an analysis, it would be very helpful to use a primary cell culture system of isolated myoblasts. In the next section, studies with primary culture of X. *laevis* larval and adult myoblasts are described in detail (Section 4).

4. In vitro characters of two types of myogenic stem cells from *Xenopus laevis*: Differential hormonal responses in cell division, cell differentiation and programmed cell death

Are the deaths of larval muscle fibers and myoblasts induced by the thyroid hormone during anuran amphibian metamorphosis? How are the cell division and differentiation for the adult muscle stem cells (i. e. adult myoblasts) regulated by thyroid hormone? To answer these questions, Shibota et al. (2000) established primary cell culture methods for adult- and larval-type myoblasts in the frog, *Xenopus laevis*, and examined the hormonal response in each case. In this study, frog leg and tadpole tail muscles were used for the source of adult and larval type myoblasts, each of which respectively has a life-or-death fate during metamorphosis. Generally, the trunk dorsal muscles (i.e. the axial muscles) of premetamorphic tadpoles should be used for the adult myoblast source in order to adjust temporal and spatial situations between the larval and adult type cells. However, the selective isolation of adult cells from tadpole dorsal muscles was found to be quite difficult. Because of this, the frog legs were employed as the source of adult type cells. The selective isolation method of adult stem cells from tadpole dorsal muscles was developed in a subsequent study (Shimizu-Nishikawa et al., 2002, see section 5).

It was found that there were several significant differences (1-7) in the nature of isolated cells between larval and adult-type myogenic stem cells as described below. (1) The cell size just after isolation in the larval-type (15 μm) was larger than that in the adult-type (5 μm). The size of spreading cells 1 day after inoculation was also larger in the larval type cells (30-50 μm) than in the adult type cells (5-10 μm) (Fig.5A, e and a). (2) Both types of cells could adhere to the plastic culture dish with different adhesion ratios (larval type=30-50%; adult type=50-60%). Most of attached cells (88% larval and 81% adult cells) were desmin-positive, showing the isolated cells to be myoblast-rich populations. (3) The timings of start of myotube-differentiation were quite different between larval and adult types: Myotube-formation by myoblasts fusion started on day 2 or 3 in larval but on day 4 in adult cell cultures (Fig.5A, f and c). (4) There was large difference in growth activity between larval and adult cells: The larval myoblasts increased only 2.5-fold over 6 days of culture but the adult ones 5.5-fold (Fig.5 B). (5) The cultured larval myoblasts responded to the metamorphic hormone, T_3, with decrease in the DNA synthetic activity (50%) (Fig.5 F). As the result, T_3 decreased the cell numbers (sum of myotubes and myoblasts) in larval cell cultures to 56% of those of control cultures over 6 days (Fig.5 E). On the other hands, T_3 did not have much influence on the total cell numbers in adult cultures (Fig.5 D). (6) T_3 promoted dose-dependently the differentiation of adult myoblasts into myotubes but diminished that of larval cells by half (Fig.6 A and B). (7) Death of differentiated myotubes was promoted by T_3 specifically in larval but not in adult cultures (Fig.5 C and Fig.6 C). In addition to the myotubes death, double staining with TUNEL and anti-desmin (a myoblast marker) antibody showed that death of myoblasts (desmin$^+$ cells) was induced by T_3 specifically in larval but not adult cells (Fig.6 D). From the differences in cell sizes, the start-timing of differentiation and cell growth activity between larval and adult myogenic stem cells (1, 3 and 4), it is conceivable that adult myoblast have a more stemness phenotype than the larval myoblasts. This is to say, adult myoblasts cannot enter the myotube-forming stage without dividing many times, but in contrast, the larval myoblasts can immediately go into myotube-differentiation. From this point of view, it would be an important issue to examine differences in gene expression of early myogenic transcription factors, such as *pax3, myf5* and *myod,* between larval- and adult-type myoblasts for further characterization of them.

Another essential difference (5-7) between larval and adult cells was found to be the difference in T_3 responses (Table 1). It was thus evident that the conversion of a larval to adult myogenic system during metamorphosis becomes possible through totally specific control of cell division, cell differentiation and programmed cell death at a precursor cell level by T_3. In studies using with a myoblast cell line (Yaoita and Nakajima, 1997) cloned from *Xenopus laevis* tadpole tail muscles, it was found that myoblast cell death was induced by T_3 with increasing level of *caspase 3* expression. Considering this and Rowe's results (Rowe et al., 2005), there must be up-regulation of caspases 3 and 9 in the primary cultures of the larval myoblasts isolated from *Xenopus* tadpole tail.

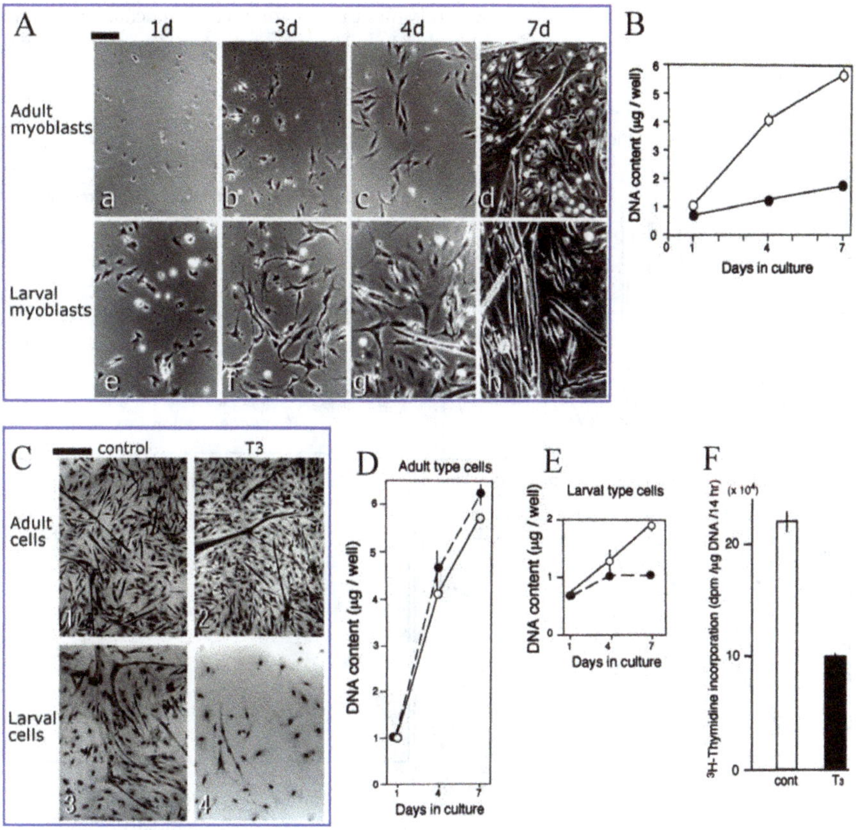

A: Phase contrast photomicrographs (Microscope: IX70, Olympus) of cultured adult (a-d) and larval myoblasts (e-h) at each culture days (1d–7d). The bar=100 µm. B: Growth curves of adult (○) and larval myoblasts (•). The cell growth activity was determined by the DNA content of total cells/culture-well. C: Photomicrographs of cultures of adult (1-2) and larval myoblasts (3-4) with (2 and 4) or without T_3 (10^{-8} M) (1 and 3). T_3 was added to the cultures at the first day of culture. Cells were fixed with 70% ethanol and stained with hematoxylin and eosin at 7th day. The bar=100 µm. D and E: The effect of T_3 on the DNA content in adult (D) and larval (E) myoblast cultures. ○: without T_3; •: with T_3 (10^{-8} M). F: The effect of T_3 on the DNA synthetic activity of larval myoblast cultures. The DNA synthetic activity was determined by measuring the ^{3}H-thymidine incorporation into cultured cells. □: without T_3; ■: with T_3 (10^{-8}M). The figures (A, B, D-F) were reformed from the original paper (Shibota et al., 2000).

Figure 5. Primary culture of adult and larval myoblasts (A and B) and the effect of thyroid hormone (T_3) on cell growth (C-F).

As noted above, there were shown to be essential differences between larval and adult type myoblasts each isolated from tadpole tail and leg muscles. Also in the trunk dorsal muscles during metamorphosis, there should be two different (larval and adult) myogenic stem cells with a life-or-death fate. Do these cells really exist within trunk dorsal muscles? If this is the case, is there a possibility that they form heterokaryon myotubes ("chimeric fate" myotubes) with different two types of cells other than myotubes consisting only of larval or adult cells? In such chimeric myotubes, which fates, a death-fate or a life-fate, possibly be selected in response to T_3? In the next section, the study about cell interaction mechanism which regulates the myotubes fate (life-or-death) in the dorsal muscle during metamorphosis of *X. laevis* was described in detail (Shimizu-Nishikawa et al., 2002).

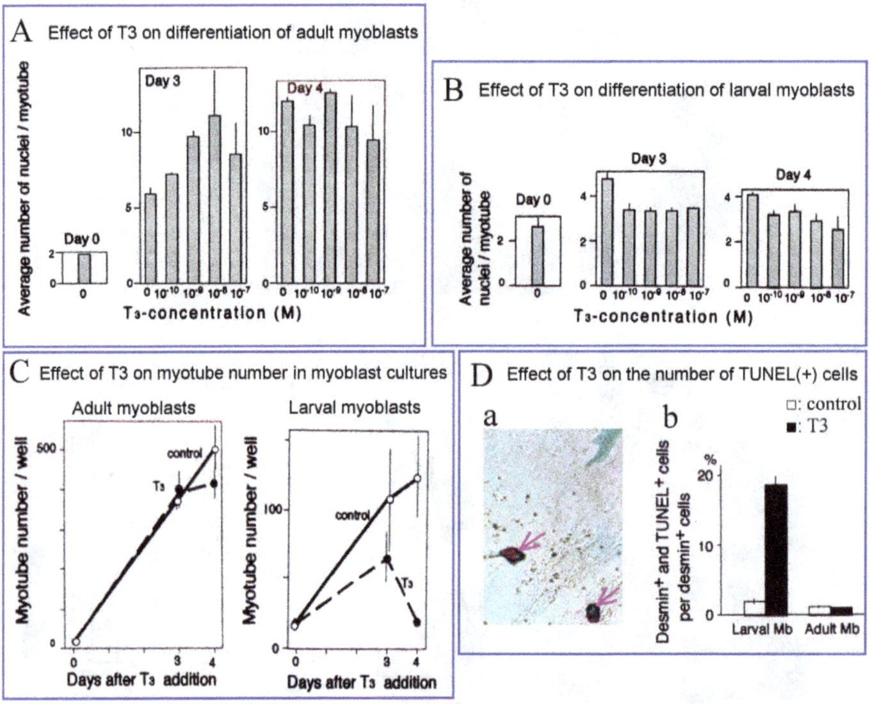

A and B: Effect of T_3 on differentiation of adult (A) and larval (B) myoblasts. Adult and larval myoblasts were cultured for 8 or 7 days, respectively. Various concentration of T_3 (10^{-10} - 10^{-7} M) were added to the cultures at 4th (A) or 3rd (B) day. Myonuclei number/myotube was determined at 0, 3 or 4th day after T_3 addition by measuring the numbers of myotubes and myonuclei within each well. The average number of myonuclei/myotube was indicated in the ordinate. C: Effect of T_3 on myotube number in adult and larval myoblast cultures. Each myoblasts were cultured for 8 or 7 days. T_3 (10^{-8} M) was added to the culture at 4th (adult cells) or 3rd (larval cells) day. The myotube number was determined at 0, 3 or 4th day after T_3 addition. ○: without T_3; •: with T_3. D: Effect of T_3 on the number of $TUNEL^+$ cells in myoblast cultures. a: Photomicrograph of $TUNEL^+$ (red) and anti-$desmin^+$ (blue) cells (arrows) in a larval cell culture. b: The larval and adult myoblasts were cultured for 7 days, fixed with formaldehyde and stained doubly with anti-desmin antibody and the TUNEL. Doubly stained cells within each well were counted and indicated in the ordinate. □: without T_3 (control); ■: with T_3 (10^{-8} M). The figures were reformed from the original paper (Shibota et al., 2000).

Figure 6. Effect of T_3 on differentiation and death of *Xenopus* adult and larval myogenic cells in vitro.

	Myogenic system	
	Larval-type	Adult-type
Replication of myoblasts	↓	n.e.
Myoblast differentiation	↓	↑
Myoblast death	↑	n.e.
Myotube death	↑	n.e.

↓: suppression; ↑: induction or promotion; n.e.: no effect. In the presence of T_3, larval type myogenic system of *Xenopus laevis* totally ceases and converts into adult type myogenic system during metamorphosis. The table was reformed from the original paper (Shibota et al., 2000).

Table 1. Comparison of T_3 responsiveness between larval and adult myogenic systems.

5. Interaction between larval- and adult-type myogenic precursor cells during metamorphosis: Regulation of cell death fate and adult muscle differentiation

5.1. Differential distribution of larval and adult myoblasts

It was expected that there should be two types of myoblasts (larval- and adult-type muscle stem cells) in trunk dorsal muscles because both larval and adult-type muscles coexist within the same regions. As a first indirect approach for proving the real existence of different two types of myogenic precursor cells in the same trunk muscle, Shimizu-Nishikawa et al. (2002) compared the enhancement of cell death activity in response to T_3 among three parts of the muscles, i.e., tadpole trunk, tail and limb muscles, each of which has a different muscle fate (Fig.7 A).

The results showed that the TUNEL$^+$ dying myoblasts were induced by T_3 strongly in the tail cells (10-fold induction; from 1.3% to 13%) and moderately in the trunk cells (2-fold; 2.5% to 5%), but not in the limb cells (1% to 1%). The value of cell death induction in trunk cells was between those of tail and limb cells, suggesting the possibility that two types of myogenic stem cells (i.e. T_3-inducible and non-inducible cells) are mixed in the trunk muscles. As the second approach, for further direct evidence, isolation of two types (larval and adult) of myoblasts from the trunk dorsal muscles was tried on the basis of their physical natures. The cells dissociated with enzyme-digestion from tadpole trunk dorsal muscles were pre-cultured for three days in high-serum (growth promoting) medium, harvested with trypsin-digestion and used for two-steps cell isolation with a percoll-density gradient centrifugation (a buoyant density-sensitive method) and an albumin-unit gravity sedimentation (a size-sensitive method)(Fig.7 B). The results showed that two different cell types, large (Lg) and small (Sm) cells, were isolated from dorsal muscle at 1:1 (Sm : Lg) ratio. For comparison, the result from the fractionation of the tail muscle cells with the same procedures showed that the two types of cells (Sm and Lg) were also obtained from the tail muscle but their ratio was 5 : 1 (Fig.7 C). Cultivation of these cells (Sm and Lg) revealed that Lg-cells could grow rapidly and T_3 decreased the number of Sm-cells but not that of Lg-cells (Fig.7 D). These results suggest that Lg-cells are the adult myoblasts and the Sm-cells are

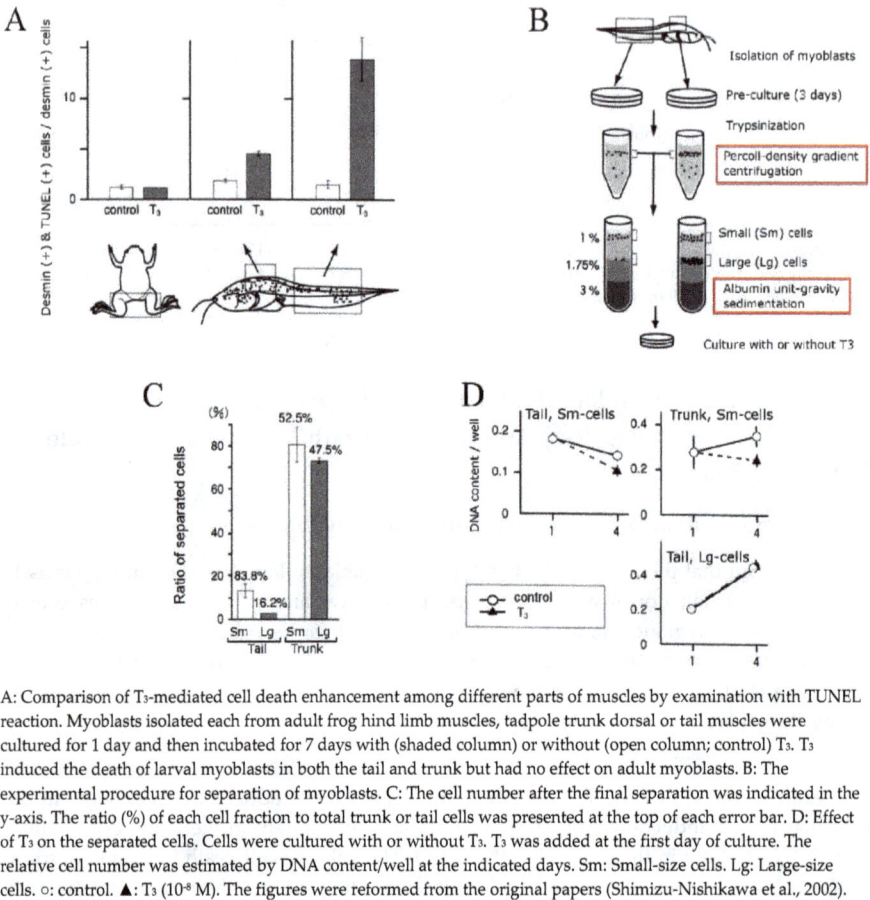

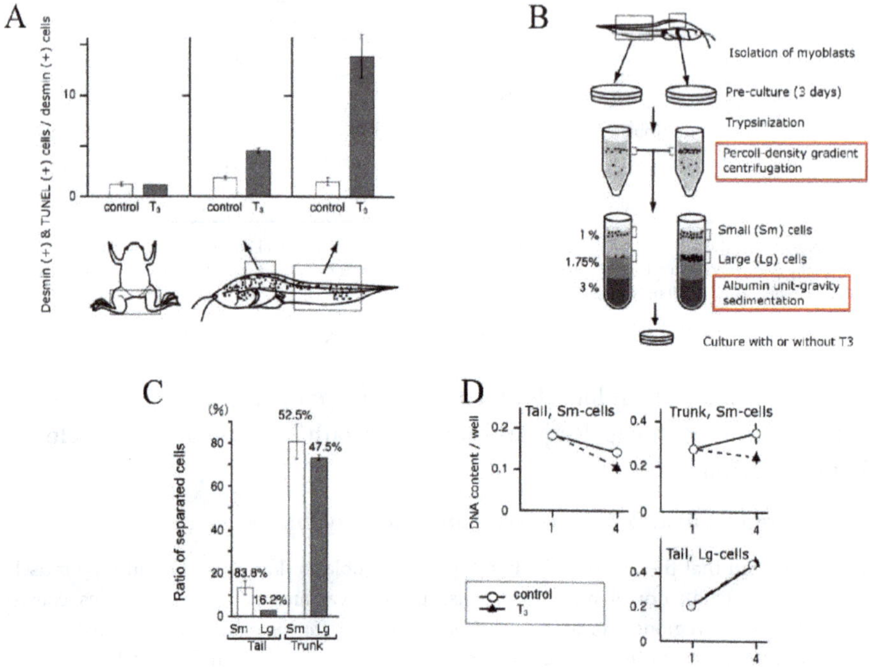

A: Comparison of T$_3$-mediated cell death enhancement among different parts of muscles by examination with TUNEL reaction. Myoblasts isolated each from adult frog hind limb muscles, tadpole trunk dorsal or tail muscles were cultured for 1 day and then incubated for 7 days with (shaded column) or without (open column; control) T$_3$. T$_3$ induced the death of larval myoblasts in both the tail and trunk but had no effect on adult myoblasts. B: The experimental procedure for separation of myoblasts. C: The cell number after the final separation was indicated in the y-axis. The ratio (%) of each cell fraction to total trunk or tail cells was presented at the top of each error bar. D: Effect of T$_3$ on the separated cells. Cells were cultured with or without T$_3$. T$_3$ was added at the first day of culture. The relative cell number was estimated by DNA content/well at the indicated days. Sm: Small-size cells. Lg: Large-size cells. ○: control. ▲: T$_3$ (10^{-8} M). The figures were reformed from the original papers (Shimizu-Nishikawa et al., 2002).

Figure 7. Cell type-dependent patterns of T$_3$-induced myoblast death.

larval myoblasts. Thus, it was shown that the trunk dorsal muscles contain almost the same number of two different (larval and adult) myoblasts. Interestingly, there were Lg-cells (adult type myoblasts) even in the tail muscle with a small ratio (1/5 of that in the dorsal muscle). In other words, there are many adult type muscle stem (AMS) cells with high growth activity in the dorsomedial (DM) portions of the trunk dorsal muscle but, on the other hands, a small number of AMS cells in the tail DM portions. These results suggest that the life-or-death fates of trunk and tail muscles are determined primarily by the differential distribution of adult myoblasts within the muscles. On the other hand, from the fact that tail muscle does not convert to adult type even though they contain AMS cells, it is conceivable that during metamorphosis the growth and differentiation of AMS cells is specifically activated in the trunk DM portions but suppressed in the tail DM portions. As to this spatial control, Yamane et al. (2011) suggested the possibility that the trunk-specific adult myogenesis is regulated by two cell-interactive mechanisms: a promotion by spinal cord (SP) cells and a suppression by notochord (Nc) cells (see section 6).

5.2. Interaction between adult and larval myoblasts

It is well known that myoblasts at first proliferate, then stop cell divisions and finally fuse among themselves to form multinuclear myotubes toward terminal differentiation. Unlike this, however, there is an exceptional uninuclear myotube formation without myoblast fusion during somitic myogenesis in *Xenopus* early embryonic development (Muntz, 1975, Boujelida and Muntz, 1987). In this uninuclear myotubes, surprisingly, the amitosis-like nuclear divisions occur twice to yield the multinuclear (4-nucleus) myotubes (Boujelida and Muntz, 1987). However, Muntz' observation (1975) showed that "satellite cells" (myogenic stem cells) occur after stage 40 and increase between stage 45 and 59, and the basement membrane becomes seen at stage 40. Furthermore, at stage 50-55, the muscle cells were shown to take on the appearance of adult skeletal muscle fibers, numerous small nuclei occupying peripheral patches of cytoplasm. From these observations, it is conceivable that the fusion of satellite cells with the myotomal muscle fibers at the onset of metamorphosis (stage 48-50) enable the further multinucleation of the tadpole trunk muscles (Muntz, 1975). Supporting this idea, the isolated myoblasts from *Xenopus* tadpole trunk muscles could actually fuse to themselves to form multinucleated myotubes in vitro (Shimizu-Nishikawa et al., 2002).

In the trunk dorsal muscles of *X. laevis* tadpoles, there are almost the same number of two kinds of different myoblasts (Fig.7 C). Then, do the two different types of myoblasts which either have a life or death fate fuse to form the heterokaryon myotubes with the "chimeric fates" during metamorphosis? If this is the case, how do the "chimeric myotubes" behave in response to T_3? Which fate do they choose death or life, under the T_3-influences? What mechanisms exist for the smooth transition from larval to adult myogenic system? From these points of view, Shimizu-Nishikawa et al. (2002) carried out the study about cell interactions between the larval and adult myogenic stem cells in *X. laevis* (Fig.8). In this study, at first, the myoblasts isolated from adult leg (adult type myoblasts) or larval tail muscles (larval type myoblasts) were implanted under the skin of tadpole tail (or dorsal region) in order to know whether each type (larval or adult) of myoblasts fuse preferentially with its own type rather than the other type (Fig.8 A). To distinguish two types (adult and larval type) of different myoblasts by means of the differential nuclear staining with quinacrine, the Xenograft system with two closely related *Xenopus* species was employed, and thus the myoblasts from *X. borealis* (intense and heterogeneous nuclear staining) were transplanted into *X. laevis* larvae (homogenous weak nuclear staining). The result showed that transplanted adult type myoblasts migrated into tadpole muscle area and fused with preexisting larval type muscles (or myoblasts) to form heterokaryon myotubes with both types of myonuclei. Thus, it was found that adult myoblasts was not committed to fuse with its own type (adult type) but could fuse also with different type (larval) myoblasts in vivo.

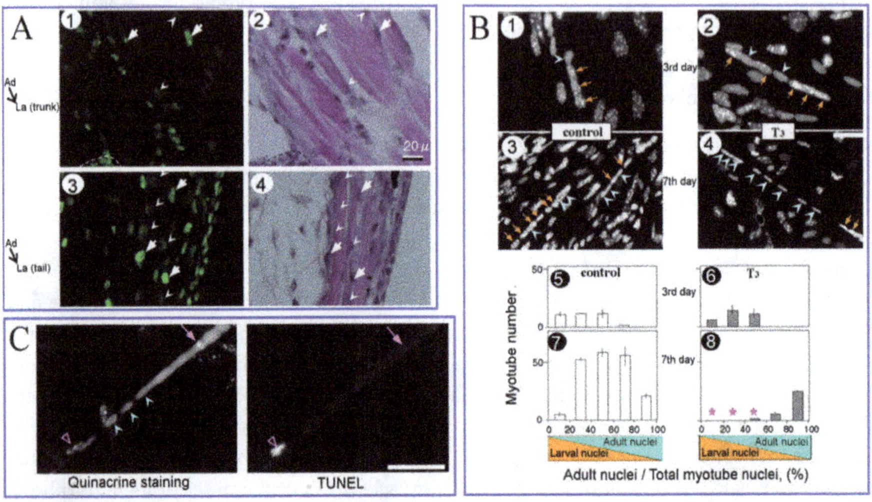

A: Formation of heterokaryon myotubes by cell transplantation. Cell aggregates (*Xenopus borealis* adult myoblasts) were transplanted subcutaneously into *X. laevis* (stage 48) trunk (1 and 2) or tail (3 and 4). Nine days after transplantation, muscle sections were made and stained with quinacrine (1 and 3) and then hematoxilin/eosin (2 and 4). The bar shows 20 μm. Nuclei of *X. borealis* show bright fluorescent spots (arrows) and can be distinguished from those of *X. laevis* (arrowheads). Dotted line (1) shows the transplanted cell aggregate. B: Heterokaryon formation in vitro and the T_3 response. Adult and larval myoblasts were co-cultured with (2 and 4: T_3) or without T_3 (1 and 3: control). 1 – 4: Fluorescent photomicrographs of the quinacrine stained nuclei of heterokaryon myotubes. The larval and adult nuclei were indicated by arrow and arrowheads, respectively. 1 and 2: Day 3. 3 and 4: Day7. The bar=50 μm. 5 – 8: Myotubes were classified by the proportion (%) of adult nuclei per total nuclei in each myotubes. The result was shown by histogram. Note that there are no heterokaryon myotubes dominated by the larval nuclei in T_3–cultures (asterisks). C: Larval nuclei do not die within the surviving heterokaryon myotubes even in the presence of T_3. Adult (*X. laevis*; arrowheads) and larval (*X. borealis*; arrows) myoblasts were co-cultured and T_3 was added 1 day after inoculation. Cells were fixed at 8th day and subjected to TUNEL reaction (left figure) and then stained with quinacrine (right figure). In adult nuclei-rich myotubes, the larval nuclei were TUNEL-negative (arrows). On the other hand, nuclei of a mononucleated larval myoblast (an open triangle) were TUNEL-positive. The bar=50 μm. The figures were reformed from the original papers (Shimizu-Nishikawa et al., 2002).

Figure 8. Formation of heterokaryon myotubes between adult and larval myoblasts.

Then, is there a possibility that the larval muscle fibers in tadpole are rescued from their T_3-mediated death by fusing with the adult type myoblasts and transform into adult muscle fibers? If this is the case, such cell-fate conversion may accelerate the speed of adult myogenesis because such tadpoles do not have to use much more energy for destroying a lot of differentiated larval muscles. So, does a fusion-mediated fate-change really occur? For the clarification of this point, it was examined using an in vitro co-culture system whether chimeric myotubes with both adult and larval myoblasts respond to T_3 to die or not (Fig.8 B). As the result, it was clarified that both larval and adult myoblasts randomly fuse to each other to make heterokaryon and the rescue from their T_3-mediated death occur only when the proportion of adult nuclei number was higher than 80 % within the myotube. Since the rescue from larval cell death thus requires incorporation of so many adult cells, the rescue of the trunk myotubes would occur at a very low rate and most of larval type cells would usually die during metamorphosis by the action of T_3. Interestingly, an apoptotic feature (DNA fragmentation) was not observed in any larval nuclei within the surviving heterokaryon myotubes (i.e. adult nuclei ratio $\geq$ 80%). This mean that the larval nuclei were protected from apoptotic death and their death fate was converted to a life fate (Fig.8 C). However, because a lot of adult cell fusion are needed for preexisting larval muscles to increase the adult-nuclear ratio up to 80%, it is reasonable in vivo situation that adult dorsal muscle conversion by the rescue of the larval myotubes seldom occurs or it occurs only at very few fibers in the anterior portion of body axis with high growth activity. Accordingly, it would never occur in the tail portion where the growth activity of myoblasts is very low (Fig.9). In essence, adult conversion of the trunk dorsal muscles is mainly carried out by the new myotubes formation rather than the old myotubes rescue.

Then, not involving the rescue mechanism, another mechanism which promotes the adult myoblasts differentiation should be needed in order to make efficiently the adult muscles in dorsal muscle region. In order to know whether such promotion of adult myogenesis involves some kinds of cell-cell interactions, experiments with "separated co-culture" of two types of (adult and larval) myoblasts were conducted (Shimizu-Nishikawa et al., 2002). In this experiment, adult (frog leg muscle) and larval (tadpole tail muscle) type myoblasts were separately inoculated in the two different areas in the same culture dish in order to avoid a direct adult-to-larval cell interaction ("separated co-culture") and their differentiation activity was compared with that in control cultures with either one of the two types by counting myotube and myotube-nuclei numbers within each areas. In this "separated co-culture" system, two types of myoblasts can communicate only through culture medium but through direct cell-to-cell interactions. The result clearly showed that differentiation of adult myoblasts into myotubes was promoted by larval myoblasts but that of larval myoblasts was not affected by adult cells (Fig.10 A).

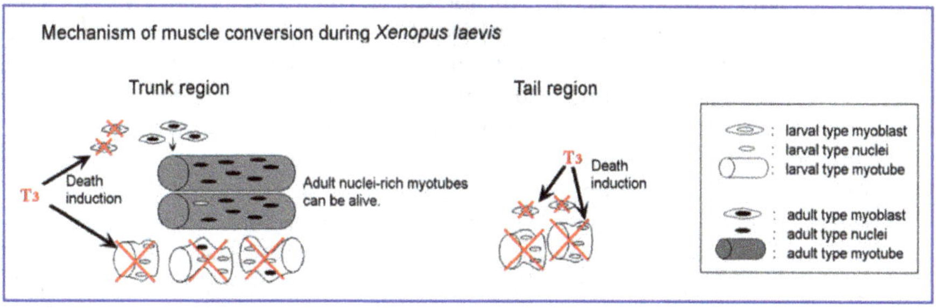

In trunk region, many adult myonuclei-rich myotubes (adult nuclei ratio< 80%) are formed and survive under the presence of T_3 while the pre-existed larval myonuclei-rich myotubes and larval myoblasts are to die by T_3. In tail region, however, adult myonuclei-rich myotubes are not formed and T_3 induces the death of larval myoblasts and larval myotubes. The figures were reformed from the original papers (Shimizu-Nishikawa et al., 2002).

Figure 9. A model for myotubes formation during *Xenopus laevis* metamorphosis.

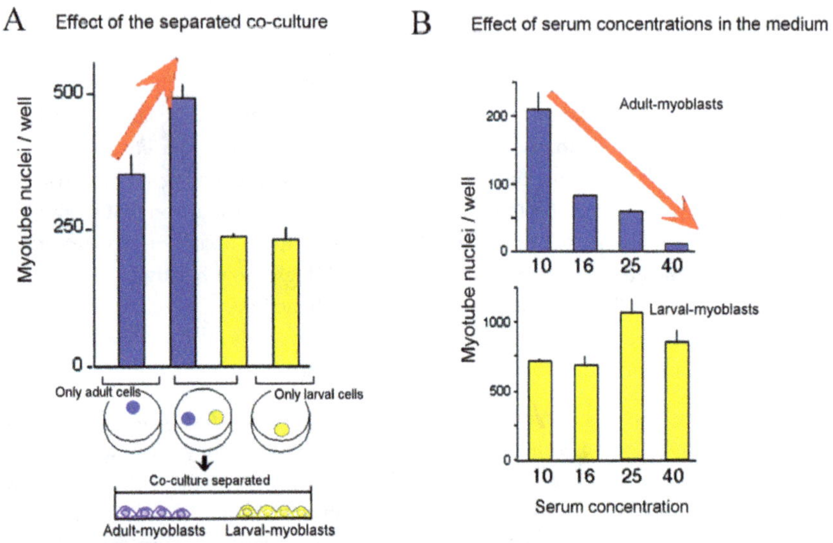

A: Effect of separated co-culture. Adult and larval myoblasts were cultured as in schematic drawings. Each type of myoblasts (3 x 10^4 cells) was inoculated into an area (5-mm diameter) in a culture plate. The number of nuclei in myotubes was analyzed at third day of culture. B: Effect of serum concentration on myoblast differentiation. Each adult and larval myoblast was cultured in culture medium with different concentration of serum. Differentiation (number of nuclei in myotubes) was analyzed at 10th (adult) or 8th (larval) day of culture. The figures were reformed from the original papers (Shimizu-Nishikawa et al., 2002).

Figure 10. Effects of separated co-culture and serum concentration on myoblast differentiation.

This effect should be caused by some humoral factors which released from larval myoblasts but not by a direct cell-to-cell contact, because it occurred at a certain distance in the "separated" areas. So, it was examined whether the activity which promotes adult differentiation was observed in conditioned medium (CM). The result indicated that the activity was found only in a larval myoblast CM but not in an adult myoblast CM, suggesting that larval cell secreted a factor(s) for adult muscle differentiation. This putative factor was found to be in the retentate (R) fraction with molecular weight (MW) more than 10,000 through ultra filtration (MW 10,000 cut-off). The ultra filtration also revealed the inverse activity that inhibits the adult myoblast differentiation in the R fraction of control culture medium, suggesting that control medium (maybe serum components in medium) intrinsically contains a factor(s) which antagonizes with the factor(s) in L-CM to inhibit the adult myoblast differentiation. Thus, in order to examine if such inhibitory factor is from serum components of the culture medium, each adult and larval myoblasts were cultured in various conditions with different fetal calf serum (FCS) concentrations and their differentiation (myotube formation) activities were measured (Fig.10 B). As a result, differentiation of adult myoblasts was found to be suppressed dose-dependently by FCS but that of larval cells not to be affected. Taken from these results, it is conceivable that adult differentiation promoting factor(s) being released from larval cells functions through antagonistic regulation of the adult differentiation-inhibitory factor(s) in the control medium (i.e. serum). It was suggested from a work with mouse myogenic cells (Cusella-DeAngelis et al., 1994) that the differentiation inhibitory factor(s) in serum could possibly be some kind of molecules related to TGF-β. Because, interestingly, TGF-β dose-dependently suppressed the differentiation of mouse fetal (but not embryonic) myoblasts in the same way as FCS dose-dependently suppressed that of *Xenopus* adult (but not larval) myoblasts. It still remains as an important question to be solved whether modulation of TGF-β activity is responsible for the specific promotion of adult myoblast differentiation in dorsal muscles during *Xenopus* metamorphosis.

In summary, it was found for the first time that the *Xenopus* larval myoblasts that have a death-fate under T_3 accept not only their death but also a big role as a cooperator for the adult myoblast differentiation via a regulation of humoral environment. The role of larval cells was thus found to be very important, however, we would not be able to fully understand the regulatory mechanism for the trunk-specific adult muscle differentiation only by analysis of the cell-cell interactions between larval and adult myoblasts. This is because, there are two types of myoblasts (adult and larval) not only in the trunk but also in the tail muscles. Then, is there any regulation from cells (or tissues) other than myoblasts (or muscles) for the trunk-specific adult myogenesis? From this point of view, Yamane et al. (2011) reported the importance of interaction between myogenic cells and non-myogenic cells (i.e. notochord and spinal cord cells) for the trunk specific adult muscle differentiation during *X. laevis* metamorphosis. This study is described in detail in the next section (section 6).

6. Interaction between adult myogenic precursor cells and axial cells

The adult muscle differentiation occurs in the trunk (but not the tail) during *Xenopus laevis* metamorphosis. Then, Yanane et al. (2011) focused on the two major tissues, the notochord and the spinal cord, which adjacent to the axial muscles (i.e. dorsal and tail muscles) in order to assess the involvement from the neighboring cells in the region-specific adult muscle differentiation. Because, both two tissues (notochord and spinal cord) are known to regulate early axial myogenesis in vitro (Munsterberg and Lassar, 1995; Stern and Hauschka, 1995) and in vivo (Blagden et al., 1997). At first, Yamane et al. (2011) examined the difference in cross-sectional areas between trunk and tail regions using histological sections of *Xenopus* tadpoles and noticed that the cross-sectional ratio of spinal cord to notochord area (SC/Nc ratio) is nearly 1:1 in the trunk but about 1:15 in the tail (Fig.11). This big difference is due to the situation that notochord area in the tail region is about 1.5 times larger than that in the trunk region and, on the contrary, spinal cord area in the tail is extremely smaller (around 1/10) than that in the trunk part. From this observation, it is conceivable that the influence of the spinal cord on myogenesis is greater in the trunk than in the tail and, in contrast, the influence of the notochord is stronger in the tail than in the trunk.

Therefore, it is reasonable to hypothesize for the trunk-specific adult muscle differentiation that adult myoblast differentiation is promoted by the spinal cord but suppressed by the notochord. So, according to this hypothesis, each of adult (from hindlimb) and larval (from tail) myoblasts was co-cultured with the spinal cord (or notochord) cells so as to compare their responses to two axial cells (i.e. spinal cord and notochord cells)(Fig.12). The result clarified the expected opposite roles of the two axial cells: The spinal cord cells increased twice the myotubes-forming activity of adult myoblasts but did not increase that of larval cells. On the other hands, the notochord cells strongly suppressed the myotube-formation by adult myoblasts but did not suppress that by larval cells (Fig.12 G-L). Thus, there is a high possibility that two contrasting mechanisms, i.e. the "spinal cord (SC)-promotion" and the "notochord (Nc)-suppression" on adult myogenesis, are involved in the trunk-specific adult muscle conversion (Fig.13).

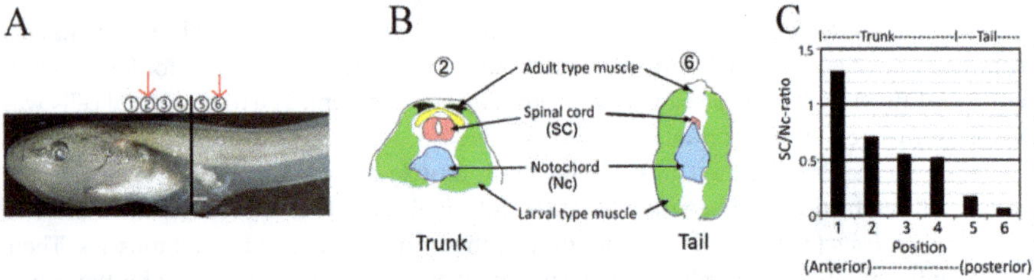

A: Side view of stage-55 tadpole.. The numbers ①-⑥ show the position of cross-sections analyzed. B: A schematic illustration of the cross-sections at positions ② and ⑥ (red arrows). C: The ratio of the spinal cord area to the notochord area (SC/Nc ratio) was shown in the ordinate and the abscissa showed the position along the anterior-posterior axis. The figures were reformed from the original papers (Yamane et al., 2011).

Figure 11. Cross-sectional proportion of spinal cord to notochord in *Xenopus laevis* tadpoles.

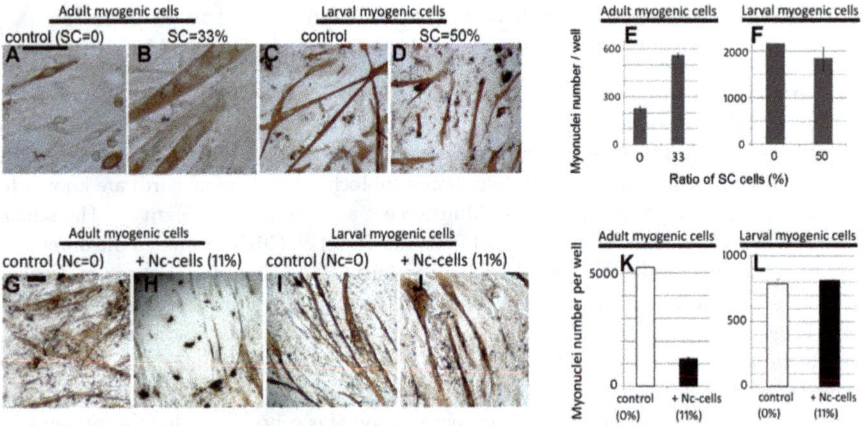

A-F: Adult (A, B and E) and larval (C, D and F) myogenic cells (1.6 x 10^4 cells) were co-cultured with (B and D) or without (A and C) SC cells (33%=8 x 103 cells or 50%=1.6 x 103 cells, respectively) for 7 days. Fixed cells were immunostained with antibodies for myogenic markers (brown color). The myo-nuclei number/well was measured and shown in E and F. G-L: Adult (G, H and K) and larval (I, J and L) myogenic cells (1.6 x 10^4 cells) were co-cultured with (H and J) or without (G and I) Nc cells (11%=2 x 10^3 cells) for 7 days. Fixed cells were immunostained with antibodies for myogenic markers (brown color). The myo-nuclei number/well was measured and shown in K and L. The bars in A and G show 50 μm. The figures were reformed from the original papers (Yamane et al., 2011).

Figure 12. Effect of spinal cord (SC) cells and notochord (Nc) cells on adult and larval myogenic cells in vitro.

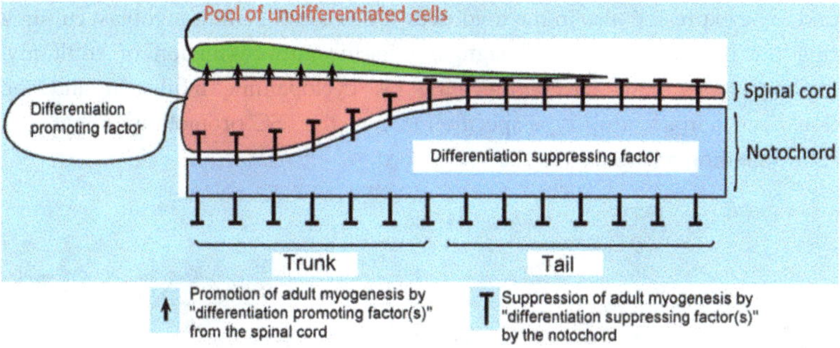

In trunk region, adult myogenesis is promoted by a "differentiation promoting factor(s)" from the spinal cord while, in the tail region, adult myogenesis is suppressed by a "differentiation suppressing factor(s)" from the notochord.

Figure 13. A model for the trunk-specific adult muscle differentiation during *Xenopus laevis* metamorphosis.

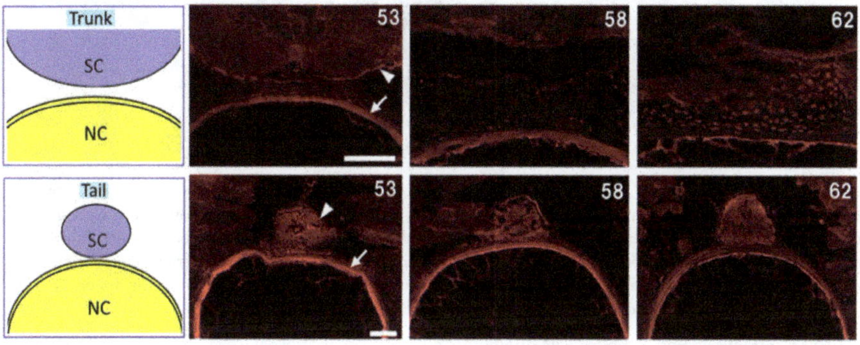

Shh expression was examined by immunostaining of tadpole sections. Left panels show the schematic illustrations of cross-sections of trunk and tail regions. SC: spinal cord. Nc: notochord. The other panels show the immunofluorescent photomicrographs with anti-Shh antibody (H160) of frozen section of tadpoles. The numbers show the stages of tadpoles. Upper panels: trunk region. Lower panels: tail region. The arrows and arrowheads show the position of notochord sheath and spinal cord, respectively. The bars show 100 μm. The figures were reformed from the original paper (Yamane et al., 2011)

Figure 14. Sonic hedgehog (Shh) distribution in notochord and spinal cord of the trunk and tail regions during *Xenopus laevis* metamorphosis.

As to the former mechanism, i.e. the "SC-promotion", the involvement of sonic hedgehog (Shh) signaling is expected, because the spinal cord and notochord expresses Shh and positively regulates the early embryonic myogenesis (Munsterberg and Lassar, 1995, Stern and Hauschka, 1995, Blagden et al., 1997). In fact, the analysis with antibody staining revealed that the N-terminal fragment (active form) of Shh proteins is present in the spinal cord and notochord regions throughout metamorphosis of *X. laevis* (Fig.14). And further analysis with a Shh inhibitor (cyclopamine) revealed that cyclopamine suppress the adult myoblast differentiation but not affect the larval one (Fig.15). In addition, the Shh protein was found to be expressed also in the myotubes formed in the adult myoblast cultures, thus suggesting the involvement of Shh signaling in the auto-regulation of adult myoblast differentiation. Since both the two events, i.e. cyclopamine inhibition and the "SC promotion", occur in an adult type-specific manner, the "SC promotion" might involve a kind of Shh signaling released from the spinal cord.

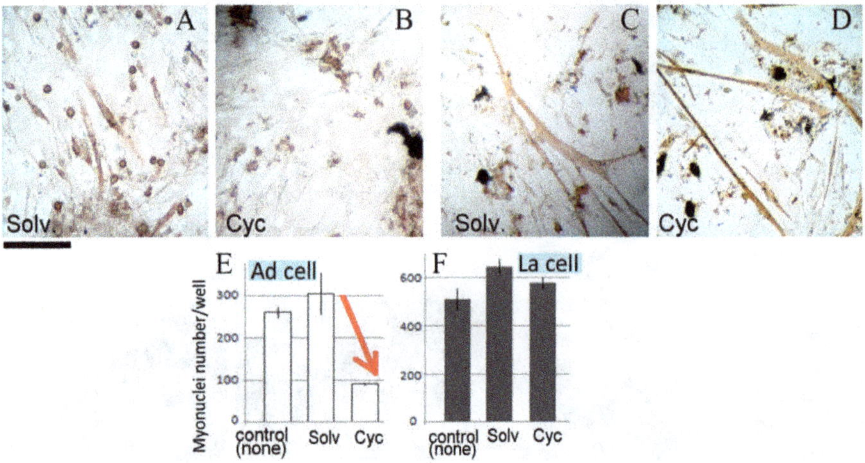

Adult (A, B and E) and larval (C, D and F) myoblasts were cultured for 6 days in the presence (B and D) or absence (A and C) of cyclopamine (1 μM). A-D: Photomicrographs of immunostained (brown color) cells at 6th day with an antibody mixture (anti-desmin and anti-myosin antibodies). The bar in A=100 μm. E and F: The numbers of nuclei in the antibody+ myotubes were counted and indicated in the ordinate. Mean values of two determinations (two wells) were shown and the bar indicate the range of the two. Solv: solvent without cyclopamone. Cyc: cyclopamine. The figures were reformed from the original paper (Yamane et al., 2011).

Figure 15. Effect of a Shh inhibitor, cyclopamine, on the differentiation of *Xenopus* adult and larval myogenic cells.

On the other hands, as to the "Nc suppression", the causing factor(s) remains to be unknown. However, another interesting feature of the notochord cells was found by a "separated co-culture" experiment. In the "separated co-culture", when the notochord cells and adult myoblasts were placed each in separated two areas on the same culture dish in order to avoid direct cell-cell interaction between them, the notochord cells lost their ability to suppress adult myogenesis but rather promoted the adult myoblast differentiation (Fig.16). Interestingly, the same effect (promotion of adult myogenesis) was also observed in a "separated co-culture" with a whole notochord tissue instead of isolated notochord cells. Thus, it was found that notochord cells have a long-distance promotive effect for adult myogenesis (i.e. the "notochord promotion") other than the "notochord suppression" effect appearing at a short distance. Since the notochord also expresses Shh throughout metamorphosis (Fig.14) and Shh is also known to positively regulate early embryonic myogenesis (Munsterberg and Lassar, 1995), this molecule is possibly be a major candidate molecule for the "notochord promotion".

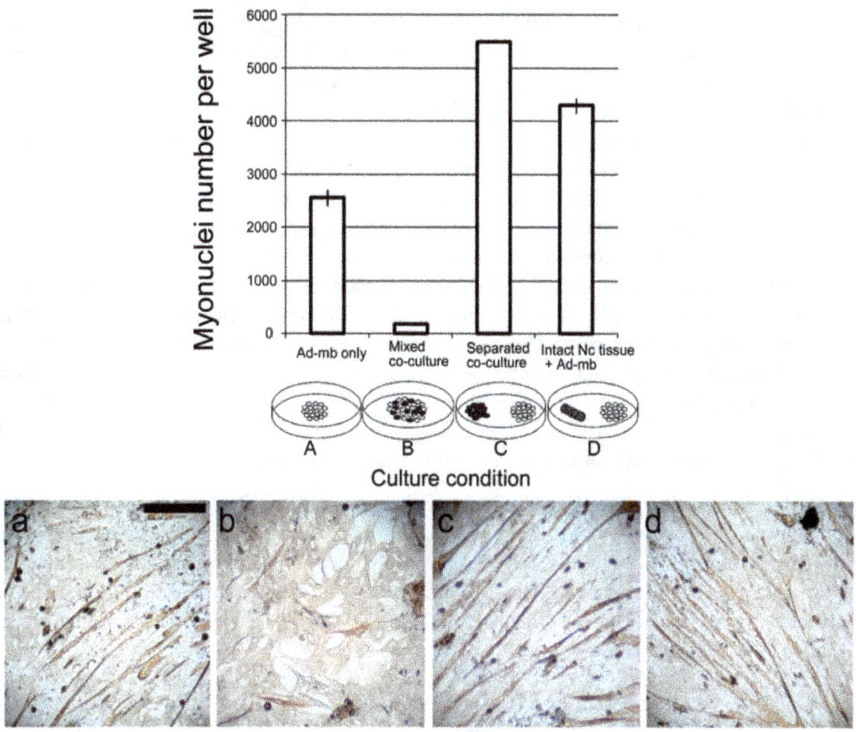

Ad-mbs were cultured for 7 days in various conditions (A-D). A: Ad-mbs only. B: Ad-mbs and Nc-cells were co-cultured within the same area of a culture dish. C: Ad-mbs and Nc-cells were separately placed in two different area of the same culture dish and cultured within the same medium. D: Separated co-culture with Ad-mbs and an intact Nc tissue. After 7 days, each culture was fixed, stained with an antibody mixture (anti-desmin and anti-myosin antibodies) and photographed (a-d for conditions A-D, respectively). The bar=200 μm. The nuclei numbers in antibody+ myogenic cells (i. e. myotubes and myoblasts) per dish were determined. The mean value for the two determinations was shown with the range of the two (vertical bar). The figures were reformed from the original paper (Yamane et al., 2011).

Figure 16. Notochord suppression on adult differentiation requires a direct cell-to-cell interaction between notochord (Nc) cells and adult myoblasts (Ad-mbs).

As described above, the multiple cell-to-cell interactions coordinately regulate in diverse ways the trunk-specific promotion and the tail-specific suppression of adult myogenesis during metamorphosis. The molecular features of such cell interactions have not fully characterized. Especially, it should be primarily emphasized as an important future work to get insight into the molecular mechanism for the "notochord suppression". Hebrok et al. (1998) reported with chick embryo that factors from notochord, such as fibroblast growth factors (FGF) and activin, suppress the prepancreatic dorsal endoderm Shh expression and thereby permit early pancreatic development. Such a FGF (or activin)-like molecule(s) might be a factor(s) responsible for the "notochord suppression". Further investigations are needed for clarifying this question.

Secondary, it is also very important issue to clarify the signaling cascade for the "SC promotion" of adult myogenesis. The "SC promotion" shows adult-specific effect on the muscle differentiation and Shh is expected to be a responsible factor for this phenomenon. As detailed above (section 4), adult myoblasts proliferate many times as undifferentiated stem cells and then after few days stop cell divisions to transit to the differentiation steps (Shibota et al., 2000). Therefore, it is reasonable that the "SC promotion" is involved in the transition step from the undifferentiated stemness stage of the adult cells to the more committed differentiation stage. Borycki et al. (1999) reported that Shh, produced by the notochord and floor plate, control epaxial muscle determination through *myf-5* activation in the mouse embryo. Gustafsson et al. (2002) also reported that *myf-5* is a direct target of long-range Shh signaling through Gli transcription factors for the specification of mouse somitic epaxial muscle progenitor cells. Therefore, also in case of *X. laevis* "SC promotion", the factor(s) (possibly Shh-related molecules) may act on commitment stages (determination and/or specification steps) of undifferentiated stem cells. If the "SC promotion" factor is indeed Shh, *myf-5* upregulation may possibly be enhanced in the adult-cells co-cultured with SC cells. These points should be clarified in the near future.

The clarification of the cell-cell interaction mechanisms and their molecular cascades for the adult muscle differentiation during *Xenopus laevis* metamorphosis would finally contribute to the accumulation of technical knowledge about how to manipulate and maintain the mammalian embryonic stem (ES) cells (Martin, 1981; Evance and Kaufman, 1981) or induced pluripotent stem (iPS) cells (Takahashi and Yamanaka, 2006) for their induction toward complete (adult-type) differentiation.

7. Conclusion

Nishikawa and Hayashi firstly examined the larval-to adult-muscle isoform transition and put forward a new model, the "cell replacement model", that clearly explains the larval-to-adult myogenic conversion during frog metamorphosis. In this model, larval-to-adult conversion of tadpole dorsal muscles was achieved through the cell replacement by both death of larval-type myogenic cells and proliferation and differentiation of adult-type myogenic cells. The death of tadpole trunk dorsal muscles was found to occur through apoptotic processes including nucleosomal DNA-fragmentation, apoptotic body formation and phagocytosis by macrophages.

In subsequent research, larval- and adult-type myogenic precursor cells (myoblasts) were isolated each from *Xenopus laevis* tadpole tail or hindlimb leg muscles and cultured in vitro to see their basic nature, developmental fates, i.e., to die or not to die, and hormonal responses. It was thus clearly demonstrated that larval-type myoblasts are to die in the presence of T_3, while adult-type myoblasts can proliferate and differentiate into myotubes in the same hormonal condition. More interestingly, cell separation using different two methods showed that not only trunk muscles but also tail muscles contain both larval- and adult-type myoblasts. In order to see cell interaction between adult- and larval-type myogenic cells, heterokaryon analyses were conducted. The results revealed the possibility of the rescue of nuclei-death of larval type cells by adult cell nuclei in the syncytium and the promoting effect by a larval cell-releasing factor(s) for the adult myoblast differentiation.

Finally, co-culture system using myogenic cells (larval and adult myoblasts) and non-myogenic axial cells (notochord and spinal cord cells) was developed to examine how adult myogenesis is promoted in the trunk muscle region but suppressed in the tail muscle region through the interactions between myogenic and non-myogenic cell or environmental signals. The results revealed the suppression of adult myogenesis by notochord cells ("notochord suppression"), promotion of adult myogenesis by spinal cord cells ("spinal cord promotion") and upregulation of adult myogenesis by sonic hedgehog (Shh)-signaling. These results present a model for the region-specific regulatory mechanism of adult myogenesis by cell-cell interactions, i.e., "spinal cord promotion" and "notochord suppression", during *X. laevis* metamorphosis.

Author details

Akio Nishikawa
Shimane University, Japan

8. References

Blagden, C. S., Currie, P. D., Ingham, P. W., & Hughes, S. M. (1997). Notochord induction of zebrafish slow muscle mediated by Sonic hedgehog. *Genes & Dev.*, 11, 2163-2175.

Borycki, A. G., Brunk, B., Tajbakhsh, S., Buckingham, M., Chiang, C., & Emerson, C. P. Jr. (1999). Sonic hedgehog controls epaxial muscle determination through Myf5 activation. *Development*, 126, 4053-4063.

Boudjelida, H., & Muntz, L. (1987). Multinucleation during myogenesis of the myotome of *Xenopus laevis*: a qualitative study. *Development*, 101, 583-590.

Brown, D. D., Cai, L., Das, B., Marsh-Armstrong, N., Schreiber, A. M., & Juste, R. (2005). Thyroid hormone controls multiple independent programs required for limb development in *Xenopus laevis* metamorphosis. *Proc. Natl. Acad. Sci. U.S.A.*, 102, 12455-12458.

Callery, E. M. & Elinson, R. P. (2000). Thyroid hormone-dependent metamorphosis in a direct developing frog. *Proc. Natl. Acad. Sci. U.S.A.*, 97, 2615-2620.

Callery, E. M. & Fang, H. & Elinson, R. P. (2001). Frogs without polliwogs: Evolution of anuran direct development. *BioEssays*, 23, 233-241.

Chino, Y., Saito, M., Yamasu, K., Suyemitsu, T., & Ishihara, K. (1994). Formation of the adult rudiment of sea urchins is influenced by thyroid hormones. *Dev. Biol.*, 161, 1-11.

Cusella-De Angelis, M. G., Molinari, S., Le Donne, A., Coletta, M., Vivarelli, E., Bouche, M., Molinaro, M., Ferrari, S., & Cossu, G. (1994). Differential response of embryonic and fetal myoblasts to TGF beta: a possible regulatory mechanism of skeletal muscle histogenesis. *Development*, 120, 925-933.

Das, B., Schreiber, A. M., Huang, H., & Brown, D. D. (2002). Multiple thyroid hormone-induced muscle growth and death programs during metamorphosis in *Xenopus laevis*. *Proc. Natl. Acad. Sci. U.S.A.*, 99, 12230-12235.

de Luze, A., Sachs, L., & Demeneix, B. (1993). Thyroid hormone-dependent transcriptional regulation of exogenous genes transferred into *Xenopus* tadpole muscle in vivo. *Proc. Natl. Acad. Sci. U.S.A.*, 90, 7322-7326.

Du Pasquier, D., Rincheval, V., Sinzelle, L., Chesneau, A., Ballagny, C., Sachs, L. M., Demeneix, B., & Mazabraud, A. (2006). Developmental cell death during *Xenopus* metamorphosis involves BID cleavage and caspase 2 and 8 activation. *Dev. Dyn.*, 235, 2083-2094.

Elinson, R. P., Remo, B., & Brown, D. D. (1999). Novel structural elements identified during tail resorption in *Xenopus laevis* metamorphosis: lessons from tailed frogs. *Dev. Biol.*, 215, 243-252.

Enari, M., Sakahira, H., Yokoyama, H., Okawa, K., Iwamatsu, A., & Nagata, S. (1998). A caspase-activated DNase that degrades DNA during apoptosis, and its inhibitor ICAD. *Nature*, 391, 43-50.

Evans, M. J., & Kaufman, M. H. (1981). Establishment in culture of pluripotential cells from mouse embryos. *Nature*, 292, 154-156.

Grimaldi, A., Tettamanti, G., Martin, B. L., Gaffield, W., Pownall, M. E., & Hughes, S. M. (2004). Hedgehog regulation of superficial slow muscle fibres in *Xenopus* and the evolution of tetrapod trunk myogenesis. *Development*, 131, 3249-3262.

Gudernatch, J. F. (1912). Feeding experiments on tadpoles. I. The influence of specific organs given as food on growth and differentiation. A contribution to the knowledge of organs with internal secretion. *Arch. Entwicklungsmech*, 35, 457-483.

Gustafsson, M. K., Pan, H., Pinney, D. F., Liu, Y., Lewandowski, A., Epstein, D. J., & Emerson, C. P. Jr. (2002). Myf5 is a direct target of long-range Shh signaling and Gli regulation for muscle specification. *Genes Dev.*, 16, 114-126.

Hebrok, M., Kim, S. K., & Melton, D. A. (1998). Notochord repression of endodermal Sonic hedgehog permits pancreas development. *Genes Dev.*, 12, 1705-1713.

Ishizuya-Oka, A., & Shi, Y. B. (2007). Regulation of adult intestinal epithelial stem cell development by thyroid hormone during *Xenopus laevis* metamorphosis. *Dev. Dyn.*, 236, 3358-3368.

Kaltenbach, J. C. (1953). Local action of thyroxin in amphibian metamorphosis. I. Local metamorphosis in R. pipiens larvae effected by thyroxin-cholesterol implants. *J. Exp. Zool.*, 122, 21-40.

Kaufmann, S. H., Mesner, P. W.,Jr, Samejima, K., Tone, S., & Earnshaw, W. C. (2000). Detection of DNA cleavage in apoptotic cells. *Methods in Enzymol.*, 322, 3-15.

Kawakami, K., Kuroda, M., & Nishikawa, A. (2009). Regulation of desmin expression in adult-type myogenesis and muscle maturation during *Xenopus laevis* metamorphosis. *Zool. Science*, 26, 389-397.

Kerr, J. F., Harmon, B., & Searle, J. (1974). An electron-microscope study of cell deletion in the anuran tadpole tail during spontaneous metamorphosis with special reference to apoptosis of striated muscle fibers. *J. Cell Sci.*, 14, 571-585.

Kitajima, C., Sato, M. & Kawanishi, M. (1967). On the effect of thyroxine to promote the metamorphosis of a conger eel-preliminary report. *Nippon Suisan Gakkaishi* (Japanese Edition), 33, 919-922.

Krieser, R. J., & Eastman, A. (1998). The cloning and expression of human deoxyribonuclease II. A possible role in apoptosis. *J. Biol. Chem.*, 273, 30909-30914.

Liu, X., Zou, H., Slaughter, C., & Wang, X. (1997). DFF, a heterodimeric protein that functions downstream of caspase-3 to trigger DNA fragmentation during apoptosis. *Cell,* 89, 175-184.

Los, M., Neubuser, D., Coy, J. F., Mozoluk, M., Poustka, A., & Schulze-Osthoff, K. (2000). Functional characterization of DNase X, a novel endonuclease expressed in muscle cells. *Biochemistry,* 39, 7365-7373.

Martin, G. R. (1981). Isolation of a pluripotent cell line from early mouse embryos cultured in medium conditioned by teratocarcinoma stem cells. *Proc. Natl. Acad. Sci. U.S.A.,* 78, 7634-7638.

Miller, L. (1996). Hormone induced changes in keratin gene expression during amphibian metamorphosis. In: *Metamorphosis: Post-Embryonic Reprogramming of Gene Expression in Amphibian and Insect Cells,* L.I. Gilbert, B.G. Atkinson, and J.R. Tata (Ed.), pp. 599-624, Academic Press, ISBN 0-12-283245-0, San Diego, California.

Miwa, S., & Inui, Y. (1987). Effects of various doses of thyroxine and triiodothyronine on the metamorphosis of flounder (Paralichthys olivaceus). *Gen. Comp. Endocrinol.,* 67, 356-363.

Munsterberg, A. E., & Lassar, A. B. (1995). Combinatorial signals from the neural tube, floor plate and notochord induce myogenic bHLH gene expression in the somite. *Development,* 121, 651-660.

Muntz, L. (1975). Myogenesis in the trunk and leg during development of the tadpole of *Xenopus laevis* (Daudin 1802). *J Emb. Exp. Morphol.,* 33, 757-774.

Nishikawa, A. (1997). Induction of cell differentiation and programmed cell death in amphibian metamorphosis. Human Cell, 10, 167-174.

Nishikawa, A., & Hayashi, H. (1994). Isoform transition of contractile proteins related to muscle remodeling with an axial gradient during metamorphosis in *Xenopus laevis. Dev. Biol.,* 165, 86-94.

Nishikawa, A., & Hayashi, H. (1995). Spatial, temporal and hormonal regulation of programmed muscle cell death during metamorphosis of the frog *Xenopus laevis. Differentiation,* 59, 207-214.

Nishikawa, A., Murata, E., Akita, M., Kaneko, K., Moriya, O., Tomita, M., & Hayashi, H. (1998). Roles of macrophages in programmed cell death and remodeling of tail and body muscle of *Xenopus laevis* during metamorphosis. *Histochem. Cell Biol.,* 109, 11-17.

Nishikawa, A., Shiokawa, D., Umemori, K., Hayashi, H. & Tanuma, S. (1997). Occurrence of DNase gamma-like apoptotic endonucleases in hematopoietic cells in *Xenopus laevis* and their relation to metamorphosis. *Biochem. Biophys. Res. Comm.,* 231, 305-308.

Parrish, J., Li, L., Klotz, K., Ledwich, D., Wang, X., & Xue, D. (2001). Mitochondrial endonuclease G is important for apoptosis in *C. elegans. Nature,* 412, 90-94.

Rauch, F., Polzar, B., Stephan, H., Zanotti, S., Paddenberg, R., & Mannherz, H. G. (1997). Androgen ablation leads to an upregulation and intranuclear accumulation of deoxyribonuclease I in rat prostate epithelial cells paralleling their apoptotic elimination. *J. Cell Biol.,* 137, 909-923.

Rowe, I., Le Blay, K., Du Pasquier, D., Palmier, K., Levi, G., Demeneix, B., & Coen, L. (2005). Apoptosis of tail muscle during amphibian metamorphosis involves a *caspase 9*-dependent mechanism. *Dev. Dyn.,* 233, 76-87.

Ryke, P. A. J. (1953). The ontogenetic development of the somatic musculature of the trunk of the aglossal anuran *Xenopus laevis* (Daudin). *Acta Zool.*, 34, 1-70.

Sachs, L. M., Abdallah, B., Hassan, A., Levi, G., De Luze, A., Reed, J. C., & Demeneix, B. A. (1997). Apoptosis in *Xenopus* tadpole tail muscles involves Bax-dependent pathways. *FASEB journal*, 11, 801-808.

Sachs, L. M., Le Mevel, S., & Demeneix, B. A. (2004). Implication of *bax* in *Xenopus laevis* tail regression at metamorphosis. *Dev. Dyn.*, 231, 671-682.

Shibota, Y., Kaneko, Y., Kuroda, M., & Nishikawa, A. (2000). Larval-to-adult conversion of a myogenic system in the frog, *Xenopus laevis*, by larval-type myoblast-specific control of cell division, cell differentiation, and programmed cell death by triiodo-L-thyronine. *Differentiation*, 66, 227-238.

Shimizu-Nishikawa, K., Shibota, Y., Takei, A., Kuroda, M., & Nishikawa, A. (2002). Regulation of specific developmental fates of larval- and adult-type muscles during metamorphosis of the frog *Xenopus*. *Dev. Biol.*, 251, 91-104.

Shiokawa, D., Hatanaka, T., Hatanaka, M., Shika, Y., Nishikawa, A., & Tanuma, S. (2006). cDNA cloning and functional characterization of *Xenopus laevis* DNase gamma. *Apoptosis*, 11, 555-562.

Shiokawa, D., Hirai, M., & Tanuma, S. (1998). cDNA cloning of human DNase gamma: chromosomal localization of its gene and enzymatic properties of recombinant protein. *Apoptosis*, 3, 89-95.

Shiokawa, D., Kobayashi, T., & Tanuma, S. (2002). Involvement of DNase gamma in apoptosis associated with myogenic differentiation of C2C12 cells. *J. Biol. Chem.*, 277, 31031-31037.

Shiokawa, D., Ohyama, H., Yamada, T., Takahashi, K., & Tanuma, S. (1994). Identification of an endonuclease responsible for apoptosis in rat thymocytes. *Eur. J. Biochem.*, 226, 23-30.

Shiokawa, D., & Tanuma, S. (2001). Characterization of human DNase I family endonucleases and activation of DNase gamma during apoptosis. *Biochemistry*, 40, 143-152.

Shiokawa, D. & Tanuma, S. (2004). Differential DNases are selectively used in neuronal apoptosis depending on the differentiation state. *Cell Death Diff.*, 11, 1112-1120.

Sonnet, C., Lafuste, P., Arnold, L., Brigitte, M., Poron, F., Authier, F. J., Chretien, F., Gherardi, R. K., & Chazaud, B. (2006). Human macrophages rescue myoblasts and myotubes from apoptosis through a set of adhesion molecular systems. *J. Cell Sci..*, 119, 2497-2507.

Stern, H. M., & Hauschka, S. D. (1995). Neural tube and notochord promote in vitro myogenesis in single somite explants. *Dev. Biol.*, 167, 87-103.

Takahashi, K., & Yamanaka, S. (2006). Induction of pluripotent stem cells from mouse embryonic and adult fibroblast cultures by defined factors. *Cell*, 126, 663-676.

Tomita, M., & Ishikawa, H. (1992). Identification of novel adhesion proteins in mouse peritoneal macrophages. *Biology of the Cell*, 76, 103-109.

Yamane, H., Ihara, S., Kuroda, M., & Nishikawa, A. (2011). Adult-type myogenesis of the frog *Xenopus laevis* specifically suppressed by notochord cells but promoted by spinal cord cells in vitro. *In Vitro Cellular & Dev. Biol. Anim.*, 47, 470-483.

Yaoita, Y., & Brown, D. D. (1990). A correlation of thyroid hormone receptor gene expression with amphibian metamorphosis. *Genes Dev.*, 4, 1917-1924.

Yaoita, Y., & Nakajima, K. (1997). Induction of apoptosis and CPP32 expression by thyroid hormone in a myoblastic cell line derived from tadpole tail. *J. Biol. Chem.*, 272, 5122-5127.

Section 2

Immune Cell Interactions

Chapter 3

Dendritic Cells Interactions with the Immune System – Implications for Vaccine Development

Aiala Salvador, Manoli Igartua, José Luis Pedraz and Rosa María Hernández

Additional information is available at the end of the chapter

1. Introduction

Dendritic cells (DCs) were firstly described by Ralph Steinman and Zanvil Cohn in 1973 [1]. These cells were identified in the spleen of mice and possessed characteristics that were used to name them; they presented uniform pseudopods, like dendrites, so they were called dendritic cells. The advances in the research regarding these cells since then have situated them as key cells that coordinate immune responses. They are distributed along the whole body but especially prevalent in peripheral tissues where antigen capturing might occur. Immature DCs play a central role in activating naïve T cells and directing the subsequent immune response towards a T helper 1 (Th1), Th2, Th17 or T regulatory (Treg) profile [2,3]. Thus, they are the main regulators of the subsequent reaction, producing the optimal response against a given antigen and developing immunity or tolerance.

Despite the increasing research and knowledge acquired in the last years, there are not effective vaccines available against certain pathogens or diseases such as malaria, HIV, hepatitis C, tuberculosis or cancer. These pathogens are intracellular, requiring the induction of strong cellular immunity, including cytotoxic responses (CTL), to remove the infected cells. The development of antibody responses can be stimulated by traditional adjuvants such as alum [4]. However, most of the currently licensed adjuvants lack the ability for inducing cellular or mixed immune responses.

The unique features of the DCs make them ideal target cells for vaccines [5]. They appear in their immature state in the peripheral tissues and once they capture an antigen, they are able to mature and become antigen-presenting cells (APCs) at the same time as they migrate to the lymph nodes. There, they are able to present antigens loaded to major histocompatibility complex (MHC) classes I and II molecules to T cells, glycolipids and glycopeptides to T cells and NKT cells as well as polypeptides to B cells [6]. In addition, it is now apparent that adjuvants are the activators of DCs. For these reasons, a better understanding of the interactions of DCs with the immune system, antigens and adjuvants is imperative to design new generation vaccines.

In this chapter we will discuss the factors that can influence the interactions of DCs with immune cells for generating an immune response and its applicability in the development of vaccines targeting DCs. Different DC subsets and the alternatives available for triggering their activation will also be explained, such as passive or active targeting using different adjuvants. Finally, the latest approaches combining multiple adjuvants will be described.

2. Dendritic cell subsets

The family of DCs is constituted of several distinct DC subsets that possess common characteristics but differ into their functions related with biasing the immune response towards the appropriate arm in each situation. In general terms, DC subsets are defined based on their expression of surface markers. Traditionally, they have been described as high in Major Histocompatibility Complex (MHC) class II molecules (HLA-DR) and lacking the lineage (Lin) surface markers characteristics of other immune or non-immune cells such as CD14 (monocytes), CD19 (B cells), CD3 (T cells), CD56 (NK cells) or CD34 (stem cells). The expression of the CD11c integrin, as well as co-stimulatory molecules (CD80, CD83, CD86 or CD40) has also been used for defining the immunobiology of DC subsets both in mice and humans. In humans, the expression of CD1c antigen and Blood Dendritic Cell Antigens (BDCA) has also been useful in their definition [7].

There are two main lineages of DCs in humans, the myeloid DCs (MDCs, also named classical or conventional DCs) and plasmacytoid DCs (PDCs). MDCs are originated from a myeloid progenitor in the bone marrow, while PDCs may come from a lymphoid progenitor in lymphoid organs [8]. The DC subsets migrate to different specific locations (tissues) or circulate in the blood. Human DCs in blood and skin are the best characterized, although DCs also can be found in other locations such as lungs or gut.

2.1. Blood dendritic cells

Different DC subsets are classified by the expression of distinct surface molecules. Thus, the PDCs are characterized by the expression of CD303 (BDCA-2), CD123 (IL-3Rα) and CD304 (BDCA-4) and MDCs by CD11c. In blood, MDCs are subdivided into three additional subsets, called CD1c$^+$ DCs, CD141$^+$ DCs and CD16$^+$ DCs (Figure 1) [9].

PDCs play a key role in the antiviral immunity derived form their unique ability to produce large quantities of type I interferon (IFN) [10], spell out IP-10, Tumoral Necrosis Factor (TNF) and IL-6 [11]. Through that secretion, PDCs support the function of other immune cells such as MDCs, NK or B cells, thus, linking the innate and adaptive arms of the immune responses. PDCs express several Toll-Like Receptors (TLRs) (TLR1, 6, 7 9, and 10) but TLR7 and TLR9, recognizing single stranded RNA and unmethylated CpG DNA respectively are the main inducers of type I IFN release.

Under steady-state conditions, PDCs are found in the bloodstream. Following inflammation, PDCs can arrive to the infection site and partake in uptake of antigen and subsequently migrate to draining lymph nodes via high endothelial venules, while most other immune cells enter though lymph vessels [12]. Human PDCs are able to present peptides onto both MHC class I and II molecules [13]. Moreover, it has been shown that mature PDCs are able to effectively develop protective immunity in mice [14].

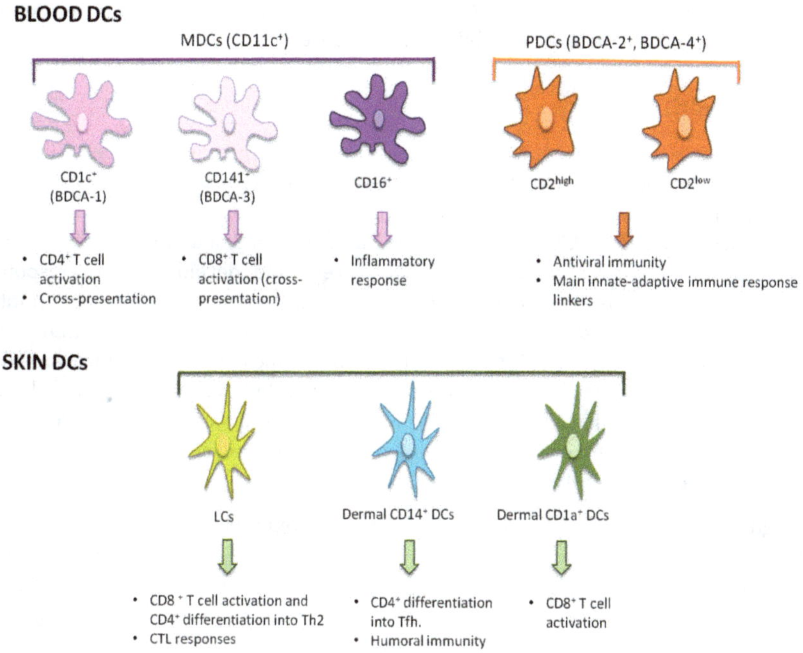

Figure 1. Schematic view of the different human DC subsets in blood and skin and their functions in immunity.

Two PDC subsets can be distinguished based on the CD2 molecule expression on their surface, which also present different functions. Matsui *et al.* [15] showed that $DC2^{high}$ PDCs highly express lysozyme and tent to be present in tonsils and tumors. In addition, they secrete higher amounts of IL12p40, express larger levels of CD80 and trigger the proliferation of naïve T cells more competently than $CD2^{low}$ PDCs.

In contrast to PDCs, MDCs are not so restricted to blood and can be also found in multiple peripheral tissues or in secondary lymphoid organs [16]. They are also superior over PDCs at antigen presentation [17,18]. Immature MDCs express a wide range of TLRs, namely TLR1, 2, 3, 4, 5, 6, 7, 8, and 10 [19,20]. It was recently shown that immature MDCs express mRNA for all TLRs except for TLR9 (they do not express it) and TLR3 in $CD16^+$ MDCs [9].

Among the different MDC subsets, CD141+ (BDCA-3) MDCs are the most specialized in cross-presentation, i.e. capturing endogenous antigens and presenting them into MHC class I molecules [21,22]. CD1c+ (BDCA-1) MDCs are CD4+ T cell stimulators, release pro-inflammatory cytokines such as IL-12 and also appear to be chemotaxis inducers [23]. In addition, they also exhibit the ability of cross-presenting antigens. In fact, although CD141+ MDCs possess higher ability, they exist in a small percentage (2-5%), which means that other DC subsets are likely also contributing overall to this function [24]. Finally, the few reports so far indicate that the CD16+ MDCs show a strong pro-inflammatory function and present a lower antigen presentation ability [9,25] (Figure 1).

2.2. Skin dendritic cells

The skin contains three types of MDCs, the epidermis resident Langerhans cells (LCs) and the dermal CD1a+ and CD14+ DCs [26], which present phenotypic and functional differences (Figure 1). The TLR panel expression is also different in the skin DC subsets. While LCs express TLR1, 3, 6 and 10, CD14+ DCs express TLR2, 4, 5, 6, 8, and 10 [16,27,28]. LCs are efficient at cross-presenting antigens to CD8+ T cells and developing CTL responses. Furthermore, they can also interact with CD4+ T cells and induce their differentiation into T helper 2 (Th2) cytokine secreting cells (IL-4, IL-5 and IL-13). Conversely, CD14+ cells stimulate the differentiation of CD4+ T cells into T helper follicular (Tfh) cells that can induce naïve B cells to change their isotype and become IgG and IgA secreting plasma cells. This ability is not shared with LCs. Thus, both humoral and cellular immune responses are achievable with CD14+ DCs and LCs respectively. CD1a+ DCs are also able to activate CD8+ T cells but in a lesser extent than LCs [29].

Although the number of LCs is not very high (they represent around 2% of the total epidermal cells), their dendritic shape helps them forming a network that increases their superficial area in contact with their surrounding space [7]. In addition, it has been recently shown that upon activation, they are able to elongate their dendrites for reaching to the antigens [30].

2.3. DCs in other mammal models

Among the animal models available, the non-human primates (NHP) are the most similar to humans. However, there exist important functional and phenotypic differences between human and NHP and also among different species of NHP, which makes difficult the establishment of a general rule for comparison. Moreover, these differences are more marked when smaller animals like mice are studied. Furthermore, although phenotypically similar, DCs from different animals may present distinct functionalities. For example, rhesus macaques and human DCs express similar TLR panel. However, signaling through TLR9 produces less IFN-α and negligible IL-12 from PDCs in macaques in comparison to the human counterparts. Furthermore, surface markers used to identify DC subsets in humans are not the same in NHP. This is exemplified by the lack of BDCA-2 and BDCA-4 markers of PDCs in NHP [31]. Skin DCs in NHP have been less studied in comparison with blood DCs but phenotypic and functional differences with the human skin DCs are also evident.

Other mammal models such as pigs [32] have also been used, but the differences with human DCs are higher in comparison with NHP. This variability among species needs to be taken into account when evaluating a vaccine in each animal model and when designing translational studies.

3. Mouse dendritic cells

In mice, DC subsets are also composed by PDCs and MDCs. Splenic MDCs include two main subsets: $CD8\alpha^+CD11b^-CD205^+$ and $CD8\alpha^-CD11b^+CD205^-$, also called lymphoid and myeloid DCs respectively. Moreover, the $CD8^-$ subset also expresses the 33D1 marker. Cells of the first subset are able to polarize $CD4^+$ T cells to a Th1 response and can also cross-present antigens to $CD8^+$ T cells. Cells of the second subset produce large amounts of IL-12 and effectively present antigens into MHC class II molecules so they are specialized in biasing the $CD4^+$ T cells toward Th2 responses [33-36]. Recently, it has been reported that there are functional commonalities between mouse $CD8\alpha^+$ DCs and the human $CD141^+$ MDCs [21,22,37]. Although the $CD141^+$ MDCs do not express CD8, they share other surface molecules such as CLEC9A, NECL2 or XCR1. More importantly, they also share the high ability of cross-presenting antigens to $CD8^+$ T cells [23].

Mouse MDCs can be also classified by the presence or absence of the CD4 surface marker. Thus, three different subsets can be distinguished, $CD8^-CD4^+$, $CD8^+CD4^-$ and $CD8^-CD4^-$ MDCs. The effector function of the $CD8^+$ MDCs appears to be clear and is based on the production of CTL responses (due to their ability of cross-presenting antigens to $CD8^+$ T cells). However, studies carried out with the $CD8^-$ subsets have shown contradictory results. It appears that $CD8^-CD4^+$ MDCs are not able to release cytokines but they effectively present antigens to $CD4^+$ T cells, while the $CD8^-CD4^-$ MDCs produce IFN-γ [38,39]. Bialecki *et al.* have recently reported that both subsets demonstrate equivalent ability to present OVA peptide or whole OVA to $CD4^+$ and $CD8^+$ T cells under steady-state conditions. Conversely, when the invariant Natural Killer T (iNKT) cell agonist α-galactosylceramide (α-GalCer) is used, $CD4^+$ and $CD4^-$ subsets noticeably differ in their ability to activate and/or polarize iNKT cells both *in vitro* and *in vivo*. Hence, $CD4^-$ DCs were more efficient in stimulating iNKT cells, an effect related to their higher ability to secrete cytokines (Figure 2) [40].

Mouse PDCs have $CD11c^+B220^+$ phenotype. As the human PDCs, they also circulate in blood and during inflammation or infection they enter to the lymphoid organs and release large amounts of type I IFNs. In fact, it has been shown that the PDC population decreases from blood 3 days post infection, mainly because they migrate to the spleen [41]. In addition, PDCs have been reported to be involved in induction of tolerance in mice [42,43]. A recent study showed that these different functions could be conducted by two distinct PDC subsets in mice. $CD9^+$Siglec-H^{low} PDCs have demonstrated to release IFN-α, activate CTLs and trigger a protective immunity when stimulated with TLR agonists. On the other hand, $CD9^-$Siglec-H^{high} PDCs produced trivial amounts of IFN-α, induced $Foxp3^+CD4^+$ T cells and were not able to produce antitumor immunity (Figure 2) [44].

With regard to the skin DCs, two main populations have been identified in mice, epidermal LCs and Langerin⁺ dermal DCs [45]. It has been suggested that these dermal Langerin⁺ DCs could be the mouse counterpart of the human CD1a⁺ dermal DCs (Figure 2) [16].

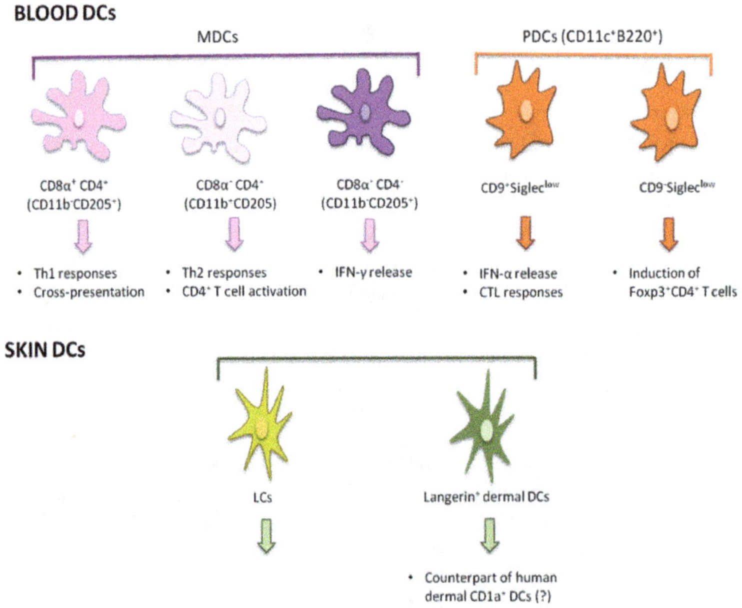

Figure 2. Schematic view of the different mouse DC subsets in blood and skin and their functions in immunity.

Therefore, the research on mouse DC subsets as a model to aid data for human vaccination should be carefully considered because of the important functional and phenotypic differences between DCs in these two species. Vaccines effective in mice could not be effective in other species or vice versa. Although some human DC subsets seem to have their mouse counterpart, translating mouse studies results to humans should be made with caution. For example, although mouse CD8⁺ DCs and human CD141⁺ share the ability for cross-presentation, there are other human DCs that have the same function, such as CD1c⁺ DCs or LCs. Other animal models such as non-human primates could represent a good alternative to mice since they present more similarities with humans. However, there exist important differences mainly in the surface molecule expression and also in maturation and some functions [46]. Thus, they can represent a better tool for preclinical studies than mice, but again, the results should be translated with caution to humans.

4. Activation of the immune response by dendritic cells

DCs are the main linkers of the innate and adaptive immune responses and depending on the context they are able to induce immunity or tolerance. They drive naïve T cells into different effector cells or into Treg cells that suppress activated T cells. These different functions of DCs require a fine orchestration of stimuli to be adequately directed.

Either circulating or in the peripheral tissues, DCs are found in an immature or resting state specialized in capturing antigens. In this state they express low levels of MHC class I and II and costimulatory molecules in their surface. They have the ability to produce MHC class II molecules but they are directed to endosomes and lysosomes. When MHC class II molecules are loaded to a peptide (from the DC itself or a captured antigen), the formed complexes can be directed to lysosomes for degradation or be stably expressed into the cell membrane. This different ways are determined by the maturation status of the DCs. When they are immature, MHC class II molecules-peptide complexes are ubiquitinated, which finally drives the complexes to the lysosomes for being degraded. When DCs are activated by microorganisms, malignant cells or by an inflammatory stimuli, the MHC class II-peptide complexes are not successfully ubiquitinated and they are guided to be expressed into the plasma membrane [47]. Those stimuli are the signals that make DCs to undergo their maturation process.

While DCs are undergoing their maturation process, they reduce the ability for capturing antigens and they also increase the expression of MHC class I molecules, costimulatory molecules (CD80, CD86) and T cell adhesion molecules (e.g. CD48, CD58). Furthermore, maturation process also involves a migration of the DCs to the T cell areas of lymphoid organs. This process becomes DCs into antigen-presenting cells (APCs). The APCs are able to present processed antigens to effector cells, undergoing to the pathogen-specific immune response. In fact, DCs can present peptides into MHC molecules I and II to T cells as well as glycopeptides and glycolipids to T cells and NKT cells, and polypeptides to B cells [6]. These processes are highly regulated and influenced by the different stimuli that interact with DCs.

The development of antibody responses is an easily achievable feature and it is produced by most of the current licensed vaccines. Antibodies exert their function by neutralizing the pathogens, binding to the complement, allowing phagocytosis after opsonization, and/or producing antibody-dependent cellular cytotoxicity [48]. However, the induction of potent cellular immune responses, including cytolytic responses (CTL), remains a challenge in vaccinology. For accomplishing a CTL response, cytolytic cells have to recognize processed peptides presented into MHC class I molecules or into the CD1-lipid complex. Thus, the $CD8^+$ T cells, $CD4^+$ Th1 cells and NKT cells become activated and release the cytolytic molecules that will destroy infected cells at the same time as they develop inflammatory reactions and some humoral responses. The presentation of antigens into MHC class I molecules is a process called cross-presentation. Although the MHC class I molecules expression is increased upon maturation, the degree of expression may vary among different DC subsets [49].

In contrast to the immune response against pathogens, the development of tolerance has traditionally thought to occur because of the absence of infection or inflammation stimuli. However, there is evidence that suggest that there are also some endogenous signals that favor its development [50,51]. Moreover, there are differences in the regulation of the self-tolerance and the environment-tolerance. The development of the central (thymic) tolerance needs the recognition of self-antigens in the thymus. Some soluble antigens can reach the thymus by the blood and be presented by resident DCs or by medullary thymic epithelial cells. In addition, PDCs have also the ability to transport self-antigens to the thymus, a process regulated by the expression of the CCR9 chemokine receptor [52]. Alteration of this balance leads to autoimmune diseases. On the other hand, the mechanisms involved in the immune tolerance to allergens are distinct, because this process requires the development of peripheral T cell tolerance, modifying the Th and Treg balance. Treg cells control the allergic responses by several mechanisms, which include the suppression of effector immune cells that act in allergic processes (DCs, Th cells, mast cells, eosinophils and basophils) and the migration of inflammatory cells to tissues. In addition they also decrease the allergen-specific IgE/IgG4 ratio, a process regulated by IL-10 and TGF-β [53]. DCs seem to have an important role on immune tolerance. In fact, both in autoimmune diseases and allergy, it has been shown that DCs are present in the site of inflammation. This process is accompanied by a decrease in the number of circulating DCs, consistent with the migration from the periphery to the inflammation site [54]. However, the role of each specific DC subset in diverse mechanisms of tolerance remains to be well established.

5. Targeting dendritic cells

In the field of vaccines against infectious diseases generation of any tolerogenic responses should be avoided. The best way for developing high immune responses tends to be the stimulation of DCs in the same way that pathogens do. DCs possess immune receptors, called pathogen recognition receptors (PRRs), specified in recognizing highly conserved pathogen associated molecular patterns (PAMPs) from the microorganisms. Currently, methods for detecting PRRs are based in the use of stained antibodies specific for each PRR [55]. There are several PRRs, including TLRs, C-type lectin receptors (CLRs), cytoplasmic retinoic acid-inducible gene-I-like receptors (RLRs), nucleotide oligomerization domain-like receptors (NLRs) and others [56]. Upon recognition of antigens, especially via PRRs DCs undergo their maturation process. However, the nature of the subsequent immune response will depend on the PRR or combination of them that are stimulated.

TLRs are the best-characterized PRRs. They are a family of transmembrane receptors. Except TLR3, they possess a major adaptor protein required for signaling, the myeloid differentiation protein 88 (MyD88). MyD88 is a cytosolic adaptor protein consisting of a N-terminal death domain (DD) and a C-terminal Toll/Interleukin-1 receptor (TIR) domain, connected by a short linker. The TIR domain is composed of three highly conserved motifs, known as box 1, 2 and 3, which play a key role in the initiation of the immune signaling upon ligand coupling to TLRs [57, 58]. Engagement of PAMPs to TLRs makes MyD88 recruit the IL-1 Receptor-associated kinases (IRAK4, IRAK1, IRAK2 and IRAK-M), which enable the induction of interferon regulatory factor (RF)-responsive genes, activation of mitogen-associated protein kinases (MAPKs), and nuclear factor κB (NF-κB). Finally, these factors modulate the production of type I IFN and inflammatory cytokines such as IL-6 and

IL-12p40 [59]. The adaptor molecule for TLR3 is called Toll/IL-1R domain-containing adaptor-inducing interferon (IFN)-α-(TRIF). Moreover, TLR4 signals through both MyD88 and TRIF pathways.

Up to date, 10 different TLRs have been identified in humans (TLR1-TLR10), while in mice there have been recognized 12 (TLR1-TLR9 and TLR11-TLR13). These signaling receptors are specialized in recognizing bacteria, viruses, fungi and protozoa components not present in mammals. TLR1, TLR2, TLR4, TLR5, TLR6 and TLR10 are expressed in the cell membrane, thus recognizing microorganism's surface molecules. In contrast, TLR3, TLR7, TLR8, and TLR9 are intracellular receptors, which recognize pathogen derived DNA and RNA [60]. In addition, an eleventh human TLR has been recently identified. TLR11 senses profilin, present in *Toxoplasma gondii.* Profilin has homologous proteins in human and also in plants. However, it has been proposed that the structure of the TLR11 in human does not allow the response to endogenous components [61].

The interactions of the TLRs with their ligands and the downstreaming effects have been elegantly reviewed by Dzopalic *et al* [58]. Briefly, TLR2 exerts its function by forming heterodimers with TLR1 and TLR6. Thus, it is the receptor with the highest spectrum of ligands, recognizing molecules such as peptidoglycan or bacterial lipoproteins. TLR3 interacts with double-stranded RNA; TLR4 recognizes LPS and its derivates; TLR5 recognizes flagellin protein; TLR7 and TLR8 are receptors for single-stranded RNA; and TLR9 interacts with unmethylated CpG DNA motifs. There is low evidence about the ligands binding TLR10. A study carried out by Govindaraj *et al.* [62] showed that PamCysPamSK4, a di-acylated peptide, might activate the human TLR10/1 hetero and TLR10 homodimer, while Pam3CSK4 might be recognized by the TLR10/2 heterodimer.

The CLRs can be presented as both soluble and transmembrane receptors and are composed of at least one carbohydrate recognition domain (CRD). Thus, these receptors interact with carbohydrate structures like mannose, fucose or glucans [63]. There are several types of CLRs such as mannose receptor, comprising CD206 and DEC-205 (CD205) [64], DC-SIGN (CD209), Langerin (CD207), dectin-1 family and DC immunoreceptor family (DCIR) [65]. Triggering CLRs produces a signaling pathway that commonly produces NK-κB. However, one of the potential problems of targeting CLRs is that some of them are able to recognize endogenous ligands, decreasing the specific response against pathogens.

NLRs and RLRs are cytoplasmic receptors that recognize components of the microorganisms. Upon PAMP stimulation, many NLRs form the inflammasome, a protein complex that activates IL-1β and IL-18. On the other hand, RLRs (RIG-I and MDA5) interact with double-stranded RNA and stimulate NK-κB and IRF3/7 signaling pathways. Stimulation of RIG-I also produces the activation of the inflammasome [66].

5.1. *In vivo* targeting of antigens and adjuvant-activation of dendritic cells

In vivo targeting of antigens to DCs has shown to be a promising approach for vaccines, particularly for enhancing the stimulation of cellular immune responses. Targeting of vaccine antigens involves the administration of an antigen linked to a specific antibody or to a specific marker expressed on DCs with the objective to more efficiently deliver the antigen to the DCs.

For an efficient vaccination, the vaccine candidate also needs to stimulate the immune system in the same way as pathogens do, i.e. signaling through PRRs. However, the immune system is not always able to respond to certain microorganism and thus, alternative ways for stimulating the immune response should be investigated in order to develop an effective vaccine. For example, unlike HIV in humans or SIV in rhesus macaques, sooty mangabeys act as a reservoir for SIV and do not develop AIDS. It has been demonstrated that those differences are derived from the distinct activation of the sooty mangabeys PDCs upon TLR7 and TLR9 activation by the virus [67]. Consequently, PDCs of the reservoir lack the characteristic chronic immune activation producing a much lower amount of IFN-α. In addition, some pathogens have the ability of modulating the expression of the TLRs, such as the hepatitis B virus, which is able to reduce the TLR9 depending IFN-α secretion by PDCs, although not the one dependent on TLR7 [68].

For developing an effective vaccination, inactivated or subunit protein based vaccine antigens need to be accompanied by adjuvants. Adjuvants are components added to the vaccine antigens that are able to modulate the bias, the intensity and the duration of the immune response interacting with DCs and also with other immune cells. Some adjuvants have been shown to directly be able to interact with receptors on APCs for regulating the immune response. Traditionally, adjuvants have been classified as immunostimulatory or carrier systems. The first ones interact with the PRRs of the APCs. The latters were supposed to form a depot that allowed the long-term release of the antigens or to passively drive antigens to DCs. However, it is nowadays clear that carrier systems are also able to directly activate the immune response [69]. For example, it has been shown that alum, which has been used in human for more than 80 years and was thought to act by a depot effect [70], interacts with NLRs and activates NALP3 inflammasome [71]. This interaction with PRR has also been shown for other molecules. Ariza *et al.* have shown that the previously known anticancer drug Bryostatin-1 exerts its activity by the interaction with TLR-4 present in MDCs, which triggers the activation of NF-κB and release of cytokines (IL-5, IL-6, and IL-10) and chemokines [72].

The first approaches to specifically deliver antigens and immune-stimulatory adjuvants to DCs were carried out directing antigens to molecules present in all DC subsets, such as CD11c or MHC class II [73,74]. However, the most recent studies have developed more specific strategies. Thus, targeting a specific PRR, particular DC subsets are being targeted and leading to different activation of the immune system. Guo *et al.* have shown an improvement of the response elicited in a porcine model by the traditional vaccine for foot-and-mouth disease incorporating a CpG-enriched recombinant plasmid into the formulation [75]. On the other hand, activation of PRRs is not a universal science, and should be studied for each pathogen. Cheng *et al.* have reported that the TLR2/TLR6 agonist Pam_2CSK_4 increases the immunogenicity of a Chlamydia trachomatis vaccine, producing strong humoral and cellular immune responses as well as better protection after challenge in mice. However, other adjuvants such as polyinosinic-polycytidylic acid (poly (I:C), TLR3), monophosphoryl lipid A (MPLA, TLR4), flagellin (TLR5), imiquimod R837 (TLR7), imidazoquinoline R848 (TRL7/8), CpG-1826 (TLR9), M-Tri-DAP (NOD1/NOD2) and muramyldipeptide (NOD2) elicited lower immune responses and were less efficient for eliciting protection upon a challenge, demonstrating that Pam_2CSK_4 is the most suitable adjuvant for further evaluation of a vaccine against *Chlamydia trachomatis* [76].

It has also been proposed that targeting specific DC receptors should be accompanied by an appropriate route of immunization. Thus, directing antigens to LCs or dermal DCs should be improved by subcutaneous or intradermal routes, and the intravenous route should be better for targeting blood circulating DCs [77]. However, it has not been clearly demonstrated.

Nowadays, it is clearer that achieving an effective immune response needs the activation of multiple PRRs. There are several adjuvants that are able to interact with more than one PRR. For example, poly(I:C), a synthetic analogue of double-stranded RNA, binds TLR3, MDA5 and RIG-I [78]. In addition, it has been shown that for an efficient immune activation, interaction with both TLR3 ad RLRs is necessary [79]. This multiple activation has been thought to be a possible reason for the development of strong immune responses.

The other alternative for efficient activation of more than one PRR at the same time is the co-administration of antigens formulated with a combination of adjuvants. Raman *et al.* have demosntrated that achieving strong and effective T cell responses against cutaneous leishmaniasis in mice required the combination of MPLA and CpG along with the antigen. However, the administration of the antigen with each adjuvant separately did not produce an optimal response [80]. In the same way, Zhu *et al.* found that mice immune response upon vaccination against HIV using envelope peptide along with macrophage-activating lipoprotein 2 (MALP2, TLR2/6), poly(I:C), CpG was importantly improved [81].

On the other hand, combination of TLR and CLR agonists appears an interesting approach [82]. Klechevsky *et al.* have evaluated the $CD8^+$ T cell responses developed in humans upon targeting antigens to DCs *in vitro* using different adjuvants. Even at low doses, antigens conjugated to antibodies against DCIR were able to trigger antigen specific $CD8^+$ T cell responses as well as memory responses by *ex-vivo* generated DCs, LCs, MDCs and PDCs. Although the use of monoclonal antibodies to target DC-SIGN did not produce such high response, the addition of the CL055 (TLR7/8) enhanced the cross-presentation and cross-priming developed by DCIR activation. This activation produced a higher $CD8^+$ T cell expansion, a high release of IFN-γ and TNF-α and production of low levels of Th2 type cytokines [83].

In addition, as discussed below, combination of immunostimulatory adjuvants with carrier adjuvants is another alternative for PRR selective targeting.

5.2. *Ex vivo* loading of antigens to dendritic cells

Ex vivo antigen-loaded and/or matured DCs have been developed starting from monocytes or $CD34^+$ precursors in the presence of GM-CSF and IL-4. Afterwards, the antigen is added to the immature DC culture, generally accompanied with various maturation stimuli (adjuvants) to ensure the ability of cross-presentation and the capacity to develop strong immune responses. Finally, these *ex vivo* matured DCs are administered to the patient.

Ex vivo loading of tumor antigens to DCs for developing anticancer vaccines has become the most important application of this approach [84]. The main objective in vaccination against cancer is to trigger potent CLT responses, able to eliminate tumoral cells and to develop memory responses capable of avoiding tumor relapses. This strategy is safe and well tolerated, as well as capable of inducing cellular immune responses, including the expansion of $CD4^+$ and $CD8^+$ tumor-specific T cells. However, most of the patients do not develop clinical responses [24].

Palucka *et al.* have addressed the most important immunological parameters that determine the effectiveness of this strategy in *ex vivo*-vaccinated patients [24]: (1) the quality of the elicited CTLs, (2) the quality of induced $CD4^+$ Th response, (3) the abrogation of Treg cells, and (4) the breakdown of the immunosupressive tumor microenvironment. In fact, the ability of *ex vivo* cultured DCs for migrate to the tumors is low, impairing the possibilities of activating T cells [85,86]. Moreover, tumors possess mechanisms able to impair the maturation of DCs, and they also have the ability of generating immune reactions that favor their own growth or development [87].

The selection of the antigen for DC-loading can affect the efficacy of the vaccination. Mainly, two kinds of antigens have been used for this approach, mutated (unique) antigens and shared non-mutated antigens. Using mutated antigens makes necessary the personalization of the treatment. In this regard, idiotype-pulsed DCs are emerging as a promising tool. The idiotype is the combination of the multiple antigenic determinants of the immunoglobulins for each individual. This approach has been evaluated in a phase III clinical trial in patients with follicular lymphoma, showing a significantly prolonged duration of the remission induced by chemotherapy treatment [88]. However, tumor-specific idiotypes have demonstrated disappointing results in some diseases such as multiple myeloma [89]. Alternatively, the use of shared non-mutated antigens will open a window for developing largely applicable vaccines, but pre-existing Tregs could impair the efficacy of these vaccines [87].

Finally, there are some limitations in the use of *ex vivo* matured DCs. The vaccines need to be prepared for each individual, which increases the time and the costs of vaccination. In addition, the manipulation of the cell cultures increases the risk of endotoxin contamination. Besides, there is a lack of universal protocols for this purpose, so the results of the vaccination may vary depending on the clinic centers.

6. Particulate adjuvants for targeting dendritic cells

Traditionally, adjuvants that were supposed to act by a depot effect were called particulate adjuvant or carriers. They were thought to sediment at the site of injection from which they released the antigen during prolonged periods of time [70]. Nowadays it is clear that, although they play a role as passive-delivery systems, their adjuvant activity is also due to their ability to interact with the immune system.

Conventional vaccines are composed of whole organisms. Although these vaccines present higher immunogenicity, they have some disadvantages that make the vaccination with antigens derived from the pathogens a better alternative [90-92]. These antigens are usually peptides and tend to be low immunogenic. The main reason is that the uptake and presentation by APCs is not well performed, so effective T cell responses are not developed. In fact, proteins and peptides are easily degraded by proteases and have limited bioavailability due to their low ability to cross biological membranes [93,94]. Carrier adjuvants present some advantages that make them very useful for the development of peptide-based vaccines. First, they protect labile antigens from the effect of hydrolytic enzymes. Secondly, they can be passively or actively delivered to APCs. Moreover, they allow the incorporation of multiple adjuvants and antigens into the same system, as well as APC-targeting molecules, leading to the development of multicomponent vaccines. Thus, finding the optimal combination between antigens and adjuvants acting by different mechanisms is the major goal of vaccination.

Although the most used adjuvant is still alum, other strategies based on particulate systems have emerged and have been licensed for human use. For example, vaccines against influenza (Inflexal®) or hepatitis A (Epaxal®) are composed of virosomes [95,96] and another vaccine against influenza (Fluad®) is formulated with MF59, a squalene-based oil in water emulsion [97]. There are also licensed vaccines composed of more than one adjuvant systems, such as Cervarix® or Gardasil®, both against human papillomavirus [98-100]. The first one is adjuvanted by a mixture of virus like particles (VLPs) and AS04, a combination of MPLA and alum. The second comprises VLPs and alum. The fact of being licensed vaccines means that this strategy is effective, and more research is needed in order to find the appropriate formulations for diseases lacking a vaccine.

6.1. Interactions of particulate adjuvants with dendritic cells

A wide variety of particulate adjuvants have been evaluated as DC-targeting vaccines, which have particular characteristics such as size, shape or composition. The antigen or additional adjuvants can be incorporated by different strategies, such as encapsulated [101], and covalently [102] or non-covalently coupled to the particles [103]. In addition, the delivery of the products can be controlled, allowing a slow and continuous release, by pulses, or triggered by factors such as pH [104], electric or magnetic fields [105], temperature [106] or ionic strength [107]. Thus, the interaction of these systems with DCs is not universal and the adjuvant effect is developed by different mechanisms.

There are two main mechanisms by which carrier adjuvants interact with APCs. i.e. passive and active targeting. Derived from their particulate nature, all of them are directed to DCs by passive targeting. Thus, endocytosis and pinocytosis by APCs is spontaneously favored. Active targeting can also be developed incorporating different ligands into the particulated system. Those ligands can be antibodies, polysaccharides, peptides and drugs that bind to receptors on DCs. Recently, we have designed an adjuvant system that includes poly lactic-co-glycolic acid based microparticles (PLGA MPs) combined with different adjuvants such as poly(I:C), MPLA, and α-galactosylceramide, co-encapsulated with bovine serum albumin (BSA). Intradermal vaccination of mice with these MPs showed higher IgG titers than those obtained by the administration of the soluble peptide mixed with the same combinations of poly(I:C), MPLA, and α-galactosylceramide used in the MPs. In contrast to the soluble peptides, MPs were able to develop cellular immune responses. Furthermore, the combination of MPLA and α-galactosylceramide triggered the higher cellular response, demonstrating the usefulness of this approach [108].

Important advances of knowledge on the mechanism of action of particulate adjuvant have been done in the last years. Recent data demonstrate that alum targets NALP3 (also known as NLRP3) inflammasome, a kind of NLR, and mediates the delivery of IL-1β and IL-18 [71], although other studies show that activation of these receptor is dispensable for the effect of alum [109], or that it can activate NALP3 inflammasome by indirect mechanisms [110]. PLGA MPs also activate NALP3 inflammasome, although it is not clear if that activation is or not directly triggered by MPs [111,112]. Other adjuvant systems include an immunomodulatory adjuvant in their structure, such as ISCOMs or ISCOMATRIX®. They are cage-like structures comprising cholesterol, phospholipids and a saponin, being the last one the main responsible of the adjuvant effect [113].

In other cases the mechanism of action is not well defined. For example, non-targeted liposomes are thought to act by passive targeting derived from their particulate nature [114], which is more easily detected by the immune systems than smaller soluble antigens and can resembles the structure of the pathogens.

6.2. Factors influencing the efficacy of particulate vaccines

There are some aspects that are crucial for this effect, such as particle size and surface charge of the particles, or their route of administration. The particle size is one of the most important parameters; the smaller the size, the higher the surface area for antigen delivery [115]. Moreover, DCs have different phagocytic ability depending on the particle size. Chua *et al.* have evaluated the differences between chitosan micro and nanoparticles (NPs) for being taken up by DCs [116]. They have found that the uptake of both micro and NPs by bone marrow derived mouse DCs depends on particle concentration. NPs were faster than MPs for reaching to the lymph nodes, although this fact did not affect the quality of the immune response, which lead to similar antibody responses than the administration of the soluble antigen emulsified in Freund's adjuvant. In addition, it has been shown that small particles can reach to the lymph nodes and be engulfed by resident DCs, a fact not achievable by larger particles (500nm-2μm) [117,118]. However, other studies show contradictory results and suggest that MPs are better for biasing the immune response towards the humoral arm while NPs can stimulate cellular responses [115]. In addition, coupling ligands to MPs and NPs for targeting DC-SIGN has shown that only NPs are able to actively target DCs, while MPs are non-specifically taken up [119]. The distinct results in these studies can also be explained because different delivery systems and protocols have been used. Thus, each case should be studied individually.

Modifying the surface charge has also showed differences in the uptake of polystyrene MPs by DCs. Covalently coupling of polyaminoacids/proteins became the particles positively charged and lead to a high increase in the uptake by monocyte derived human blood DCs [120].

The evaluation of different routes of administration has also lead to contradictory results. We have previously reported that PLGA MPs encapsulating the SPf66 malarial peptide produced higher humoral and cellular immune responses when they were administered by the intradermal route rather than when the subcutaneous route was used. Furthermore, the intradermal route allowed a 10 fold reduction of the dose without compromising the immune response [121]. A study by Mohanan *et al.* was designed to address the differences in the immune response of mice vaccinated with ovalbumin-loaded liposomes, N-trimethyl-chitosan NPs and PLGA MPs by subcutaneous, intradermal, intramuscular and intralymphatic routes. They found that only Th1 immune responses were sensitive to the route of administration, but not the Th2 responses [117]. On the other hand, Speiser *et al.* have carried out a phase I/II clinical trial where they have not found significant differences upon subcutaneous or intradermal vaccination with a vaccine against melanoma comprising VLPs and CpG [122].

7. Conclusion

DCs are the cells that join the innate and adaptive immune responses. The interaction of these cells with T or B cells regulates the immunity against pathogens. Moreover, the interaction of antigens and/or adjuvants with DCs also influences the immune response. These unique characteristics of DCs make them an ideal target for vaccination purposes. However, there are several fundamental concerns when designing vaccines directed towards DCs. Studying the human immune system and understanding its complexity is essential for this purpose. Animal models play an important role in this field, but the results obtained in those studies have to be carefully interpreted and translated to humans. The fact of having similar immune effector functions is not enough for accepting any specie as a good model. For example, both human and mouse PDCs have an important role on antiviral immunity by releasing high amounts of type I IFN. Nevertheless, the phenotypic differences between human and mice are going to condition the response. CpG motifs can be used for selective activation of human PDCs with TLRs agonist, but in mice MDCs would be also activated because they do express TLR9. Differences in the expression of uptake receptors such as BDCA-2 will also limit the comparison of studies that target antigens to BDCA-2, as mouse DCs do not present this C-type lectin. Furthermore, it is also important to take into account that mice and humans have distinct susceptibilities to some diseases (for example mice do not develop HIV and they do not have a similar counterpart like SIV for monkeys), impairing the translation of studies addressing immunological responses against pathogens.

In conclusion, there are a lot of factors influencing the interaction of DCs with the immune systems for activating an effective immune response. Consequently, the use of distinct approaches for stimulating DCs would lead to a different activation of the immune system. Thus, the study and understanding of these relations is essential for developing useful vaccines against malaria, HIV, or tuberculosis, or to treat allergies, autoimmunity disorders or rejection of transplanted organs.

Author details

Aiala Salvador, Manoli Igartua, José Luis Pedraz and Rosa María Hernández
NanoBioCel Group, Laboratory of Pharmaceutics, University of the Basque Country, School of Pharmacy; Biomedical Research Networking Center in Bioengineering, Biomaterials and Nanomedicine (CIBER-BBN), Vitoria, Spain

Acknowledgement

This project was partially supported by the "Ministerio de Ciencia e Innovación" (SAF2007-66115), "University of the Basque Country (UPV/EHU)" (UFI 11/32) and FEDER funds. A. Salvador thanks the "University of the Basque Country" for the Fellowship Grant.

8. References

[1] Steinman RM, Cohn ZA (1973) Identification of a novel cell type in peripheral lymphoid organs of mice. I. Morphology, quantitation, tissue distribution. J Exp Med. 137: 1142-1146.

[2] de Jong EC, Smits HH, Kapsenberg ML (2005) Dendritic cell-mediated T cell polarization. Springer Semin Immunopathol. 26(3): 289-307.

[3] Stockinger B, Veldhoen M, Martin B (2007) Th17 T cells: Linking innate and adaptive immunity. Semin Immunol. 19: 353-361.

[4] Tritto E, Mosca F, De Gregorio E (2009) Mechanism of action of licensed vaccine adjuvants. Vaccine. 27: 3331-3334.

[5] Salvador A, Igartua M, Hernández RM, Pedraz JL (2011) An overview on the field of micro- and nanotechnologies for synthetic peptide-based vaccines. Journal of Drug Delivery. 2011: 2011: 181646.

[6] Blanco P, Palucka AK, Pascual V, Banchereau J (2008). Dendritic cells and cytokines in human inflammatory and autoimmune diseases. Cytokine Growth Factor Rev. 19: 41-52.

[7] Teunissen MB, Haniffa M, Collin MP (2012) Insight into the immunobiology of human skin and functional specialization of skin dendritic cell subsets to innovate intradermal vaccination design. Curr Top Microbiol Immunol. 351: 25-76.

[8] Liu Y (2001) Dendritic cell subsets and lineages, and their functions in innate and adaptive immunity. Cell. 106: 259-262.

[9] Mittag D, Proietto AI, Loudovaris T, Mannering SI, Vremec D, Shortman K, *et al* (2011) Human dendritic cell subsets from spleen and blood are similar in phenotype and function but modified by donor health status. J Immunol. 186: 6207-6211.

[10] Siegal FP, Kadowaki N, Shodell M, Fitzgerald-Bocarsly PA, Shah K, Ho S, *et al* (1999) The nature of the principal type 1 interferon-producing cells in human blood. Science. 284: 1835-1837.

[11] Palucka AK, Blanck JP, Bennett L, Pascual V, Banchereau J (2005) Cross-regulation of TNF and IFN-alpha in autoimmune diseases. Proc Natl Acad Sci USA. 102: 3372-3377.

[12] Conrad C, Meller S, Gilliet M (2009) Plasmacytoid dendritic cells in the skin: To sense or not to sense nucleic acids. Semin Immunol. 21: 101-109.

[13] Lui G, Manches O, Angel J, Molens JP, Chaperot L, Plumas J (2009) Plasmacytoid dendritic cells capture and cross-present viral antigens from influenza-virus exposed cells. PLoS One. 4: e7111.

[14] Remer KA, Apetrei C, Schwarz T, Linden C, Moll H (2007) Vaccination with plasmacytoid dendritic cells induces protection against infection with leishmania major in mice. Eur J Immunol. 37: 2463-2467.

[15] Matsui T, Connolly JE, Michnevitz M, Chaussabel D, Yu CI, Glaser C, *et al* (2009) CD2 distinguishes two subsets of human plasmacytoid dendritic cells with distinct phenotype and functions. J Immunol. 182: 6815-6823.

[16] Klechevsky E, Liu M, Morita R, Banchereau R, Thompson-Snipes L, Palucka AK, *et al* (2009) Understanding human myeloid dendritic cell subsets for the rational design of novel vaccines. Hum Immunol. 70: 281-288.

[17] Lore K, Adams WC, Havenga MJ, Precopio ML, Holterman L, Goudsmit J, *et al* (2007) Myeloid and plasmacytoid dendritic cells are susceptible to recombinant adenovirus vectors and stimulate polyfunctional memory T cell responses. J Immunol. 179: 1721-1729.

[18] Jones L, McDonald D, Canaday DH. Rapid MHC-II antigen presentation of HIV type 1 by human dendritic cells (2007) AIDS Res Hum Retroviruses. 23: 812-816.

[19] Kadowaki N, Ho S, Antonenko S, Malefyt RW, Kastelein RA, Bazan F, *et al* (2001) Subsets of human dendritic cell precursors express different toll-like receptors and respond to different microbial antigens. J Exp Med. 194: 863-869.

[20] Jarrossay D, Napolitani G, Colonna M, Sallusto F, Lanzavecchia A (2001) Specialization and complementarity in microbial molecule recognition by human myeloid and plasmacytoid dendritic cells. Eur J Immunol. 31: 3388-3393.

[21] Jongbloed SL, Kassianos AJ, McDonald KJ, Clark GJ, Ju X, Angel CE, *et al* (2010) Human CD141+ (BDCA-3)+ dendritic cells (DCs) represent a unique myeloid DC subset that cross-presents necrotic cell antigens. J Exp Med. 207: 1247-1260.

[22] Bachem A, Guttler S, Hartung E, Ebstein F, Schaefer M, Tannert A, *et al* (2010) Superior antigen cross-presentation and XCR1 expression define human CD11c+CD141+ cells as homologues of mouse CD8+ dendritic cells. J Exp Med. 207: 1273-1281.

[23] Ueno H, Klechevsky E, Schmitt N, Ni L, Flamar AL, Zurawski S, et al (2011). Targeting human dendritic cell subsets for improved vaccines. Semin Immunol. 23: 21-27.

[24] Palucka K, Banchereau J, Mellman I (2010) Designing vaccines based on biology of human dendritic cell subsets. Immunity. 33: 464-478.

[25] Piccioli D, Tavarini S, Borgogni E, Steri V, Nuti S, Sammicheli C, *et al* (2007) Functional specialization of human circulating CD16 and CD1c myeloid dendritic-cell subsets. Blood. 109: 5371-5379.

[26] Valladeau J, Saeland S (2005) Cutaneous dendritic cells. Semin Immunol. 17: 273-278.

[27] van der Aar AM, Sylva-Steenland RM, Bos JD, Kapsenberg ML, de Jong EC, Teunissen MB (2007). Loss of TLR2, TLR4, and TLR5 on langerhans cells abolishes bacterial recognition. J Immunol. 178: 1986-1990.

[28] Flacher V, Bouschbacher M, Verronese E, Massacrier C, Sisirak V, Berthier-Vergnes O, *et al* (2006) Human langerhans cells express a specific TLR profile and differentially respond to viruses and gram-positive bacteria. J Immunol. 177: 7959-7967.

[29] Klechevsky E, Morita R, Liu M, Cao Y, Coquery S, Thompson-Snipes L, *et al* (2008). Functional specializations of human epidermal langerhans cells and CD14+ dermal dendritic cells. Immunity. 29: 497-510.

[30] Kubo A, Nagao K, Yokouchi M, Sasaki H, Amagai M (2009) External antigen uptake by langerhans cells with reorganization of epidermal tight junction barriers. J Exp Med. 206: 2937-2946.

[31] Jesudason S, Collins MG, Rogers NM, Kireta S, Coates PTH (2012) Non-human prmate dendritic cells. J of Leukocyte Biol. 91: 217-228.

[32] Bertho N, Marquet F, Pascale F, Kang C, Schwartz-Cornil I (2011) Steady state pig dendritic cells migrating in skin draining pseudo-afferent lymph are semi mature. Vet Immunol Immunopathol. 144: 430-436.

[33] Dudziak D, Kamphorst AO, Heidkamp GF, Buchholz VR, Trumpfheller C, Yamazaki S, *et al* (2007) Differential antigen processing by dendritic cell subsets in vivo. Science. 315: 107-111.

[34] Hildner K, Edelson BT, Purtha WE, Diamond M, Matsushita H, Kohyama M, *et al* (2008) Batf3 deficiency reveals a critical role for CD8alpha+ dendritic cells in cytotoxic T cell immunity. Science. 322: 1097-1100.

[35] den Haan JM, Lehar SM, Bevan MJ (2000) CD8(+) but not CD8(-) dendritic cells cross-prime cytotoxic T cells in vivo. J Exp Med. 192: 1685-1696.

[36] Schnorrer P, Behrens GM, Wilson NS, Pooley JL, Smith CM, El-Sukkari D, *et al* (2006) The dominant role of CD8+ dendritic cells in cross-presentation is not dictated by antigen capture. Proc Natl Acad Sci USA. 103: 10729-10734.

[37] Poulin LF, Salio M, Griessinger E, Anjos-Afonso F, Craciun L, Chen JL, *et al* (2010) Characterization of human DNGR-1+ BDCA3+ leukocytes as putative equivalents of mouse CD8alpha+ dendritic cells. J Exp Med. 207: 1261-1267.

[38] Liu K, Nussenzweig MC. Development and homeostasis of dendritic cells (2010) Eur J Immunol. 40: 2099-2102.

[39] Steinman RM, Banchereau J (2007). Taking dendritic cells into medicine. Nature. 449: 419-426.

[40] Bialecki E, Macho Fernandez E, Ivanov S, Paget C, Fontaine J, Rodriguez F, *et al* (2011) Spleen-resident CD4+ and CD4- CD8alpha- dendritic cell subsets differ in their ability to prime invariant natural killer T lymphocytes. PLoS One. 6: e26919.

[41] Langellotti C, Quattrocchi V, Alvarez C, Ostrowski M, Gnazzo V, Zamorano P, *et al* (2012) Foot-and-mouth disease virus causes a decrease in spleen dendritic cells and the early release of IFN-alpha in the plasma of mice. Differences between infectious and inactivated virus. Antiviral Res. 94: 62-71.

[42] Tsuchida T, Matsuse H, Fukahori S, Kawano T, Tomari S, Fukushima C, *et al* (2012) Effect of respiratory syncytial virus infection on plasmacytoid dendritic cell regulation of allergic airway inflammation. Int Arch Allergy Immunol. 157: 21-30.

[43] Goubier A, Dubois B, Gheit H, Joubert G, Villard-Truc F, Asselin-Paturel C, *et al* (2008) Plasmacytoid dendritic cells mediate oral tolerance. Immunity. 29: 464-475.

[44] Bjorck P, Leong HX, Engleman EG (2011) Plasmacytoid dendritic cell dichotomy: Identification of IFN-alpha producing cells as a phenotypically and functionally distinct subset. J Immunol. 186: 1477-1485.

[45] Bursch LS, Wang L, Igyarto B, Kissenpfennig A, Malissen B, Kaplan DH, et al (2007) Identification of a novel population of langerin+ dendritic cells. J Exp Med. 204: 3147-3156.

[46] Jesudason S, Collins MG, Rogers NM, Kireta S, Coates PT (2012) Non-human primate dendritic cells. J Leukoc Biol. 91: 217-228.

[47] van Niel G, Wubbolts R, Stoorvogel W (2008) Endosomal sorting of MHC class II determines antigen presentation by dendritic cells. Curr Opin Cell Biol. 20: 437-444.

[48] Pulendran B, Ahmed R (2011) Immunological mechanisms of vaccination. Nat Immunol. 12: 509-17.

[49] Delamarre L, Mellman I (2011) Harnessing dendritic cells for immunotherapy. Semin Imunol. 23: 2-11.

[50] Jiang A, Bloom O, Ono S, Cui W, Unternaehrer J, Jiang S, *et al* (2007) Disruption of E-cadherin-mediated adhesion induces a functionally distinct pathway of dendritic cell maturation. Immunity. 27: 610-624.

[51] Manicassamy S, Reizis B, Ravindran R, Nakaya H, Salazar-Gonzalez RM, Wang YC, *et al* (2010) Activation of beta-catenin in dendritic cells regulates immunity versus tolerance in the intestine. Science. 329: 849-853.

[52] Hadeiba H, Lahl K, Edalati A, Oderup C, Habtezion A, Pachynski R, et al (2012) Plasmacytoid dendritic cells transport peripheral antigens to the thymus to promote central tolerance. Immunity. 36: 438-450.

[53] Fujita H, Meyer N, Akdis M, Akdis CA (2012) Mechanisms of immune tolerance to allergens. Chem Immunol Allergy. 96: 30-38.

[54] Cools N, Petrizzo A, Smits E, Buonaguro FM, Tornesello ML, Berneman Z, *et al* (2011) Dendritic cells in the pathogenesis and treatment of human diseases: A Janus Bifrons? Immunotherapy. 3: 1203-1222.

[55] RampeyAM, Lathers DMR, Woodworth BA, Schlosser RJ (2007) Immunolocalization of dendritic cells and pattern recognition receptors in chronic rhinosinusitis. Am J Rhinol. 21: 117-121.

[56] Dzopalic T, Rajkovic I, Dragicevic A, Colic M. The response of human dendritic cells to co-ligation of pattern-recognition receptors. Immunol Res. 2012 Mar 6.

[57] Li C, Zienkieviwicz J, Hawiger (2005) Interactive Sites in the MyD88 Toll/Interleukin (IL) 1 Receptor Domain Responsible for Coupling to the IL1β Signaling Pathway. J Biol Chem. 280: 25152-26159.

[58] Ohnishi H, Tochio H, Kato Z, Orii KE, Li A, Kimura T, Hiroaki H, Kondo N, Shirakawa M (2009) Structural basis for the multiple interactions of the MyD88 TIR domain in TLR4 signaling. Proc NAtl Acad Sci USA. 106: 10260-10265.

[59] Kawai T, Akira S (2010) The role of pattern-recognition receptors in innate immunity: Update on toll-like receptors. Nat Immunol. 11: 373-384.

[60] Montero Vega MT (2008) A new era for innate immunity. Allergol Immunopathol. 36: 164-175.

[61] Balenga NA, Balenga NA (2007) Human TLR11 gene is repressed due to its probable interaction with profilin expressed in human. Med Hypotheses. 68: 456.

[62] Govindaraj RG, Manavalan B, Lee G, Choi S (2010) Molecular modeling-based evaluation of hTLR10 and identification of potential ligands in toll-like receptor signaling. PLoS One. 5: e12713.

[63] Zelensky AN, Gready JE (2005) The C-type lectin-like domain superfamily. FEBS Journal. 272: 6179-6217.

[64] Figdor CG, van Kooyk Y, Adema GJ (2002) C-type lectin receptors on dendritic cells and langerhans cells. Nat Rev Immunol. 2: 77-84.

[65] van den Berg LM, Gringhuis SI, Geijtenbeek TBH (2012) An evolutionary perspective on C-type lectins in infection and immunity. Ann N Y Acad Sci. *In press.*

[66] O'Neill LAJ, Bowie AG (2010) Sensing and signaling in antiviral innate immunity. Current Biology. 20: R328-333.

[67] Mandl JN, Akondy R, Lawson B, Kozyr N, Staprans SI, Ahmed R, *et al* (2011) Distinctive TLR7 signaling, type I IFN production, and attenuated innate and adaptive immune responses to yellow fever virus in a primate reservoir host. The Journal of Immunology. 186: 6406-6401.

[68] Vincent IE, Zannetti C, Lucifora J, Norder H, Protzer U, Hainaut P, *et al* (2011) Hepatitis B virus impairs TLR9 expression and function in plasmacytoid dendritic cells. PLoS One. 6: e26315.

[69] De Gregorio E, D'Oro U, Wack A (2009) Immunology of TLR-independent vaccine adjuvants. Curr Opin Immunol 21: 339-345.

[70] Mannhalter JW, Neychev HO, Zlabinger GJ, Ahmad R, Eibl MM (1985) Modulation of the human immune response by the non-toxic and non-pyrogenic adjuvant aluminium hydroxide: effect on antigen uptake and antigen presentation. Clin Exp Immunol 61: 143-151.

[71] Franchi L, Núñez G (2008) The Nlrp3 inflammasome is critical for aluminium hydroxide-mediated IL-1b secretion but dispensable for adjuvant activity. Eur J Immunol. 38: 2085-2089.

[72] Ariza ME, Ramakrishnan R, Singh NP, Chauhan A, Nagarkatti PS, Nagarkatti M (2011) Bryostatin-1, a naturally occurring antineoplastic agent, acts as a Toll-like receptor 4

(TLR-4) ligand and induces unique cytokines and chemokines in dendritic cells. J Biol Chem. 286: 24-34.

[73] Castro FVV, Tutt AL, White AL, Teeling JL, James S, French RR, *et al* (2008) CD11c provides an effective immunotarget for the generation of both CD4 and CD8 T cell responses. Eur J Immunol. 38: 2263-2273.

[74] Carayanniotis G, Barber BH (1987) Adjuvant-free IgG responses induced with antigen coupled to antibodies against class II MHC. Nature. 327: 59-61.

[75] Guo X, Jia H, Zhang Q, Yuan W, Zhu G, Xin T, *et al* (2012) CpG-enriched plasmid enhances the efficacy of the traditional foot-and-mouth disease killed vaccine. Microbiol Immunol. *In press.*

[76] Cheng C, Jain P, Bettahi I, Pal S, Tifrea D, de la Maza LM (2011) A TLR2 agonist is a more effective adjuvant for a Chlamydia major outer membrane protein vaccine than ligands to other TLR and NOD receptors. Vaccine. 29: 6641-6649.

[77] Caminschi I, Shortman K (2012) Boosting antibody responses by targeting antigens to dendritic cells. Trends Immunol. 33: 71-77.

[78] Perrot I, Deauvieau F, Massacrier C, Hughes N, Garrone P, Durand I, et al (2010) TLR3 and Rig-Like Receptor on Myeloid Dendritic Cells and Rig-Like Receptor on Human NK Cells Are Both Mandatory for Production of IFN-γ in Response to Double-Stranded RNA. J Immunol 185: 2080-2088.

[79] Kumar H, Koyama S, Ishii KJ, Kawai T, Akira S (2008) Cutting Edge: Cooperation of IPS-1- and TRIF-Dependent Pathways in Poly IC-Enhanced Antibody Production and Cytotoxic T Cell Responses. J immunol 180: 683-687.

[80] Raman VS, Bhatia A, Picone A, Whittle J, Bailor HR, O'Donnell J, *et al* (2010) Applying TLR synergy in immunotherapy: implications in cutaneous leishmaniasis. J Immunol. 185: 1701-1710.

[81] Zhu Q, Egelston C, Gagnon S, Sui Y, Belyakov IM, Klinman DM, *et al* (2010) Using 3 TLR ligands as a combination adjuvant induces qualitative changes in T cell responses needed for antiviral protection in mice. J Clin Invest. 120: 607-616.

[82] Lang R, Schoenen H, Desel C (2011) Targeting Syk-Card9-activating C-type lectin receptors by vaccine adjuvants: Findings, implications and open questions. Immunobiology. 216: 1184-1191.

[83] 7883 Klechevsky E, Flamar AL, Cao Y, Blanck JP, Liu M, O'Bar A, et al (2010) Cross-priming CD8+ T cells by targeting antigens to human dendritic cells through DCIR. Blood. 116: 1685-1697.

[84] Ueno H, Schmitt N, Klechevsky E, Pedroza-Gonzalez A, Matsui T, Zurawski G, *et al* (2010) Harnessing human dendritic cell subsets for medicine. Immunol Rev. 234: 199-212.

[85] Appay V, Douek DC, Price DA (2008) CD8+ T cell efficacy in vaccination and disease. Nat Med Jun;14(6):623-628.

[86] Harlin H, Meng Y, Peterson AC, Zha Y, Tretiakova M, Slingluff C, *et al* (2009) Chemokine Expression in Melanoma Metastases Associated with CD8+ T-Cell Recruitment. Cancer Res. 69: 3077-3085.

[87] Palucka K, Banchereau J (2012) Cancer immunotherapy via dendritic cells. Nat Rev Cancer. 12: 265-277.

[88] Schuster SJ, Neelapu SS, Gause BL, Muggia FM, Gockerman JP, Sotomayor EM, *et al* (2009) Idiotype vaccine therapy (BiovaxID) in follicular lymphoma in first complete remission: Phase III clinical trial results. ASCO Meeting Abstracts. 27: 2.

[89] Nguyen-Pham TN, Lee YK, Lee HJ, Kim MH, Yang DH, Kim HJ, *et al* (2012) Cellular immunotherapy using dendritic cells against multiple myeloma. Korean J Hematol. 47: 17-27.

[90] Look M, Bandyopadhyay A, Blum JS, Fahmy TM (2010) Application of nanotechnologies for improved immune response against infectious diseases in the developing world. 62: 378-393.

[91] Peek LJ, Middaugh CR, Berkland C (2008) Nanotechnology in vaccine delivery. Adv Drug Deliv Rev. 60: 915-928.

[92] Aguilar JC, Rodriguez EG (2007) Vaccine adjuvants revisited. Vaccine 25: 3752-3762.

[93] Chadwick S, Kriegel C, Amiji M (2010) Nanotechnology solutions for mucosal immunization. Adv Drug Deliv Rev. 62: 394-407.

[94] Azad N, Rojanasakul Y (2006) Nanobiotechnology in Drug Delivery. American Journal of Drug Delivery. 4: 79-88.

[95] Thoelen S, De Clercq N, Tornieporth N. A prophylactic hepatitis B vaccine with a novel adjuvant system. 2001;19:2400-2403.

[96] Tregnaghi MW, Voelker R, Santos-Lima E, Zambrano B (2010) Immunogenicity and safety of a novel yeast Hansenula polymorpha-derived recombinant Hepatitis B candidate vaccine in healthy adolescents and adults aged 10-45 years. Vaccine. 28: 3595-3601.

[97] FDA, United States Food and Drug Administration (2010) Vaccines licensed for immunization and distribution in the US with supporting documents. Available at: http://www.fda.gov/BiologicsBloodVaccines/Vaccines/ApprovedProducts/ucm093833.htm. Accessed 2012 April 15.

[98] Jones T (2009) GSK's novel split-virus adjuvanted vaccines for the prevention of the H5N1 strain of avian influenza infection. Curr Opin Mol Ther. 11: 337-345.

[99] Bovier PA (2008) Epaxal: a virosomal vaccine to prevent hepatitis A infection.Expert Rev Vaccines. 7: 1141-1150.

[100] Herzog C, Hartmann K, Künzi V, Kürsteiner O, Mischler R, Lazar H, *et al* (2009) Eleven years of Inflexal V-a virosomal adjuvanted influenza vaccine. Vaccine 27: 4381-4387.

[101] Mata E, Igartua M, Patarroyo ME, Pedraz JL, Hernández RM (2011) Enhancing immunogenicity to PLGA microparticulate systems by incorporation of alginate and RGD-modified alginate. Eur J Pharm Sci. 44: 32-40.

[102] Carrillo-Conde B, Song EH, Chavez-Santoscoy A, Phanse Y, Ramer-Tait AE, Pohl NL, *et al* (2011) Mannose-functionalized "pathogen-like" polyanhydride nanoparticles target C-type lectin receptors on dendritic cells. Mol Pharm. 8: 1877-1886.

[103] Saini V, Jain V, Sudheesh MS, Jaganathan KS, Murthy PK, Kohli DV (2011) Comparison of humoral and cell-mediated immune responses to cationic PLGA microspheres containing recombinant hepatitis B antigen. Int J Pharm. 408: 50-57.

[104] Makhlof A, Tozuka Y, Takeuchi H (2009) pH-Sensitive nanospheres for colon-specific drug delivery in experimentally induced colitis rat model. Eur J Pharm Biopharm. 72: 1-8.

[105] Butoescu N, Seemayer CA, Foti M, Jordan O, Doelker E (2009) Dexamethasone-containing PLGA superparamagnetic microparticles as carriers for the local treatment of arthritis. Biomaterials. 30: 1772-1780.

[106] Tang Y, Singh J (2009) Biodegradable and biocompatible thermosensitive polymer based injectable implant for controlled release of protein. Int J Pharm. 365: 34-43.

[107] Yang A, Yang L, Liu W, Li Z, Xu H, Yang X (2007) Tumor necrosis factor alpha blocking peptide loaded PEG-PLGA nanoparticles: Preparation and in vitro evaluation. Int J Pharm. 331: 123-132.

[108] Salvador A, Igartua M, Hernández RM, Pedraz JL (2012) Combination of immune stimulating adjuvants with poly(lactide-co-glycolide) microspheres enhances the immune response of vaccines. Vaccine. 30: 589-596.

[109] Kool M, Petrilli V, De Smedt T, Rolaz A, Hammad H, van Nimwegen M, *et al* (2008) Cutting Edge: Alum Adjuvant Stimulates Inflammatory Dendritic Cells through Activation of the NALP3 Inflammasome. J Immunol. 181: 3755-3759.

[110] Kool M, Soullie T, van Nimwegen M, Willart MAM, Muskens F, Jung S, *et al* (2008) Alum adjuvant boosts adaptive immunity by inducing uric acid and activating inflammatory dendritic cells. J Exp Med. 205: 869-882.

[111] Sharp FA, Ruane D, Claass B, Creagh E, Harris J, Malyala P, *et al* (2009) Uptake of particulate vaccine adjuvants by dendritic cells activates the NALP3 inflammasome. Proc Natl Acad Sci USA. 106: 870-875.

[112] Demento SL, Eisenbarth SC, Foellmer HG, Platt C, Caplan MJ, Mark Saltzman W, *et al* (2009) Inflammasome-activating nanoparticles as modular systems for optimizing vaccine efficacy. Vaccine. 27: 3013-3021.

[113] Sun H, Xie Y, Ye Y (2009) ISCOMs and ISCOMATRIX(TM). Vaccine. 27: 4388-4401.

[114] Altin JG, Parish CR (2006) Liposomal vaccines--targeting the delivery of antigen. Methods. 40: 39-52.

[115] Kanchan V, Panda AK (2007) Interactions of antigen-loaded polylactide particles with macrophages and their correlation with the immune response. Biomaterials. 28: 5344-5357.

[116] Chua BY, Al Kobaisi M, Zeng W, Mainwaring D, Jackson DC (2012) Chitosan microparticles and nanoparticles as biocompatible delivery vehicles for peptide and protein-based immunocontraceptive vaccines. Mol Pharm. 9: 81-90.

[117] Mohanan D, Slütter B, Henriksen-Lacey M, Jiskoot W, Bouwstra JA, Perrie Y, *et al* (2010) Administration routes affect the quality of immune responses: A cross-sectional evaluation of particulate antigen-delivery systems. J Contol Release. 147: 342-349.

[118] Manolova V, Flace A, Bauer M, Schwarz K, Saudan P, Bachmann MF (2008) Nanoparticles target distinct dendritic cell populations according to their size. Eur J Immunol 38: 1404-1413.

[119] Cruz LJ, Tacken PJ, Fokkink R, Joosten B, Stuart MC, Albericio F, et al (2010) Targeted PLGA nano- but not microparticles specifically deliver antigen to human dendritic cells via DC-SIGN in vitro. J Contol Release. 144: 118-126.

[120] Foged C, Brodin B, Frokjaer S, Sundblad A (2005) Particle size and surface charge affect particle uptake by human dendritic cells in an in vitro model. Int J Pharm 298: 315-322.

[121] Carcaboso Á, Hernández R, Igartua M, Rosas J, Patarroyo M, Pedraz J (2004) Enhancing Immunogenicity and Reducing Dose of Microparticulated Synthetic Vaccines: Single Intradermal Administration. Pharm Res. 21: 121-126.

[122] Speiser DE, Schwarz K, Baumgaertner P, Manolova V, Devevre E, Sterry W, *et al* (2010) Memory and effector CD8 T-cell responses after nanoparticle vaccination of melanoma patients. J Immunother. 33: 848-858.

Chapter 4

Stress-Induced Molecules in Regulation of NK Cell Activity

Elena Kovalenko, Leonid Kanevskiy, Anna Klinkova, Anastasiya Kuchukova, Maria Streltsova, William Telford and Alexander Sapozhnikov

Additional information is available at the end of the chapter

1. Introduction

Natural killer (NK) cells are a heterogeneous multifunctional population of immune cells traditionally classified as cytotoxic lymphocytes of the innate immune system [1]. NK cells possess several key characteristics of innate immune cells. They are capable of rapid effector response without prior sensitization and express a range of germline-encoded receptors that do not need any additional rearrangement. On the other hand NK cells share some phenotypic and functional characteristics with cytotoxic T cells: they express common surface markers (CD2, CD7, CD8), recognize infected and tumor-transformed cells of host organism and also mediate the similar mechanism of granule-dependent cytolysis. NK cells are able to interact with major histocompatibility complex class I (MHC-I) self-molecules as distinct from other innate immune cells. Moreover, recent findings show that NK cells possess a form of immunological memory [2]. Thus, NK cells represent evolutionary and functional link connecting innate and adaptive immunity.

In the organism NK cells fulfill two main functions: they provide constitutive cytotoxic activity against infected, malignant and other damaged cells and regulate innate and adaptive immune reactions by production of chemokines and cytokines. At early stage of immune response NK cells are the main sources of IFN-γ, triggering the adaptive immune response [3]. NK cells interact with other cells of immune system, participating in contact-dependent costimulation. Primed dendritic cells (DCs) and macrophages produce a range of cytokines including IL-12, IL-15 and IL-18 which activate NK cells. Contact interaction of NK cells with these cells also induces macrophage and DC maturation [4, 5]. NK cells express receptors for many cytokines and chemokines and therefore might be influenced by other cells producing them [6].

NK cells are classically identified in humans as lineage-negative $CD3^-CD14^-CD19^-CD56^+$ leukocytes circulating in peripheral blood and residing in various organs and tissues of the body. Two main subpopulations of NK cells based on the expression levels of CD16 and CD56 have been described [7]. Subpopulation $CD56^{dim}CD16^{+}$ is prevalent in peripheral blood. These cells express receptors of killer-cell immunoglobulin-like receptor (KIR) family and possess high levels of cytotoxic activity. Because they express CD16 (FcγRIII), $CD56^{dim}$ NK cells are able to recognize and lyze cells opsonized with IgG antibodies [8, 9]. $CD56^{bright}CD16^{-}$ NK cells consist of approximately 5-10% of peripheral blood NK cells; they reside mostly in secondary lymphoid organs, liver and uterus. $CD56^{bright}$ cells express CCR7 and CD62L (L-selectin) which allow them to migrate to the lymph nodes. They are considered to be a regulatory NK cell subpopulation because they are able to contact with DCs, have low cytotoxicity and high levels of cytokine production compared with $CD56^{dim}$ cells. Due to constitutitive expression of CD25 these cells can easily proliferate in response to IL-2 stimulation.

In general, functional activity of NK cells is regulated by fine dynamic balance between the activating and inhibitory signals providing by the interaction of receptors and co-stimulatory molecules with surface ligands of potential target cells, and depends also on the soluble factors produced by cellular environment [10, 11]. The density of interacting molecules on both effector and target cell surface is crucial for NK cell activity. Low or absence of inhibitory signals leads predominantly to NK cell activation through activating receptors followed by target cell lysis and cytokine release. On the other hand, high expression of activating ligands on target cells may trigger NK cell cytolytic activity, despite normal expression of MHC-I molecules [12]. The repertoire of activating and inhibitory receptors expressed on single NK cells is developed randomly during its maturation and differentiation [13]. The result of this expression is a diverse set of NK cell clones in the body. NK cells also express various adhesion and co-stimulatory molecules that participate in cell interaction and signaling. Diversity of NK cells allows them to identify damaged cells with distinct phenotypic characteristics and display functional response adapted to microenvironment.

NK cells express multiple inhibitory and activating receptors regulating their cytolytic activity. Receptors of the KIR family recognize MHC-I molecules expressed by most normal cells in the body. Interaction of inhibitory ITIM-bearing KIRs with MHC-I results in abrogation of activation. MHC-I expression on the surface of virally infected and tumor cells have often been shown to decrease significantly, making these cells "invisible" to cytotoxic T lymphocyte recognition [14, 15]. In contrast, NK cells are able to recognize and lyse such cells due to diminished inhibitory signals. It is important to note that KIRs are able to distinguish between groups of MHC alleles. In the case of transplantation, graft cells with MHC-I molecules differing from the host proteins can be rejected by host NK cells. Various activating receptors for the most part serve to recognize cell surface molecules whose expression indicates viral infection, tumorigenesis or cell damage caused by cellular stress. Activating receptors of natural cytotoxicity receptor (NCR) family including NKp46, NKp44 and NKp30 drive natural cytotoxicity against many tumor and virally infected cells upon signaling through linked ITAM-bearing CD3ζ, FcRγ, or DAP12 molecules [16, 17]. It is believed that these receptors have the ability to recognize molecules not expressed on the surface of normal cells. In recent years, most of the ligands recognized by these receptors have been identified. NKp46 and NKp44 bind virus-derived hemagglutinins, and NKp30 is able to recognize human cytomegalovirus-encoded pp65 protein [18, 19]. The normally intracellular proteins vimentin and leukocyte antigen-B-associated transcript 3 (BAT3) are recognized by NKp46 and NKp30, respectively [20]. NKp30 binds a B7 family homolog (B7-H6) expressed by some tumor but not normal cells [21]. A group of heparan sulfate/heparin molecules has been also proposed as ligands for NCR receptors [22]. Another important activating receptor NKG2D recognizes self-molecules which are usually absent or expressed on low level on the surface of normal cells but become expressed in tumor transformation, infection or in conditions of cell stress. A number of highly polymorphic "self" stress-induced proteins such as MICA, MICB and a group of UL16 binding proteins (ULBPs) are ligands of NKG2D [23-25]. Appearance of these molecules on the cell surface serves for detection by NK cells of cell damaging alterations induced by oxidative stress, heat shock, hypoxia, genotoxic stress, virus infection or tumor transformation.

Heat shock proteins (HSPs) represent another group of endogenous stress-induced proteins recognized by NK cells on the surface of target cells. HSPs, normally presented in cytoplasm, can be found on cell surface, in particular on the surface of tumor cells, virus-infected cells and cells subjected to stress. Tumor and stressed cells can also release HSPs in the extracellular milieu. It has been shown that expression of HSPs, in particular HSP70, on the surface of tumor cells increased cytotoxic activity of NK cells towards the targets. However, the activating receptors of NK cells for these molecules have not been identified. Membrane associated HSPs and their extracellular pool may constitute danger-associated molecular patterns (DAMPs) in the context of the "Danger model" proposed by P. Matzinger [26, 27]. Detection of stress-induced molecules as "danger signals" in conjunction with self MHC-I recognition by multiple sensor system possessed by NK cells underlies the complex process of NK cell activity regulation, although a full range of activating ligands recognized by NK cells still remains unknown. Reactive oxygen species (ROS) produced by activated monocytes and neutrophils also affect on NK cell functionality. ROS action leads to lowering of NKG2D and NKp46 expression by NK cells, that should decrease their cytolytic ability. Histamine was shown to prevent this effect of ROS on NK cells. Interesting, that $CD56^{bright}$ NK cells are more resistant to ROS action: NKG2D expression is unchanged on these cells [28, 29].

Simultaneous interactions of various surface molecules of target cells with NK cells lead to the integration of different intracellular signals, which together dictate the quality and intensity of effector NK cell response. Small changes in the target cell surface molecular profile may significantly influence the susceptibility of the cells to NK cell action. Cell stress induces multiple changes in protein synthesis and intracellular localization, and influences oligosaccharide expression. Intracellular perturbations caused by some types of cell stress might influence cell death and survival processes and are in close connection with the process of tumor cell transformation.

Activity of NK cells may be influenced by both contact interactions with target cells and various soluble factors produced by surrounding cells under stress conditions, in particular by extracellular pool of HSPs and MICA/B molecules. Besides stimulation of stress-induced protein expression, cell stress significantly modifies surface carbohydrate phenotype affecting NK cell effector functions. Membrane-associated oligosaccharides participate in NK cell – target cell interactions. Some of them have been identified to decrease NK cell cytolytic activity, others displayed stimulating effects on NK cells [30]. In this work we discuss our results obtained in the study of surface oligosaccharide influence on NK cell activation utilizing polymeric lipophilic neoglycoconjugates. We also present a study of involvement of surface and extracellular pools of HSP70 and MICA/B in regulation of human NK cells activity. The obtained results demonstrate that surface and soluble HSP70, as well as some fragments of this protein, activate NK cells in contrast to extracellular forms of MICA/B which inhibit cytotoxicity of NK cells. In addition, some characteristics of cell surface MICA/B expression in a model of ethanol-induced cell stress are discussed.

2. Oligosaccharides in regulation of NK cell activity

Oligosaccharides exposed on the plasma membrane of mammalian cells are an essential component of intercellular contacts inhibiting or cooperating with certain molecular interactions. Carbohydrate-carrying molecules also play a role in cell adhesion and signaling and contribute to immune cell activation or inhibition. Considerable changes in cell surface carbohydrate expression can be evoked by malignant transformation, infection or other unfavorable conditions and can be detected by NK cell recognizing system. Membrane-associated glycans by their structural diversity might provide fine tuning of NK cell – target cell interactions and regulate NK cell activity. Positive and negative effects of various carbohydrate compounds including sialo-oligosaccharides, glycosphingolipids, complex and high mannose-type N-glycans, and others on NK cell-mediated effector functions have also been reported [31-33]. However only in rare cases ligand-receptor signaling pathways underlying the saccharide effects were identified. Recently, inhibitory receptor Siglec-7 expressed by NK cells has been revealed to recognize specifically disaccharide Neu5Acα2-8Neu5Ac [34]. It has therefore been suggested that the NK cell recognition of the sialic residues contributes to their tolerance to self-tissues.

In our study we tested synthetic glycoconjugates designed to mimic the biological activities of native oligosaccharides for investigation of the effects of an array of oligosaccharides on NK cell activity. The lipophilic glycoconjugates were incorporated in target cells modifying carbohydrate phenotype of the plasma membrane, and NK cell cytotoxicity and cytokine production were assessed (Fig. 1).

We found that glycoconjugates containing Le^x, HSO_3Le^x and Le^y structural motifs enhanced NK cell-mediated cytolysis of K562 cells as measured by analysis of NK cell-mediated caspase 6 activation in these cells (Fig. 2). The same oligosaccharides exposed on the cell surface displayed a potency to stimulate IFN-γ production with the strongest effect revealed for Le^y-containing glycoconjugate. Only polymeric and not monomeric glycoconjugates displayed the abilities to increase NK cell activity. Le^x antigen expressed as a part of glycoproteins and glycosphingolipids on many hematopoietic cells is mainly involved in intercellular adhesion. Both Le^x and Le^y are also considered to be tumor-associated antigens [35]. Not all donors exhibited NK cell sensitivity to these glycoconjugates. This suggests a role for individual genetic polymorphism in glycan receptor expression, and also other mechanisms where the expression of molecules interacted with glycans might be regulated in the human organism throughout life. Polymeric glycoconjugates also had more biological activity than monomeric constructs in this system, suggesting that appropriate presentation is critical for carbohydrate recognition and subsequent biological effects. The unique mode of clustered oligosaccharide presentation provided by polymeric glycoconjugates may effectively mimic carbohydrate recognition found in nature. Most carbohydrate-lectin and carbohydrate-carbohydrate bonds require multivalent carbohydrate interactions for proper adhesion and signaling. The possible scenarios of carbohydrate-mediated interaction NK cells with target cells include the intercellular adhesion strengthening or/and the specific NK cell recognition of the saccharide determinants. Interestingly, only polymeric glycoconjugates presented on the cell surface in appropriate microenvironment affected NK cell activity. Glycoconjugates presented alone had no effect.

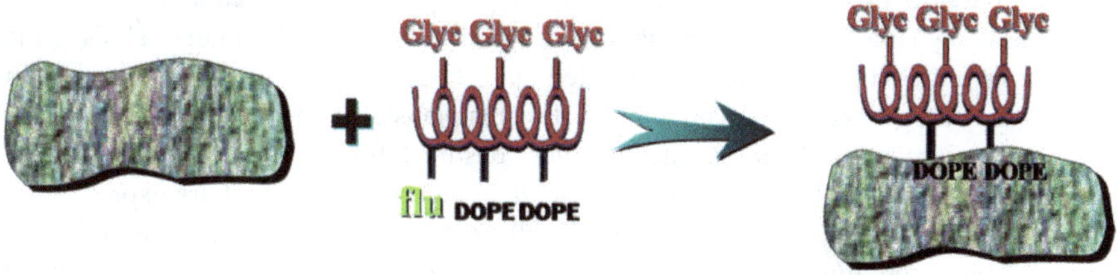

Figure 1. Modification of K562 cells targeting by NK cells with polymeric glycoconjugates.

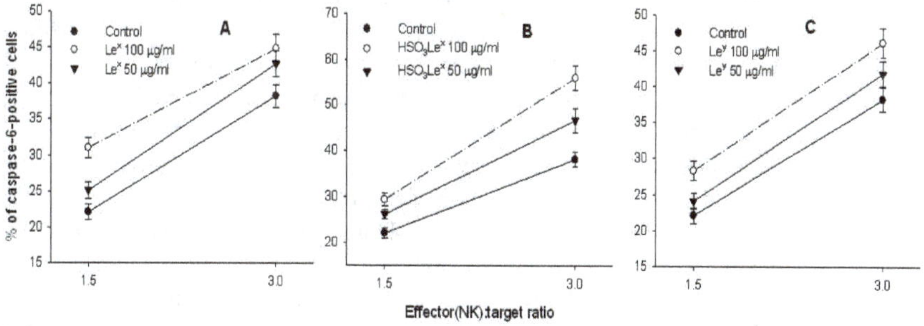

Figure 2. Effects of glycoconjugates containing (A) Le^x, (B) sulfated Le^x or (C) Le^y incorporated in target cell membrane on NK cell-mediated cytotoxicity.

Thus, a proper presentation mode of the oligosaccharides was an essential requirement for biological effects of glycans connected with carbohydrate adhesion and cell recognition. Recently suggested model of a "glycosynapse" involves membrane assembly of different types of glycosylated molecules interacting with carbohydrate partners [36]. According to this model, oligosaccharide effects depend on their clustering and proper orientation of "glycotopes", and may result in a synergistic effect with protein-dependent interactions. Further investigations may discover additional physiological functions of glycans consisting of certain glycoproteins and glycolipids expressed in specific cell types.

3. Surface and exogenous HSP70 in NK cell – Target cell interaction

3.1. Involvement of membrane-associated HSP70 in NK cell – Target cell interaction

As was mentioned above, stress-induced proteins exposed on surface of cells interacting with NK cells can be recognized by NK activating receptors followed by cytotoxic attack toward target cells. This fact highlights an important role of cell stress in the process of NK cell – target cell interaction. It is known that cellular stress response to any form of damaging factor results in a wide range of protective intracellular processes directed toward cell survival. Of critical importance among these counteracting cell reactions is the induction of high levels of HSP expression. HSPs are highly conserved and ubiquitously expressed proteins that are predominantly located in the cytoplasm and function as molecular chaperones [37]. However, evidence has now accumulated that translocation of HSPs to the cell surface may occur in stressed, infected and transformed cells [38-40]. Although mechanisms of translocation remain unclear, it seems plausible that surface HSPs may play a role as molecular markers of "altered" cells, destined to be eliminated by cytotoxic lymphocytes [39, 40]. Some tumors produce large amounts of HSPs that are expressed on the cell surface and can potentially attract NK cells [41, 42]. Surface expression of inducible form of HSP70 on human lung carcinoma cells was associated with their increased sensitivity to lysis by NK cells [43]. Also, IL-2 activated human NK cells were shown to recognize HSP70 on the surface of K562 erythroleukemia and human sarcoma cells subjected to heat shock [44-46]. Mechanisms by which surface HSPs may mediate recognition and elimination of tumor by cytotoxic lymphocytes remain poorly understood. One possibility is that immunogenic peptides expressed in association with HSP70 and HSP90 on tumor cells can be recognized by T lymphocytes, as well as NK cells [44, 48]. Another possibility is that HSPs themselves, e.g. HSP60 and HSP70 on the surface of tumor cells can be recognized by $\gamma\delta$-T cells in a non-MHC restricted manner [49]. Thus, the available evidence strongly suggests that surface HSPs, in particular HSP70, may be important in mediating recognition and/or elimination of tumor cells by host lymphocytes. Previously, we have shown that culture-adapted EL-4 mouse lymphoma lines express surface HSPs [37]. This fact was accounted for selection of EL-4 cell line in our *in vitro* mouse model aimed at investigation of a role of target cell surface HSP70 in tumor recognition and activation of cytotoxic effector cells. In this study we used EL-4 cells to analyze involvement of surface HSP70 of the target lymphoma cells in interaction with cytotoxic immune effectors.

For these experiments C57BL/6 (H-2^b), CBA (H-2^k) and Balb/c (H-2^d) mice were used. Cytotoxic effector cells were obtained from mouse spleen using standard methods [50]. The cytotoxic effectors were obtained either from intact mice or mice inoculated with

EL-4 lymphoma. In the latter case, C57BL/6 mice were inoculated intraperitoneally with $1x10^6$ EL-4 cells, and splenic effector cells were isolated 10 days after inoculation. Mouse EL-4 (H-2^b) lymphoma cells were maintained *in vitro* by passages in RPMI-1640 medium supplemented with 10% fetal calf serum (FCS), 4 mM L-glutamine, 20 mM HEPES, 10^{-5} M 2-ME and 50 μg/ml gentamicin. For flow cytometric analysis FITC- or PE-conjugated anti-CD8, anti-CD4, anti-CD3, anti-γδTCR and anti-NK-1.1 mAb were used for cell staining. Fluorescent and laser scanning confocal microscopy was used to confirm surface localization of HSP70 in EL-4 cells stained with anti-HSP70 antibody (clone BRM22). MTT- and LDH viability assays [51, 52] were applied to evaluate cytotoxic effect of the effectors in the *in vitro* models. To determine whether expression of MHC class I molecules and HSP70 on the surface of EL-4 cells plays a role in cytotoxic response against these targets, we performed antibody blocking assay using following mAb: anti-MHC class I (clones HB11 and HB51), anti-HSP70 (clone BRM22), or anti-Thy-1.2, as a control (clone 53-2). To remove $CD8^+$ cytotoxic T lymphocytes from the effector cell population a magnetic bead separation method was used (Dynal, Germany).

At the first stage of the study the surface localization of HSP70 on EL-4 cells registered previously by flow cytometry was confirmed with confocal microscopy (Fig. 3).

Splenic effector cells from normal C57BL/6 mice demonstrated strong cytotoxicity against EL-4 cells. By MTT assay, after 6 h incubation of EL-4 with effector cells, percent cytotoxicity ranged from 6 ± 1% to 28 ± 3%, at E:T ratios 5 and 20, respectively. Similar levels were detected by LDH release assay (Fig. 4A), thus confirming the validity of the MTT method. After 24 h incubation, cytotoxicity increased up to 18 ± 2% and 61 ± 15%, at E:T ratios 5 and 20, respectively (Fig. 4A). Effector cells from allogeneic mice (CBA and Balb/c) demonstrated similar levels of cytotoxicity against EL-4 cells (data not shown). Splenic effector cells from C57BL/6 mice inoculated with EL-4 cells 10 days prior to the experiment, showed negligible cytotoxicity after 6 h incubation. However, they exhibited a pronounced cytotoxic effect against EL-4 cells after 24 h of incubation with the targets (Fig. 4B).

To determine whether surface expression of HSP70 on EL-4 cells is important for their recognition and elimination by cytotoxic lymphocytes, we first incubated EL-4 cells with anti-HSP70 antibody and then used these cells as targets in a 6 hour cytotoxicity assay. Splenic effector cells from normal C57BL/6 mice showed markedly decreased cytotoxicity against target EL-4 cells pretreated with anti-HSP70 mAb (Fig. 5A). Inhibitory effect of the antibody was also observed in allogeneic co-culture, i.e. with effector cells from CBA (Fig. 5C) and Balb/C mice (data not shown). Similar result was obtained using splenic effector cells from C57BL/6 mice inoculated with EL-4 tumor: percent cytotoxicity 21 ± 5% (target EL-4 cells pretreated with anti-HSP70 antibody) versus 53 ± 11% (untreated EL-4 cells), at E:T ratio 20 and 24 h incubation. In contrast, pretreatment of target EL-4 cells with anti-H-2^b antibody resulted in their enhanced killing by syngeneic effector cells (Fig. 5B). A control anti-Thy1.2 antibody was without effect (Fig. 5D). It should be mentioned that treatment with antibodies alone (anti-H-2^b, anti-HSP70) for up to 24 h, did not affect viability of either target EL-4 cells or splenic effector cells (data not shown).

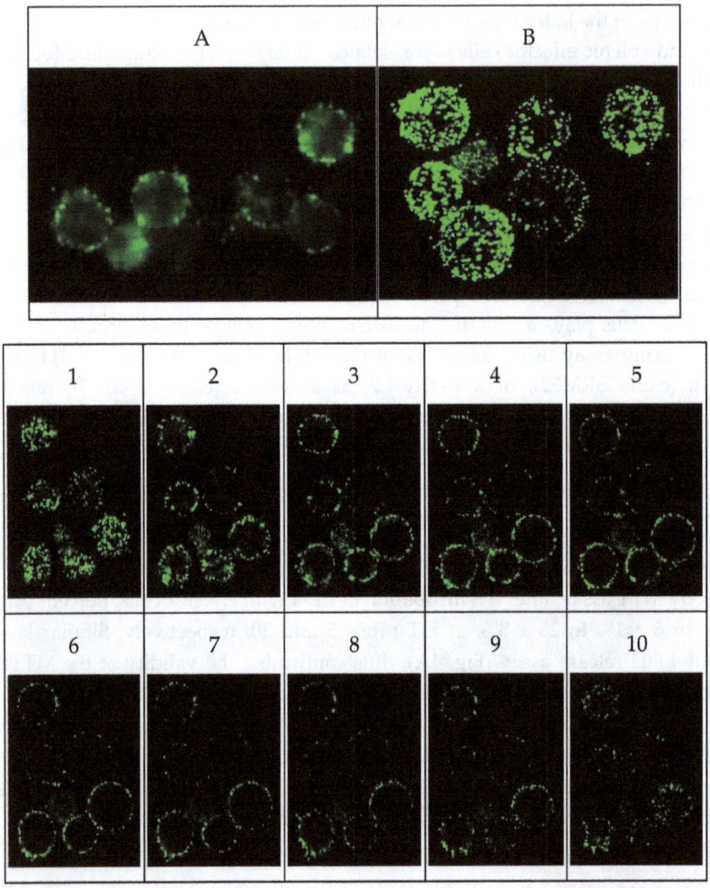

Figure 3. Visualization of cell surface expression of HSP70 in *in vitro* cultures of EL-4 cells performed by using fluorescent (A) and laser scanning confocal (B) microscopy (images were recorded using a Nikon TE-2000 confocal microscope; excitation was at 488 nm argon-ion laser) ; fluorescence emission was detected in the 500-550 nm for Alexa Fluor 488. Cells were imaged in a Z-stack and both sections and the reconstructed images examined for HSP localization. These images clearly demonstrate the cell surface localization of HSP70.

Next, we addressed the question of whether killing of EL-4 cells in our model was mediated by NK cells or cytotoxic T lymphocytes. To this end, we depleted $CD8^+$ cells from the splenic effector cell population obtained from intact C57BL/6 mice. Typically, effector cell population contained 5% sIg^+, 36% $CD8^+$, 48% $CD4^+$, 88% $CD3^+$, 2% $\gamma\delta TCR^+$, and 10% $NK1.1^+$ cells. After depletion of $CD8^+$ cells, the population contained 7% sIg^+, 60% $CD4^+$, 57% $CD3^+$, 3% $\gamma\delta TCR^+$, and 25% $NK1.1^+$ cells. Thus, it was relatively enriched for NK cells. As is shown in Fig. 6, cytotoxicity against EL-4 cells increased, rather than decreased, after depletion of $CD8^+$ cells from the effector cell population. Moreover, cytotoxicity was significantly

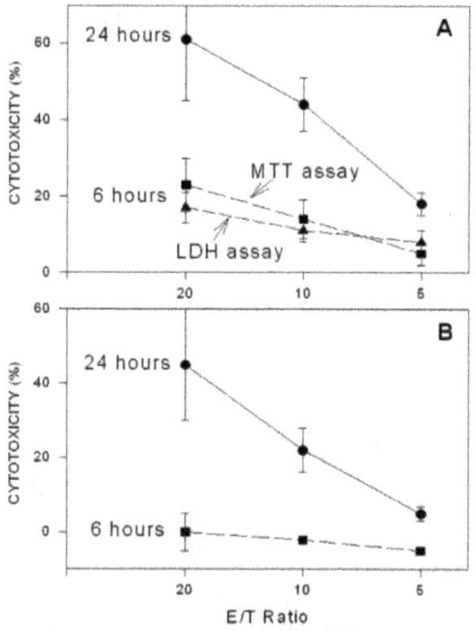

Figure 4. Cytotoxicity of C57BL/6 splenic effector cells after 6 and 24 h of incubation with the target EL-4 cells measured by MTT and LDH assays. (**A**) Cytotoxic effector cells were isolated from spleens of intact mice. (**B**) Cytotoxic effector cells were isolated from spleens of mice inoculated with EL-4 cells 10 days prior to experiment. Results are the mean of triplicate cultures ±S.E.M.

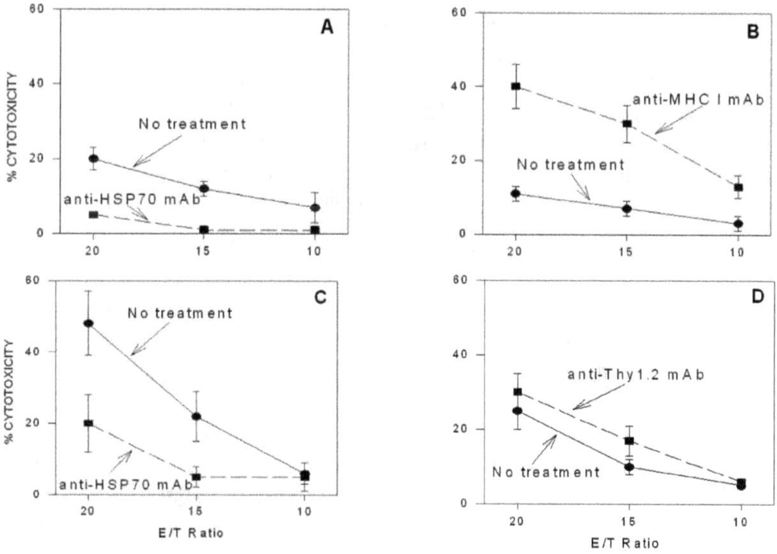

Figure 5. The effect of treatment of EL-4 target cells with antibodies to HSP70, MHC class I molecules and Thy1.2 on their sensitivity to killing by splenic effector cells. Cytotoxic effector cells were isolated from syngeneic C57BL/6 mice (**A**, **B**, and **D**) or allogeneic CBA mice (**C**). Results are the mean of triplicate cultures ±S.E.M.

inhibited by pretreatment of target EL-4 cells with anti-HSP70 mAb. Hence, lysis of EL-4 cells in this system seems to be predominantly mediated by $CD8^-$ cells, possibly NK cells, and may involve recognition of surface HSP70 present on EL-4 target cells.

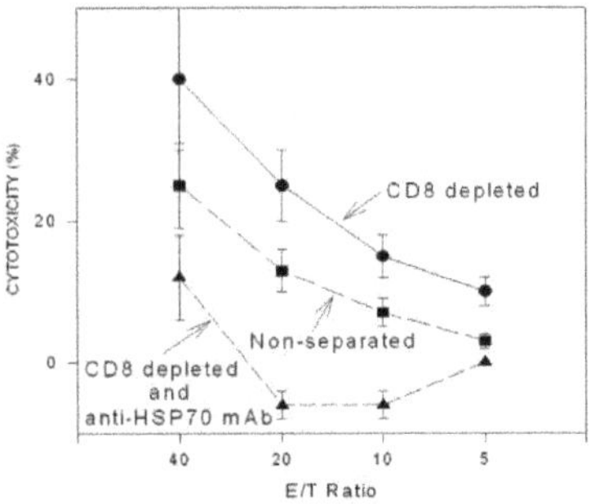

Figure 6. The effect of depletion of $CD8^+$ cells from splenic effector cell population obtained from intact C57BL/6 mice on cytotoxicity against EL-4 target cells, treated or untreated with anti-HSP70 antibody. Results are the mean of triplicate cultures ±S.E.M.

3.2. Exogenous HSP70 as mediators of NK cell activation

As stated above HSPs are intracellular proteins functioning as molecular chaperones; at the same time evidence has now accumulated that HSPs can localize on cell surface and can exit to the extracellular space, including blood. These two forms of HSPs – membrane-associated and extracellular soluble proteins – have immunomodulatory capabilities realizing by different mechanisms. Translocation of HSPs to plasma membrane has been observed in stressed, infected and transformed cells. It seems plausible that surface HSPs may represent molecular targets for recognition and elimination of 'altered' cells by cytotoxic lymphocytes, in particular NK cells. In this report we describe some models enabling to study mechanisms of interaction of NK cells with the target tumor cells bearing HSP70 on their surface.

Extracellular HSPs act as soluble immunostimulators affecting different immune cells such as DCs, macrophages, T and NK cells. HSPs can bind extracellular peptides from dead cells. Such HSP-peptide complexes become internalized by professional antigen-presenting cells – DCs or macrophages – where peptides release from complexes with HSP and become reassociated with MHC molecules. This process is called "cross-presentation"; HSPs act here as extracellular chaperones able to bind peptides of different structure [53-55]. On the other hand HSPs can also directly stimulate immune cells. For example, HSP70 added to DCs and macrophages induces their maturation and cytokine secretion. Exogenous HSP70 also stimulates proliferation and cytotoxicity of human NK cells, moreover, NK cells were shown to move by the gradient of HSP70 concentration [56, 57]. Thus, HSPs can act on immune cells similar to cytokines or chemokines. Due to such dualism extracellular HSPs were named "chaperokines", i.e. chaperones plus cytokines [58].

To clearly elucidate any immunostimulatory action of HSP70 on NK cells, we tested whether exogenous HSP70 influences NK cell cytokine secretion. In our experiments we analyzed effects of HSP70 on IFN-γ production by human NK cells. NK cells were isolated from peripheral blood samples by centrifugation on Ficoll density gradient with subsequent magnetic separation of mononuclear cell population using MACS NK cell isolation kit (Miltenyi Biotec, Germany) or cell sorting using fluorescent labeled CD3, CD16 and CD56 antibodies. Recombinant human HSP70 expressed in *E.coli* (#NSP-555, Stressgen, Canada) was used in experiments. HSP70 was added to isolated NK cells without or with 500 U/ml IL-2 (Hoffmann-La-Roche, Switzerland). After incubation for 18 h in 37°C, 5% CO_2, supernatants were collected and analyzed for IFN-γ content using Human IFN-γ ELISA kit (Pierce Endogen, USA). Effect of HSP70 on IFN-γ production by NK cells was also analyzed by flow cytometry. NK cells incubated with HSP70 and IL-2 were collected and fixed by paraformaldehyde. Cells were then labeled intracellularly with fluorescent conjugated antibody to human IFN-γ (BD Pharmingen, USA). Samples were then analyzed by flow cytometry (BD FACScan, BD Biosciences, San Jose, CA USA) and the percentage of NK cells producing IFN-γ was measured.

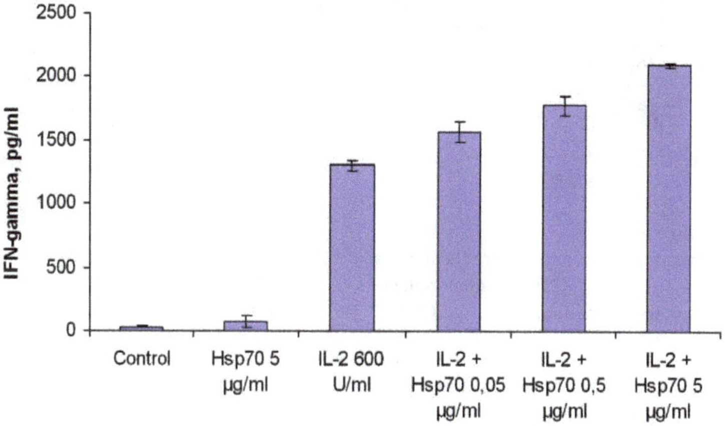

Figure 7. Recombinant human HSP70 stimulated IFN-γ production in NK cells activated by IL-2. IFN-γ production was measured by ELISA. Similar results were obtained with intracellular IFN-γ measurement by flow cytometry (data not shown).

HSP70 in dose 5 μg/ml did not induce IFN-γ production in native NK cells (Fig. 7). However HSP70, when added together with IL-2, did markedly stimulate it. Similar data were obtained with intracellular staining of NK cells. Thus HSP70 was shown to stimulate cytokine production in IL-2 activated NK cells. This fact is very consistent with the assumed role of HSPs as danger-associated molecular patterns (DAMPs) able to activate a range of cellular mediators of innate immunity. It seems HSP70 exerts a cytokine-like action on NK cells; however, it does not work by itself. It needs in initial activation by cytokines like IL-2 (similar results were obtained using IL-12 and IL-15, data not shown). It is important to note that no receptors for HSP70 on NK cells were found. Based on presented data we can speculate about a putative HSP70 receptor which expression is triggered under cytokine-mediated NK cell activation. There is also a possibility that some NK cell subpopulations are more sensitive to HSP70 action. To test this we investigated whether HSP70 induced IFN-γ production in sorted $CD56^{bright}$ and $CD56^{dim}$ human NK cells.

Cell sorting of $CD3^{-}CD16^{-}CD56^{+}$ and $CD3^{-}CD16^{+}CD56^{+}$ NK cells was performed on a fluorescence-activated cell sorter (FACSVantage DiVa, BD Biosciences, USA), with final purities at 99%. HSP70 was added to sorted NK cells simultaneously with IL-2 (600 U/ml). After 18 h incubation supernatants were collected and analyzed by ELISA as described above.

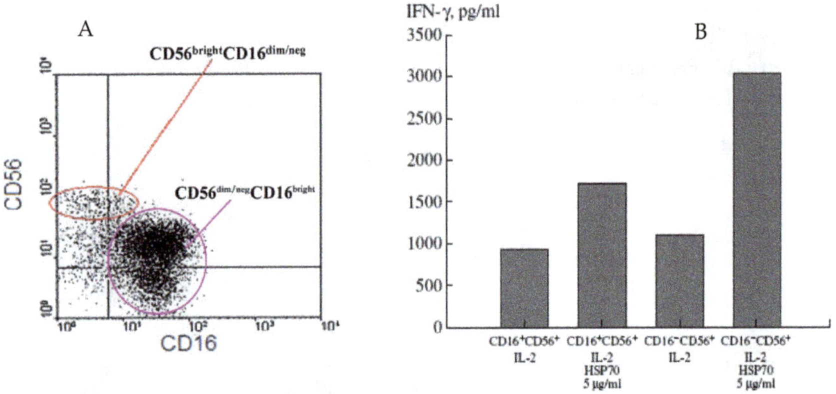

Figure 8. Effect of HSP70 on IFN-γ production of two major NK cell subpopulations. (A) Flow cytometric analysis of sorted NK cell subpopulations. (B) Recombinant human HSP70 stimulates IFN-γ production in both NK cell subpopulations $CD16^{-}CD56^{+}$ and $CD16^{+}CD56^{+}$ in presence of IL-2 (600 U/ml).

HSP70 enhanced IFN-γ production in both NK cell subpopulations (Fig. 8), and $CD16^{-}CD56^{+}$ cells produced larger amount of that cytokine. It is known $CD56^{bright}$ NK cells produce markedly higher amount of cytokines than $CD56^{dim}$. So, possibly the higher effect of HSP70 on $CD56^{bright}$ cells was due to properties of this NK cell subpopulation.

The next step of our work was the elucidation of HSP70 fragments which are responsible for apparent effects of that protein. Previously Botzler and coworkers demonstrated that substrate-binding domain of HSP70 is responsible for stimulating HSP70 effect on NK cells [44]. Then they had shown NK cells could be activated by a 14 aa fragment of HSP70 with sequence TKDNNLLGRFELSG. This fragment named TKD-peptide corresponds to 450-463 aa residues belonging to substrate-binding domain of HSP70 [59].

Using a special *in silico* approach we analyzed five peptide fragments belonging to substrate-binding domain in comparison with TKD-peptide for their ability to activate NK cells. The method of analysis of informational structure (ANIS) proposed by A.N. Nekrasov [60, 61] was used for search of potentially active fragments of HSP70 sequence. Five peptides were chosen and synthesized (Fig. 9). All of them were tested for ability to induce IFN-γ production in human NK cells isolated from peripheral blood by magnetic separation. Peptides were added to NK cells together with IL-2 in concentration 2 μg/ml. Cells were incubated for 18 h and then intracellular staining with antibody to IFN-γ was performed as described above.

Most peptides including TKD-peptide developed by G. Multhoff did not significantly enhance the amount of IFN-γ producing cells. Only peptide comprising residues 526-543 of HSP70 (P2) markedly induced IFN-γ production. Its effect was even greater than HSP70 full-length protein. It can be explained by difference in molar concentrations: 2 μg/ml solution of such peptide corresponds to the greater molar concentration than 5 μg/ml solution of HSP70 protein. Similar data were obtained from experiments with cytotoxicity of NK cells against K562 target cells measured using assay for caspase 6 activity (Oncoimmunin, USA). Only P2 peptide significantly stimulated lysis of K562 by NK cells.

399-408	P1	LSLGLETAGG	
411-424	P2	TALIKRNSTIPTKQ	
450-463	P3	TKDNNLLGRFELSG	TKD-peptide
461-470	P4	LSGIPPAPRG	
509-515	P5	RLSKEEI	
526-543	P6	KAEDEVQRERVSAKNALE	active peptide fragment

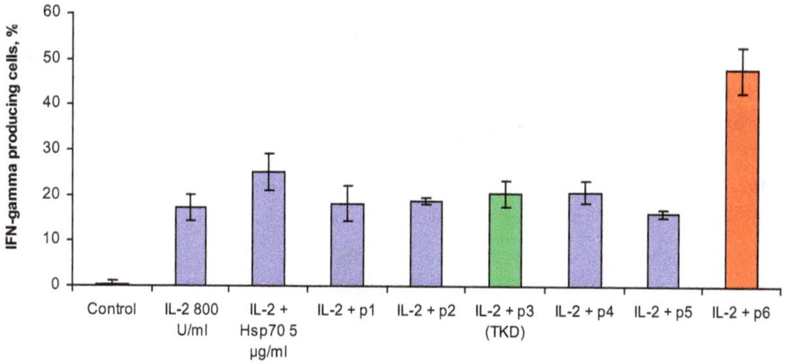

Figure 9. Testing of HSP70 peptide fragments for an ability to stimulate IFN-γ production in NK cells. Peptides were added in concentration 2 μg/ml. Amount of IFN-γ producing NK cells was determined by flow cytometry using intracellular staining with anti-IFN-γ antibody. Testing of the peptides in cytotoxicity assay provided similar results (data not shown).

In human peripheral blood there are many cells which have shown to be activated by heat shock proteins, including monocytes and blood DCs. Even small numbers of such cells can exert an influence on the apparent effect of HSP70 on NK cells. We therefore analyzed whether the presence of additional cell populations in NK cell preparations can affect HSP70 induced IFN-γ production. For that purpose NK cells and DCs were isolated by fluorescence activated cell sorting. PBMCs were stained with antibodies CD3-TC, CD11c-APC (Caltag, USA), CD16-FITC and CD56-PE (Beckman Coulter, USA). PBMCs were sorted into CD3$^-$CD11c$^-$CD16$^+$CD56$^+$ (NK cell) and CD3$^-$CD11c$^+$CD16$^-$CD56– (myeloid DC) fractions using FACSVantage DiVa. Purity was 97-98%. NK cells were then mixed with DCs in different proportions in presence of suboptimal dose of IL-2 (300 U/ml) and HSP70 was added. After 18 h incubation supernatants were collected and analyzed for IFN-γ content by ELISA (Pierce Endogen, USA).

Low doses of IL-2 did not markedly induce IFN-γ production (Fig. 10A). Mixtures of NK cells with DCs produced larger amounts of IFN-γ proportional to the numbers of DCs. This is fully consistent with current view on NK-DC interactions. It seems that DCs in the mixture with NK cells provided a stimulating signal which is additional to suboptimal dose of IL-2. Effect of HSP70 was more noticeable if it was added to the cell mixtures and increased proportionally to the elevation of basal level of IL-2 induced IFN-γ production. Thus, the amount of DCs in NK cell preparations was very important factor affecting NK cell functions, though in our experimental conditions DCs did not influence HSP70-mediated induction of NK cell activation.

Other important problem linked to HSP70 physiological action is its potential association with bacterial lipopolysaccharides (LPS). LPS can activate a range of cells of immune system via Toll-like receptor 4 (TLR4) [62]. Many reports describe LPS contamination causing apparent HSP effects on different immune cells [63-65]. HSP70 protein used in our experiments was recombinant, expressed in *E.coli,* so we might assume it was contaminated by LPS. To investigate it we have performed amoebocyte Limulus lysate (LAL) assays for LPS activity in HSP70 samples (E-Toxate kit, Sigma-Aldrich, USA). HSP70 samples were shown to contain more than 10000 U/mg of LPS. Thus, additional experiments were necessary to investigate the role of LPS in NK cell activation. NK cells isolated by magnetic separation were incubated with LPS, HSP70 known to be LPS-negative ((#ESP-555, Stressgen, Canada) and IL-2 for 18 h. Cells were collected, fixed and stained with fluorescent labeled antibody to human IFN-γ (BD Pharmingen, USA).

As with HSP70, LPS did not markedly stimulate IFN-γ production in native NK cells (Fig. 10B). However it did lead to a significant increase of IFN-γ producing cells then added together with IL-2. Thus, LPS stimulated NK cells only in cooperation with IL-2. Interestingly, HSP70 purified from LPS (HSP70 low endotoxin, HSP70-LE) added together with IL-2 did not stimulate IFN-γ production in contrast to non-purified protein. Moreover, HSP70-LE added together with LPS inhibited LPS-induced IFN-γ production showing effect opposite to LPS. One explanation for the inhibitory effect of HSP70-LE is that LPS and HSP70 compete for the same receptors on the cell surface. However it does not explain the activating effects of p6 and TKD peptides on NK cells. Additional experiments are required to characterize the regulatory role of exogenous HSP70 in NK cell functioning.

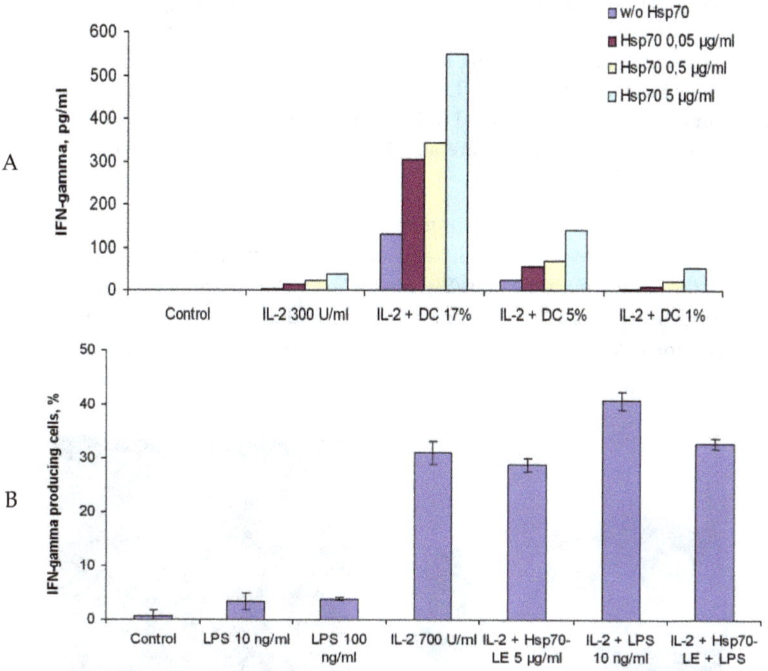

Figure 10. (A) Admixture of blood DCs affects IFN-γ production by NK cells. DCs and NK cells were isolated by cell sorting and mixed in different proportions with IL-2 in suboptimal dose, HSP70 was added. IFN-γ concentration in supernatants was analyzed by ELISA. (B) LPS stimulates IFN-γ production by NK cells in presence of IL-2. LPS, IL-2 and HSP70-LE were added to isolated NK cells, amount of IFN-γ producing NK cells was determined by flow cytometry using intracellular staining with anti-IFN-γ antibody.

Thus, multiple data support that activity of NK cells can be modulated by HSP70 both bound to the target cell surface and extracellular soluble HSP70. However, exact mechanisms of HSP70 reception by NK cells and HSP70-induced signaling are still not known. Moreover, a role of LPS in NK cell activation remains to be investigated.

4. Role of proteins MICA/B in regulation of NK cell activity

Another type of stress-induced proteins that play an important role in NK cell regulation is represented by molecules reacting with the activating receptor NKG2D. NKG2D is expressed by most NK cells; it recognizes two groups of ligands – ULBPs and MIC proteins distinct by structure and membrane anchoring [23, 25]. MICA and MICB, the most extensively studied NKG2D ligands, are highly glycosylated transmembrane proteins with one site of S-acylation necessary for anchoring in plasma membrane. They share high homology with MHC-I but are not associated with β_2-microglobulin and do not bind antigenic peptide. Like MHC-I, they consist of three extracellular domains ($\alpha 1$, $\alpha 2$ and $\alpha 3$), with two of them, α1 and α2, forming a site for NKG2D binding [66]. Oppositely, ULBPs have less homology with MHC-I molecules. They have neither an extracellular α3-domain nor a transmembrane domain, anchoring to the plasma membrane through GPI. This feature determines constitutive localization of ULBPs in lipid rafts of the plasma membrane [25, 67]. MIC proteins in their turn are also found in lipid rafts, however, their physical link to these membrane domains is less tight: they are able to transfer between rafts and the detergent-sensitive, "usual" bilipid zone [67].

Previous data demonstrates that in humans the MICA-NKG2D pair participates in regulatory interactions between NK cells and other cells involving in immune response. Interactions of membrane bound MICA/B with NKG2D lead to activation of NK cells and elimination of aberrant cells. MICA/B proteins exposed on the cell surface serve as a "danger signal" for cytotoxic lymphocytes and mark the cells that should be eliminated (Fig. 11).

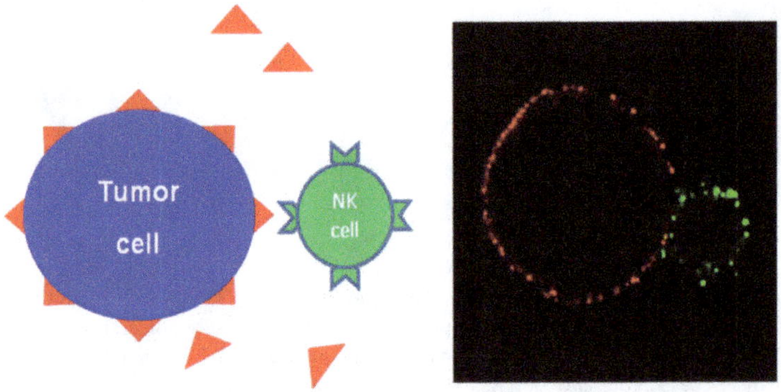

Figure 11. Receptor NKG2D recognizes proteins MICA/B. NK cells and K562 target cells were labeled with anti-NKG2D-Alexa Fluor 488 antibody and primary anti-MICA/B and secondary antibody conjugated with Alexa Fluor 555, respectively.

In our experiments we found a role for MICA/B in NK cell-mediated cytotoxicity using both intact human NK cells and cells of NK-cell-like line NKL. NK cells were isolated from human peripheral mononuclears by magnetic separation. Cytotoxicity was measured by flow cytometry method based on estimation of caspase 6 activity in target cells. Antibodies against both NKG2D and MICA/B significantly inhibited cytotoxic activity of effector cells against erythroleukemia cell line K562 and B-lymphocytic cell line 721 (Fig. 12). The values of cytotoxicity and levels of the inhibition varied for different target cell lines not depending directly on MICA/B expression. The results suggested that NKG2D-MICA/B interactions were not the only way of NK cell recognition of these target cells. Expression of NKG2D and cytotoxic activity of NK cells increased remarkably following their incubation with IL-2. Finding ways of additional induction of surface expression of MICA/B on the cell surface might be useful for developing new approaches for increase of tumor cell sensitivity to cytotoxic lymphocyte action.

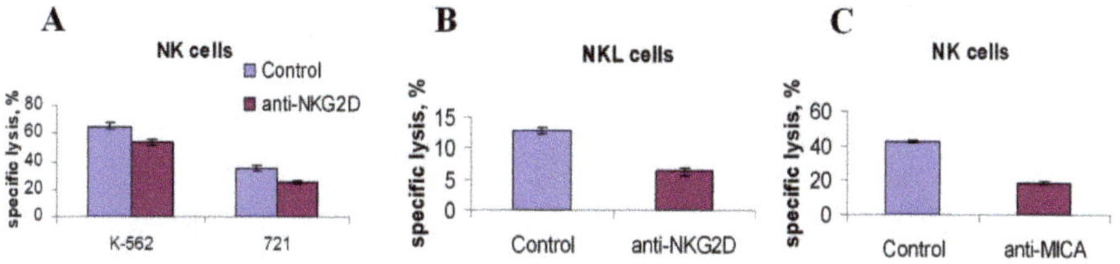

Figure 12. Cytolytic action of NK cells is mediated by NKG2D-MICA interaction. (A) Measurement of cytotoxicity of NK cells against K562 and 721 target cells in presence of antibody to NKG2D. (B) Measurement of cytotoxicity of NKL cells against 721 target cells in presence of antibody to NKG2D. (C) Cytotoxicity of NK cells against K562 cells treated with antibody to MICA.

MIC proteins can be released from the cell surface into the extracellular space. At least two pathways of release were documented: proteolytic shedding of extracellular part of MIC molecule and formation of exosomes containing MICA/B [68, 69]. Both mechanisms result in accumulation of soluble form of MIC (sMIC) in the extracellular space. Proteolytically cleaved sMICA/B circulate in human blood and can be found at remarkable levels in sera of tumor patients. This type of sMIC is also generated during viral infections. In contrast to membrane-associated MIC proteins, circulating forms of MICA/B are able to inhibit activation of NK cells and lead to anergy. It is hypothesized that tumor cells can use this mechanism to evade immune surveillance [25]. Proteolysis of MIC proteins occurs by a number of metalloproteinases, at least three of them were recently identified: two proteins of ADAM (A Disintegrin And Metalloproteinase) family, ADAM10 and ADAM17, and matrix metalloproteinase MMP14 [70, 71]. Metalloproteinases cleave MICA in a linker sequence between the transmembrane domain and the $\alpha3$ domain. There is no definite site of hydrolysis, but sufficient length of the linker region is essential [72]. MICA is expressed on the cell surface in complex with proteins originally found in endoplasmic reticulum (ER): ERp5, GRP78, and two proteins of 47 and 48 kDa of thioredoxin family. ERp5 reduces a unique disulfide bond in $\alpha3$ domain of MICA and induces conformation changes of the domain rendering the MICA molecule accessible for metalloproteinases. The roles of GRP78 and two other proteins associated with MICA on cell surface in sMICA shedding remain to be elucidated [73].

We estimated the amounts of sMIC in peripheral blood sera of patients with lymphoproliferative diseases. Levels of sMIC were measured in seven groups of patients with different types of lymphoproliferative diseases consisted of from 5 to 20 patients in a group (Fig. 13). sMICA/B concentrations were measured by ELISA (R&D Systems). The analysis of the data in every group was performed using a nonparametric Mann-Whitney test. The mean value of sMIC in the control group equaled 40 pg/ml and was not greater than 250 pg/ml. In all studied groups with the exception of Burkitt-like lymphoma group the mean sMIC serum levels exceeded significantly the levels in the group of normal donors. The highest levels of sMIC content were found in patients with T-cell anaplastic lymphomas.

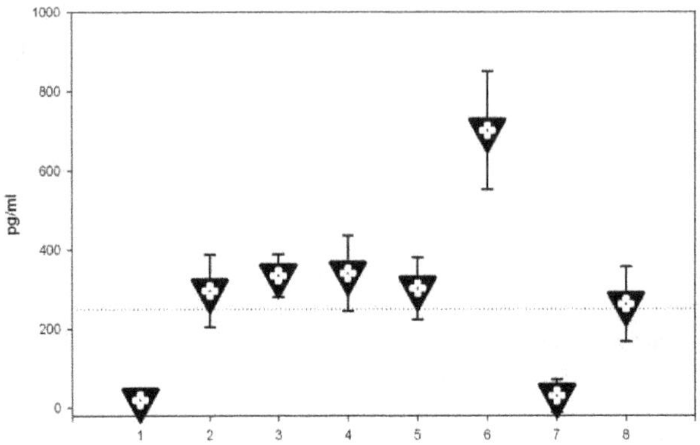

Figure 13. sMICA/B content determined in sera of patient with lymphoproliferative diseases. 1 – healthy donors, 2 – chronic lymphocytic leukemia, 3 – diffuse large B-cell lymphoma, 4 – lymphatic plasmacytoma, 5 – follicular lymphoma, 6 – T-cell anaplastic lymphoma, 7 – Burkitt-like lymphoma, 8 – lymphocytoma.

4.1. MICA/B expression under ethanol-induced cell stress conditions

The mechanisms of MICA/B regulation are still not completely clear. Expression of *MICA/B* genes on the mRNA level is found in virtually all tissues of the organism. However considerable constitutive expression of membrane-associated proteins was detected only on intestine epithelial cells [74]. It is suggested that MIC proteins participate in maintaining the integrity of the gut epithelium promoting NKG2D-dependent recognition and prompt elimination of damaged epithelial cells by intraepithelial lymphocytes [75]. There are evidence that some cellular miRNA can impair translation of $mRNA^{MICA/B}$, setting up a threshold for MICA/B expression. In stress conditions as well as in cell transformation, viral infection and surgical procedures including transplantation, *MIC* gene transcription markedly increased, specific miRNA was degraded and MIC proteins appeared on the cell surface [76, 77]. Transcription factors NFκB, Sp1, Sp3 were shown to stimulate MIC surface expression. *MIC* expression induced by heat shock is dependent on HSF1 binding to heat shock response element (HSRE) located in promoter region of *MIC* genes [25]. Genotoxic stress causes MIC surface expression via ATM/ATR kinases, sensors of DNA damages, triggering signaling cascade which leads to cell cycle arrest, induction of DNA reparation and expression of NKG2D ligands [78]. In many cases mechanisms of MIC expression triggering remain unclear.

In our work we produced a reliable model of induction of MICA/B surface expression based on cell treatment with ethanol. Using this model we analyzed MICA/B expression in transcriptional, surface and extracellular levels of MICA expression.

It is well known that ethanol can exert damaging effects on many types of cells. It influences membrane fluidity and induces oxidative stress in cells [79]. Ethanol can also affect immune response, particularly by triggering apoptosis in lymphoid and myeloid cells [80]. All these data characterize ethanol as a good inductor of cell stress conditions. We applied ethanol to induce cell stress in a range of hemopoietic cell lines in order to analyze its influence on MICA/B expression. Surface MIC expression was analyzed by flow cytometry (FACScan, BD Biosciences) and confocal microscopy (Nikon TE-2000 confocal microscope). Monoclonal antibodies to MICA/B (R&D Systems) were used for protein visualization. Expression of *MICA/B* genes was registered by RT-PCR. Percentages of apoptotic and necrotic cells were measured by cell staining with fluorescent-labeled annexin V and propidium iodide using flow cytometry. Soluble MICA/B forms were detected by ELISA.

For the study we chose several human hemopoietic cell lines of different origin: acute T cell leukemia Jurkat, erythroblastoid leukemia K562, monocytic leukemia THP-1 and melanoma cell line MelJuSo. The cell line C1R-MICA transfected with MICA gene was used as a positive control. The cell lines differed in spontaneous MICA/B surface expression (Fig. 14). The maximal MICA/B expression comparable with expression in C1R-MICA was registered in THP-1 cells. Even those cells, which had no spontaneous MICA/B expression on the cell surface had high levels of mRNA^{MIC} (data not shown).

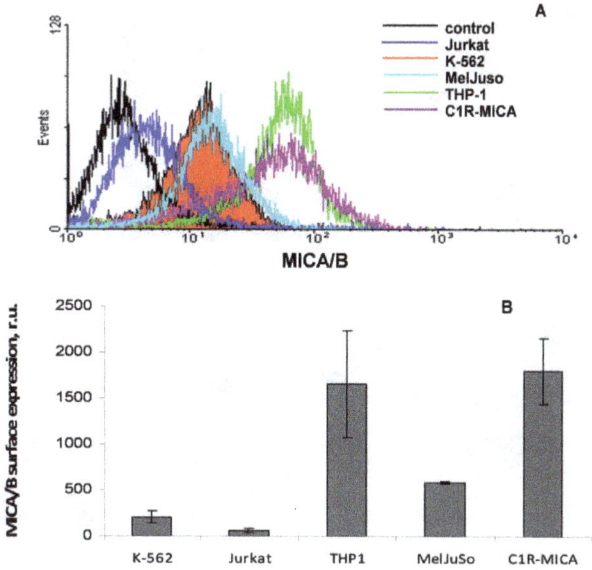

Figure 14. Spontaneous MICA/B surface expression in different hemopoietic tumor cell lines: K562 – human erythroblastoid leukemia, Jurkat – human T cell leukemia, THP-1 – human acute monocytic leukemia, MelJuSo - human melanoma , C1R-MICA – transfected cells (positive control).

We found that treatment of cells with ethanol modulated MICA/B cell expression. Short term cell treatment with ethanol (1 h) leaded to partial elimination of MICA/B from the cell surface. In the same time incubation of cells in the presence of ethanol in concentration range 0.25-3% for 18 h resulted in increased expression of MICA/B on the cell surface (Fig. 15).

Ethanol at high concentrations resulted in a reduction of MICA/B surface expression and an elevation of apoptosis and necrosis in cell culture (Fig. 16). Ethanol-induced increase of MICA/B surface expression preceded cell death caused by higher ethanol concentration. We can assume that the exposure of MICA/B proteins on the cell surface under cell stress conditions allows cytotoxic lymphocytes expressing receptor NKG2D to identify stressed cells in which the program of cell death has not yet been activated.

We also checked expression of MICA/B at the mRNA level in the hemopoietic cell lines and normal human leukocytes. $mRNA^{MICA/B}$ was easily detected in all cells tested without cell stress conditions. Increase of $mRNA^{MICA/B}$ level in cells was registered at 3 h of ethanol treatment (Fig. 17). One of the mechanisms of damaging effects of ethanol on cells is the induction of oxidative stress in cells that can cause DNA damage. We found out that ethanol-induced MICA/B upregulation did not depend on ATM/ATR kinase activity, suggesting that genotoxic effects were not the mechanism of this process (data not shown).

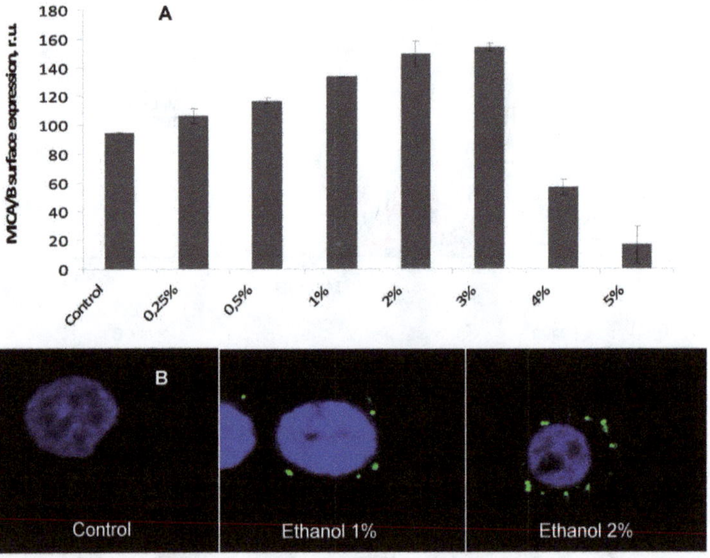

Figure 15. MICA/B surface expression alterations in Jurkat cells followed ethanol treatment for 18 hours. Incubation of cells ($5x10^5$ cells/ml) was performed in RPMI medium containing 10% FCS. Images were collected with a Nikon TE-2000 confocal microscope equipped with 405, 488 and 543 nm lasers and standard set of detectors. A. MICA/B surface expression measured by flow cytometry. Cells were labeled with primary MICA/B antibody and secondary antibody conjugated with AlexaFluor488. Nuclei were identified with Hoechst 33342. MICA/B expression levels were calculated as ((exp/2*ab*)-1)*100%, where exp – mean of fluorescence intensity of experimental sample, 2ab – mean of fluorescence intensity of control sample labeled only with secondary antibody. B – MICA/B surface expression after ethanol treatment analyzed by confocal microscopy.

Another mechanism of MICA/B surface expression increase under ethanol action was revealed in K562 cells. With confocal microscopy, it was shown that under the influence of ethanol-induced cellular stress the inner pool of MICA/B decreased and the surface pool increased as compared with the control cells indicating that under the alcohol influence MICA/B moved from the cytoplasm to the cell surface (Fig. 18).

Thus, ethanol affects MICA/B expression in cells at several levels resulting in induction of MICA/B *de novo* expression, translocation of intracellular protein pool to the cell surface and release of MICA/B proteins from the cells. MICA/B exposure on the cell surfaces and shedding of these proteins with formation of the extracellular pool are the latest stages of MICA/B cell expression. Both surface and extracellular forms of MICA/B proteins may take part in immune regulation acting on NKG2D-expressing cytotoxic lymphocytes.

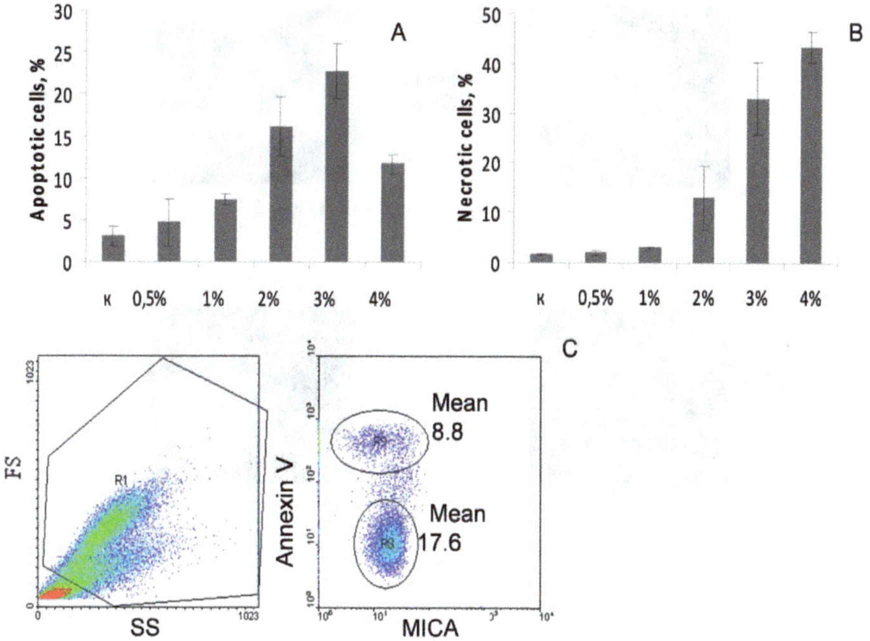

Figure 16. Ethanol influence on cell viability of THP-1 cells. **A** - Percentage of apoptotic cells in cell culture treated with ethanol for 18 h. **B** - Percentage of necrotic cells in cell culture treated with ethanol for 18 h. **C** - Flow cytometry analysis of MICA/B surface expression. THP-1 cells were stained with anti-MICA/B antibodies with FITC-labeled secondary antibodies, Alexa Fluor 647-labeled annexin V and propidium iodide.

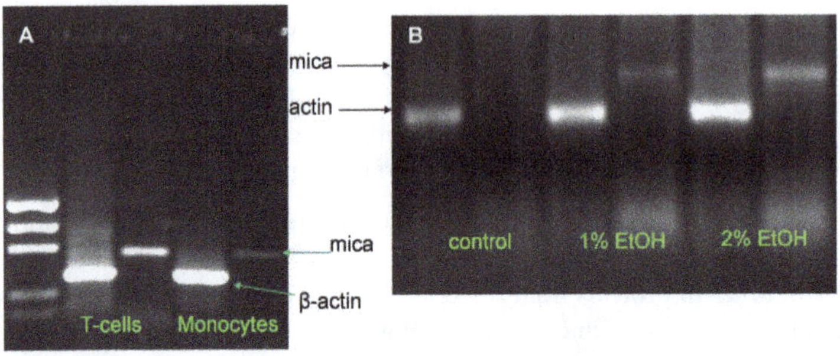

Figure 17. mRNA$^{MICA/B}$ expression estimated by RT-PCR in different kinds of cells. A. mRNA$^{MICA/B}$ expression in human T cells and monocytes. B. mRNA$^{MICA/B}$ expression in Jurkat cells subjected to ethanol treatment.

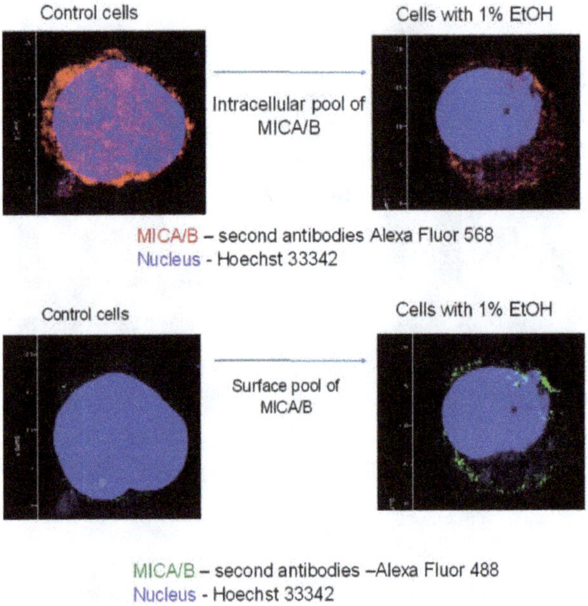

Figure 18. Intracellular and surface pool of MICA/B proteins in control K562 cells and the cells after ethanol treatment. MICA/B expression was detected with secondary antibodies labeled with Alexa Fluor 488 and Alexa Fluor 568 for surface and intracellular proteins, respectively. Images were recorded with Nikon TE-2000 confocal microscope equipped with 405, 488 and 543 nm lasers and filters appropriate for Alexa Fluor 488, Alexa Fluor 568 and Hoechst 33342.

5. Conclusion

These experiments confirmed an important role of stress-induced modification of target cell surface for the process of NK cell – tumor cell interaction. Our data showed that this interaction is modulated by both alteration of cancer cell surface molecular pattern (including surface oligosaccharide profile) and extracellular pool of stress-induced proteins released by the target cells. Noteworthily, our results show that surface and soluble HSP70 activate cytolytic NK activity in contrast to extracellular form of MICA/B which inhibits cytotoxicity of NK cells. We hypothesize that stress-induced HSP70 and MICA/B protect target cells from NK-mediated effects on two levels: HSP70 serves as protective molecule for key cell components (in the cell space) and MICA/B provide a distant resistance inhibiting NK cell cytotoxic activity before the interaction with a target.

An important point of the obtained results is connected with HSP70 fragments identified by a special *in silico* approach. Testing of synthetic analogs of the fragments demonstrated in our models an essential activity of some peptides, similar to the effect of whole molecule. This data has important implications for future application of stress-induced proteins in cancer diagnostic and therapy.

Author details

Elena Kovalenko*, Leonid Kanevskiy, Anna Klinkova, Anastasiya Kuchukova, Maria Streltsova and Alexander Sapozhnikov
Laboratory of Cell Interactions, Shemyakin-Ovchinnikov Institute of Bioorganic Chemistry, Moscow, Russia

William Telford
Experimental Transplantation and Immunology Branch, National Cancer Institute, National Institutes of Health, Bethesda, MD, USA

Acknowledgement

We thank A. Nekrasov for the help in HSP70 peptide identification. This work is supported by Ministry of Education and Science of the Russian Federation (grant #16.740.11.0200) and Molecular and Cellular Biology Program of Russian Academy of Sciences.

6. References

[1] Trinchieri G (1989) Biology of Natural Killer Cells. Adv. Immunol. 47: 187-376.
[2] Sun JC, Beilke JN, Lanier LL (2009) Adaptive Immune Features of Natural Killer Cells. Nature 457(7229): 557-61.

---

* Corresponding Author

[3] Whiteside TL, Herberman RB (1994) Role of Human Natural Killer Cells in Health and Disease. Clin. diagn. lab. immunol. 1(2): 125-33.

[4] Walzer T, Dalod M, Robbins SH, Zitvogel L, Vivier E (2005) Natural-Killer Cells and Dendritic Cells: "L'Union Fait la Force". Blood 106(7): 2252-8.

[5] Nedvetzki S, Sowinski S, Eagle RA, Harris J, Vély F, Pende D, Trowsdale J, Vivier E, Gordon S, Davis DM (2007) Reciprocal Regulation of Human Natural Killer Cells and Macrophages Associated with Distinct Immune Synapses. Blood. 109(9): 3776-85.

[6] Cooper MA, Fehniger TA, Fuchs A, Colonna M, Caligiuri MA (2004) NK Cell and DC Interactions. Trends immunol. 25(1): 47-52.

[7] Cooper MA, Fehniger TA, Caligiuri MA (2001) The Biology of Human Natural Killer-Cell Subsets. Trends immunol. 22(11): 633-640.

[8] Caligiuri M (2008) Human Natural Killer Cells. Blood 112: 461-469.

[9] O`Connor GM, Hart OM, Gardiner CM (2005) Putting the Natural Killer Cells in its Place. Immunology 117: 1-10.

[10] Bottino C, Castriconi R, Moretta L, Moretta A (2005) Cellular Ligands of Activating NK Receptors. Trends immunol. 26: 221.

[11] Long EO (2008) Negative Signaling by Inhibitory Receptors: the NK Cell Paradigm. Immunol. rev. 224: 70–84.

[12] Ljunggren HG, Karre K (1990) In Search of the 'Missing Self': MHC Molecules and NK Cell Recognition. Immunol. today 11: 237-244.

[13] Bjorkstrom NK, Riese P, Heuts F, Andersson S, Fauriat C, Ivarsson MA, Björklund AT, Flodström-Tullberg M, Michaëlsson J, Rottenberg ME, Guzmán CA, Ljunggren HG, Malmberg KJ (2010) Expression Patterns of NKG2A, KIR, and CD57 Define a Process of CD56dim NK-cell Differentiation Uncoupled from NK-cell Education. Blood 116(19): 3853 - 3864.

[14] Jamil KM, Khakoo SI (2011) KIR/HLA Interactions and Pathogen Immunity. J. biomed. biotechnol. 2011: 298348.

[15] Caligiuri M (2008) Human Natural Killer Cells. Blood 112: 461-469.

[16] Bryceson YT, March ME, Ljunggren HG, Long EO (2006) Synergy among Receptors on Resting NK Cells for the Activation of Natural Cytotoxicity and Cytokine Secretion. Blood 107: 159-165.

[17] Blery M, Olcese L, Vivier E (2000) Early Signaling via Inhibitory and Activating Receptors. Hum. immunol. 61: 51-64.

[18] Vales-Gomes M, Reyburn H, Strominger J (2000) Interaction Between the Human NK Receptors and their Ligands. Crit. rev. immunol. 20: 223-244.

[19] Garg A, Barnes PF, Porgador A, Roy S, Wu S, Nanda JS, Griffith DE, Girard WM, Rawal N, Shetty S, Vankayalapati R (2008) Vimentin Expressed on Mycobacterium Tuberculosis - Infected Human Monocytes is Involved in Binding to the NKp46 Receptor. J immunol. 177: 6192-8.

[20] Pogge von Strandmann E, Simhadri VR, von Tresckow B, Sasse S, Reiners KS, Hansen HP, Rothe A, Böll B, Simhadri VL, Borchmann P, McKinnon PJ, Hallek M, Engert A

(2007) Human Leukocyte Antigen-B-Associated Transcript 3 is Released from Tumor Cells and Engages the NKp30 Receptor on Natural Killer Cells. Immunity 27(6): 965-74.

[21] Brandt CS, Baratin M, Yi EC, Kennedy J, Gao Z, Fox B, Haldeman B, Ostrander CD, Kaifu T, Chabannon C, Moretta A, West R, Xu W, Vivier E, Levin SD (2009) The B7 Family Member B7-H6 is a Tumor Cell Ligand for the Activating Natural Killer Cell Receptor NKp30 in Humans. J. exp. med. 206(7): 1495-503.

[22] Hecht ML, Rosental B, Horlacher T, Hershkovitz O, De Paz JL, Noti C, Schauer S, Porgador A, Seeberger PH (2009) Natural Cytotoxicity Receptors NKp30, NKp44 and NKp46 Bind to Different Heparan Sulfate/Heparin Sequences. J. proteome res. 8(2): 712-20.

[23] Burgess SJ, Maasho K, Masilamani M, Narayanan S, Borrego F, Coligan JE (2008) The NKG2D Receptor: Immunobiology and Clinical Implications. Immunol. res. 40: 18–34.

[24] Diefenbach A, Jensen ER, Jamieson AM, Raulet DH (2001) Rae1 and H60 Ligands of the NKG2D Receptor Stimulate Tumour Immunity. Nature 413 (6852): 165-71.

[25] Gonzalez S, Groh V, Spies T (2006) Immunobiology of Human NKG2D and its Ligands. CTMI 298: 121-138.

[26] Matzinger P (1994) Tolerance, Danger and the Extended Family. Annu. rev. immunol. 12: 991-1045.

[27] Matzinger P (2002) The Danger Model: a Renewed Sense of Self. Science 296: 301-305.

[28] Romero AI, Thorén FB, Brune M, Hellstrand K (2006) NKp46 and NKG2D Receptor Expression in NK Cells with CD56dim and CD56bright Phenotype: Regulation by Histamine and Reactive Oxygen Species. Br. j. haematol. 132(1):91-8.

[29] Peraldi MN, Berrou J, Dulphy N, Seidowsky A, Haas P, Boissel N, Metivier F, Randoux C, Kossari N, Guérin A, Geffroy S, Delavaud G, Marin-Esteban V, Glotz D, Charron D, Toubert A (2009) Oxidative Stress Mediates a Reduced Expression of the Activating Receptor NKG2D in NK Cells from End-stage Renal Disease Patients. J immunol. 182(3):1696-705.

[30] Crocker PR, Paulson JC, Varki A (2007) Siglecs and their Roles in the Immune System. Nat. rev. immunol. 7(4): 255-266.

[31] Ogata S, Maimonis PJ, Itzkowitz SH (1992) Mucins Bearing the Cancer-Associated Sialosyl-Tn Antigen Mediate Inhibition of Natural Killer Cell Cytotoxicity. Cancer res. 52: 4741-4746.

[32] Yoshimura M, Ihara Y, Ohnishi A, Ijuhin N, Nishiura T, Kanakura Y, Matsuzawa Y, Taniguchi N (1996) Bisecting N-acetylglucosamine on K562 Cells Suppresses Natural Killer Cytotoxicity and Promotes Spleen Colonization. Cancer res. 56: 412-418.

[33] Sol MA, Vacaresse N, Lule J, Davrinche C, Gabriel B, Teissie J, Ziegler A, Thomsen M, Benoist H (1999) N-linked Oligosaccharides Can Protect Target Cells from the Lysis Mediated by NK Cells but not by Cytotoxic T Lymphocytes: Role of NKG2-A. Tissue antigens 54: 113-121.

[34] Nicoll G, Avril T, Lock K, Furukawa K, Bovin N, Crocker PR (2003) Ganglioside GD3 Expession on Target Cells Can Modulate NK Cell Cytotoxicity via Siglec-7-Dependent and -Independent Mechanisms. Eur. j. immunol. 33: 1642-1648.

[35] Zarcone D, Tilden AB, Friedman HM, Grossi CE (1987) Human Leukemia-Derived Cell Lines and Clones as Models for Mechanistic Analysis of Natural Killer Cell-Mediated Cytotoxicity. Cancer res. 47: 2674–2682.

[36] Hakomori S (2002) The Glycosynapse. Proc. natl. acad. sci. USA 99:225-232.

[37] Smith DF, Whitesell L, and Katsanis E (1998) Molecular Chaperones: Biology and Prospects for Pharmacological Intervention. Pharmacological reviews 4: 493-513.

[38] Sapozhnikov AM, Ponomarev ED, Tarasenko TN and Telford GW (1999) Spontaneous Apoptosis and Expression of Cell Surface Heat Shock Proteins in Cultured EL-4 Lymphoma Cells. Cell proliferation 32: 363-378.

[39] van Eden W, van der Zee R, Paul AG, Prakken BJ, Wendling U, Anderton SM and Wauben MH (1998) Do Heat Shock Proteins Control the Balance of T-cell Regulation in Inflammatory Diseases? Immunology today 19: 303-307.

[40] Poccia F, Piselli S, Vendetti S, Bach S, Amendola A, Placido R and Collizzi V (1996) Heat-Shock Protein Expression on Membrane of T Cells Undergoing Apoptosis. Immunology 88: 6-12.

[41] Hightower LE and Hendershot LM (1997) Molecular Chaperones and the Heat Shock Response at Cold Spring Harbor. Cell Stress and Chaperones 2: 1-11.

[42] Kurosawa S, Matsuzaki G, Harada M, Ando T and Nomoto K (1993) Early Appearance and Activation of Natural Killer Cells in Tumor-Infiltrating Lymphoid Cells During Tumor Development. Eur. j. immunol. 23: 1029-1033.

[43] Botzler C, Issels R and Multhoff G (1996) Heat-Shock Protein 72 Cell-Surface Expression on Human Lung Carcinoma Cells in Associated with an Increased Sensitivity to Lysis Mediated by Adherent Natural Killer Cells. Cancer immunol. immunother. 43: 226-230.

[44] Botzer C, Li G, Issels RD and Multhoff G (1998) Definition of Extracellular Localized Epitopes of Hsp70 Involved in an NK Immune Response. Cell stress chaperones 3: 6-11.

[45] Multhoff G, Botzler C, Wiesnet M, Eissner G and Issels R (1995) CD3- Large Granular Lymphocytes Recognize a Heat-Inducible Immunogenic Determinant Associated with the 72-kD Heat Shock Protein on Human Sarcoma Cells. Blood 86: 1374-1382.

[46] Multhoff G (1997) Heat Shock Protein 72 (HSP72), a Hyperthermia-Inducible Immunogenic Determinant on Leukemic K562 and Ewing's Sarcoma Cells. International Journal of Hyperthermia 13: 39-48.

[47]Blom DJ, De Waard-Siebinga I, Apte RS, Luyten GP, Niederkorn JY and Jager MJ (1997) Effect of Hyperthermia on Expression of Histocompatibility Antigens and Heat-Shock Protein Molecules on Three Human Ocular Melanoma Cell Lines. Melanoma res. 7: 103-109.

[48] Przepiorka D and Srivastava PK (1998) Heat Shock Protein-Peptide Complexes as Immunotherapy for Human Cancer. Mol. med. today 478: 478-484.

[49] Schild H, Arnold-Schild D, Lammert E and Rammensee H (1999) Stress Proteins and Immunity Mediated by Cytotoxic T Lymphocytes. Curr. opin. immunol. 11: 109-113.

[50]Mage MG, McHugh LL and Rothstein TL (1977) Mouse Lymphocytes with and without Surface Immunoglobulin: Preparative Scale Separation in Polystyrene Tissue Culture Dishes Coated with Specifically Purified Anti-Immunoglobulin. J. immunol. methods 15: 47-56.

[51] van de Loosdrecht AA, Nennie E, Ossenkoppele GJ, Beelen RH and Langenhuijsen MM (1991) Cell Mediated Cytotoxicity against U 937 Cells by Human Monocytes and Macrophages in a Modified Colorimetric MTT Assay. A Methodological Study. J. immunol. methods 141: 15-22.

[52] Decker T and Lohmann-Matthes M (1988) A Quick and Simple Method for the Quantitation of Lactate Dehydrogenase Release in Measurements of Cellular Cytotoxicity and Tumor Necrosis Factor (TNF) Activity. J. immunol. methods 15: 61-69.

[53] Arnold-Schild D, Hanau D, Spehner D, Schmid C, Rammensee HG, de la Salle H, Schild H (1999) Receptor-Mediated Endocytosis of Heat Shock Proteins by Professional Antigen-Presenting Cells. J. immunol. 162: 3757–3760.

[54] Creswell P, Ackerman AL, Giodini A, Peaper DR, Wearsch PA (2005) Mechanisms of MHC Class I-restricted Antigen Processing and Cross-presentation. Immunol. rev. 207: 145-157.

[55] Rock KL, Shen L (2005) Cross-presentation: Underlying Mechanisms and Role in Immune Surveillance. Immunol. rev. 207: 166-183.

[56] Multhoff G (2002) Activation of Natural Killer Cells by Heat Shock Protein 70. Int. j. hyperthermia 18(6): 576-85.

[57] Srivastava P (2002) Roles of Heat-Shock Proteins in Innate and Adaptive Immunity. Nat. rev. immunol. 2(3):185-94.

[58] Asea A (2006) Initiation of the Immune Response by Extracellular Hsp72: Chaperokine Activity of Hsp72. Curr. immunol. rev. 2(3):209-215.

[59] Multhoff G, Pfister K, Gehrmann M, Hantschel M, Gross C, Hafner M, Hiddemann W (2001) A 14-mer Hsp70 Peptide Stimulates Natural Killer (NK) Cell Activity. Cell stress chaperones 6(4): 337-44.

[60] Nekrasov AN (2002) Entropy of Protein Sequences: an Integral Approach. J. biomol. struct. dyn. 20(1): 87-92

[61] Nekrasov AN (2004) Analysis of the Information Structure of Protein Sequences: a New Method for Analyzing the Domain Organization of Proteins. J. biomol. struct. dyn. 21(5): 615-24.

[62] Palsson-Mcdermott EM, O`Neill LA (2004) Signal Transduction by the Lipopolysaccharide Receptor, Toll-like Receptor-4. Immunology 113: 153-162.

[63] Bausinger H, Lipsker D, Ziylan U, Manié S, Briand JP, Cazenave JP, Muller S, Haeuw JF, Ravanat C, de la Salle H, Hanau D (2002) Endotoxin-Free Heat-Shock Protein 70 Fails to Induce APC Activation. Eur. j. immunol. 32(12): 3708-13.

[64] Gao B, Tsan MF (2003) Endotoxin Contamination in Recombinant Human Heat Shock Protein 70 (Hsp70) Preparation is Responsible for the Induction of Tumor Necrosis Factor Alpha Release by Murine Macrophages. J. biol. chem. 278(1): 174-9.

[65] Tsan MF, Gao B (2009) Heat Shock Proteins and Immune System. J. leukoc. biol. 85(6): 905-10.

[66] Li P, Morris DL, Willcox BE, Steinle A, Spies T and Strong RK (2001) Complex Structure of the Activating Immunoreceptor NKG2D and its MHC Class I-like Ligand MICA. Nat. immunol. 2 (5): 443-451.

[67] Eleme K, Taner SB, Onfelt B, Collinson LM, McCann FE, Chalupny NJ, Cosman D, Hopkins C, Magee AI and Davis DM (2004) Cell Surface Organization of Stress-Inducible Proteins ULBP and MICA that Stimulate Human NK Cells and T Cells via NKG2D. J. exp. med. 199 (7): 1005–1010.
[68] Ashiru O, Boutet P, Fernández-Messina L, Agüera-González S, Skepper JN, Valés-Gómez M, Reyburn HT (2010) Natural Killer Cell Cytotoxicity is Suppressed by Exposure to the Human NKG2D Ligand MICA*008 that is Shed by Tumor Cells in Exosomes. Cancer res. 70(2): 481-9.
[69] Kaiser BK, Yim D, Chow IT, Gonzalez S, Dai Z, Mann HH, Strong RK, Groh V, Spies T (2007) Disulphide-Isomerase-Enabled Shedding of Tumour-Associated NKG2D Ligands. Nature 447(7143): 482-6.
[70] Waldhauer I, Goehlsdorf D, Gieseke F, Weinschenk T, Wittenbrink M, Ludwig A, Stevanovic S, Rammensee HG, Steinle A (2008) Tumor-Associated MICA is Shed by ADAM Proteases. Cancer res. 68(15): 6368-76.
[71] Liu G, Atteridge CL, Wang X, Lundgren AD, and Wu J (2010) The Membrane Type Matrix Metalloproteinase MMP14 Mediates Constitutive Shedding of MICA Independent of ADAMs. J. immunol. 184 (7): 3346–3350.
[72] Wang X, Lundgren AD, Singh P, Goodlett DR, Plymate SR, Wu JD (2009) An Six-Amino Acid Motif in the Alpha3 Domain of MICA is the Cancer Therapeutic Target to Inhibit Shedding. Biochem. biophys. res. commun. 387(3): 476-81.
[73] Huergo-Zapico L, Gonzalez-Rodriguez AP, Contesti J, Gonzalez E, Lopez-Soto A, Fernandez-Guizan A, Acebes-Huerta A, de los Toyos JR, Lopez-Larrea C, Groh V, Spies T, Gonzalez S (2012) Expression of ERp5 and GRP78 on the Membrane of Chronic Lymphocytic Leukemia Cells: Association with Soluble MICA Shedding. Cancer immunol. immunother., epub. ahead of print.
[74] Suemizu H, Radosavljevic M, Kimura M, Sadahiro S, Yoshimura S, Bahram S and Inoko H (2002) A Basolateral Sorting Motif in the MICA Cytoplasmic Tail. PNAS 99 (5): 2971-2976.
[75] Perera L, Shao L, Patel A, Evans K, Meresse B, Blumberg R, Geraghty D, Groh V, Spies T, Jabri B, Mayer L (2007) Expression of Nonclassical Class I Molecules by Intestinal Epithelial Cells. Inflamm bowel dis 13: 298 –307.
[76] Nachmani D, Stern-Ginossar N, Sarid R and Mandelboim O (2009) Diverse Herpesvirus microRNAs Target the Stress-Induced Immune Ligand MICB to Escape Recognition by Natural Killer Cells. Cell host & microbe 5: 376–385.
[77] Yadav D, Ngolab J, Seung-Hwan Lim R, Krishnamurthy S and Bui JD (2009) Cutting Edge: Down-Regulation of Major Histocompatibility Complex Class I-related Chain A (MICA) on Tumor Cells by IFNγ-Induced microRNA. J. immunol. 182 (1): 39–43.
[78] Gasser S and Raulet DH (2006) The DNA Damage Response Arouses the Immune System. Cancer res. 66(8): 3959-62.
[79] Koch OR, Pani G, Borrello S, Colavitti R, Cravero A, Farrè S, Galeotti T (2004) Oxidative Stress and Antioxidant Defenses in Ethanol-Induced Cell Injury. Mol. aspects med. 25(1-2): 191-8.
[80] Szabo G, Mandrekar P (2009) A Recent Perspective on Alcohol, Immunity, and Host Defense. Alcohol clin. exp. res. 33(2): 220-32.

Section 3

Neural Cell Interactions

Chapter 5

Cells, Molecules and Mechanisms Involved in the Neuro-Immune Interaction

Rodrigo Pacheco, Francisco Contreras and Carolina Prado

Additional information is available at the end of the chapter

1. Introduction

The existence of a surveillance and protection system against pathogens and malignant cells has evolved until reach, in the most complex organisms, a very sophisticated immune system composed of an innate arm and an adaptive arm. The innate immune system constitutes the first defense barrier against foreign organisms but it comprises relatively poor diversity of antigen (Ag)-specific recognition elements. In contrast, the adaptive immune system shows a delayed response, and involves numerous Ag-specific recognition elements and highly efficient mechanisms both to eliminate foreign pathogens and tumor cells, and to maintain tolerance to self constituents. T cells are the central players in the adaptive immune response. These cells direct and regulate the function of several immune system cells, thus orchestrating efficient elimination of threats and, at the same time, promoting tolerance to healthy self tissues. Growing evidence has shown that some cells from the adaptive immune system, specially the T cell compartment, constitute an important link between the nervous system and the immune system. Due to their pivotal role directing immune responses and tolerance, T cells become strategic target cells to be regulated by the nervous system. On the other hand, evidences point toward T cells not only play an important role driving the elimination of pathogens and tumors and maintaining tolerance to self constituents, but also regulate some nervous system functions such as acquisition of memory and behaviour. This chapter is geared toward analyze and discuss the current knowledge and growing evidences about the interactions between the nervous system and the immune system of the complex organisms. Because mammals present the most complex and sophisticated immune system among the organisms, and most evidences described in the literature correspond to studies of neuro-immune interactions analyzed in these animals, the discussion in this chapter is bounded to mechanisms operating in mammals. The bidirectional interaction between the nervous system and immune system will be analyzed at the level of cellular and molecular mechanisms. In the first half of this chapter, interactions between immune and nervous systems will be analyzed in both directions: the immune system mediated regulation of nervous system and the nervous system mediated regulation of immunity. Both kinds of interactions will be analyzed separately for simplification and better understanding. The second half of this chapter will be geared to analyze how immune cells can communicate between them using neurotransmitters and how nervous system cells can also interact with each other using cytokines. Despite this second part of the chapter does not analyze direct neuro-immune interactions, it contributes to understand how both, nervous system and immune system use the same mediators and receptors, which therefore may participate in regulation of immunity, nervous system regulation and also in neuro-immune interactions.

2. The neuro-immune interaction

Neuroimmunology studies the interaction between the nervous and the immune system, a relation that can be analyzed in two directions: the nervous system-mediated regulation of immunity and the immune system-mediated regulation of nervous system function. Both kinds of interactions are mediated by complex mechanisms in which participate the same groups of cells and the same kind of molecules (figure 1). These two kinds of neuro-immune interactions will be separately analyzed in this section.

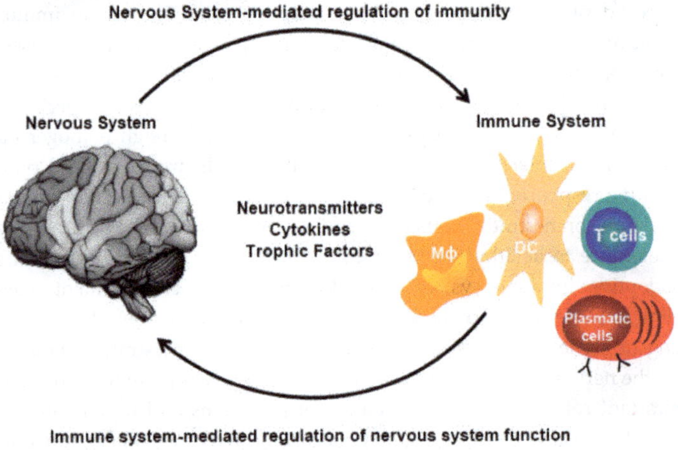

Figure 1. The Neuroimmune interactions. An extensive bidirectional communication takes place between nervous and immune system in both health and disease. The same molecules, including cytokines, neurotransmitters and trophic factors, participate as mediators in both directions.

The first part of this section (section 2.1) will start with a brief introduction about the adaptive immune response. The central role of T cells in orchestrating the development of immune responses will be highlighted. The development and the role of distinct phenotypic effector T cells and regulatory T cells generated during immune response will be summarized. Afterward, in the second part of this section (section 2.2) neurotransmitter receptors expressed on immune cells will be described. Circuits by which peripheral nervous system can modulate the function of immune cells will be discussed.

Several studies have shown that immune cells, and mainly CD4$^+$ T cell, contribute significantly to the development and regulation of nervous system processes, such as learning and acquisition of memory. These and other immune cell-mediated mechanisms regulating behaviour in healthy animals will be discussed in the third part of this section (section 2.3). In addition, several studies have shown a key role of T cell-mediated responses in the development and progression of neurodegenerative diseases and others pathologic scenarios occurring in the central nervous system. Examples of these mechanisms occurring during Parkinson's disease, multiple sclerosis and other pathologies will be also discussed in the third part of this section.

2.1. The central role of T cells and dendritic cells during adaptive immune response

The immune system has evolved to recognize and eliminate a large variety of Ags that could be potentially harmful for the host. Ag recognition is a highly specific and tightly regulated process, that promotes the destruction of dangerous agents without causing a significant damage to self-constituents. Such a sensitivity and specificity for Ag recognition is mediated by the adaptive arm of the immune system, which is constituted by B and T cells. These cells express Ag-specific receptors on their surfaces that bind to Ags in distinct forms and play very different roles during adaptive immune responses. B cells recognize Ags as soluble molecules and have the ability to differentiate and secrete a soluble form of the B cell receptor called antibody, upon appropriate Ag-stimulation. In contrast, T cells do not recognize soluble Ags, but require a specialized molecular system for Ag-stimulation. Such a system involves intracellular proteolytic processing of protein Ags into small peptides by Ag-presenting-cells (APCs) and the subsequent loading of these peptides on major histocompatibility complex (MHC) molecules, which are expressed on the surface of APCs. Antigenic peptide-MHC molecule complexes (pMHC) expressed on the APC surface are the ligands for the Ag-receptor of T cells (TCR). In addition to specific pMHC recognition, T cell activation can be positively or negatively regulated by a number of non-Ag-specific intercellular interactions that are mediated by surface-bound and soluble molecules, which bind to their respective receptors on the T cell surface. In response to the simultaneous presentation of pathogen or tumor-derived Ags on MHC molecules and co-stimulatory molecules on the APC surface, Ag-specific naïve T cells are activated. As a result, depending on the nature of the stimuli provided by the APC, T cells undergo massive expansion and differentiate into effector T cells, which can acquire several effector mechanisms to promote the elimination of pathogens (Wong & Pamer, 2003) or tumors (Amedei et al., 2011; Park et al., 2011).

Adaptive immune response against foreign proteins requires the activation of Ag-specific T cells. Help from T cells contributes to B cell activation and differentiation into antibody secreting plasma cells. T cell help is mediated by both soluble and membrane-bound molecules

(Parker, 1993). In addition, T cell-derived cytokines can also promote or potentiate effector functions of other immune cells, such as macrophages, dendritic cells (DCs), and natural killer (NK) cells. Thus, T cells play a central role coordinating adaptive immune responses (figure 2).

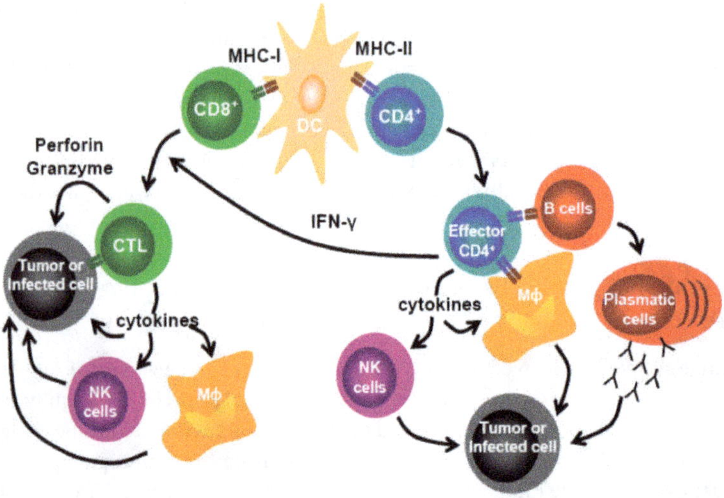

Figure 2. The central role of T cells and DCs in the adaptive immune response. DCs promote the activation and differentiation of naïve $CD8^+$ T cells to Cytotoxic T Lymphocytes (CTL) that recognize and kill tumor or infected cells by secreting cytotoxic granules containing perforin and granzymes. In addition, by secreting IFN-γ, cytotoxic $CD8^+$ T cells may also potentiate function of macrophages and NK cells. On the other hand, DCs activate naïve $CD4^+$ T cells and promote their differentiation toward effector $CD4^+$ T cells, which contribute to an efficient activation of $CD8^+$ T cells and B cells, and also regulate function of innate immune cells as macrophages and NK cells.

Two main T cell subsets have been described, which may be phenotypically differentiated by expression on the cell surface of CD4 or CD8, which work as co-receptors for MHC molecules. $CD8^+$ and $CD4^+$ T cells recognize Ags as peptides bound to class I and class II MHC molecules, respectively. Effector $CD8^+$ T cells may directly recognize tumor or infected cells expressing foreign Ags as surface pMHC complexes and subsequently mediate killing of those cells by secreting cytotoxic granules containing toxic proteins, such as perforin and granzyme (Schoenborn & Wilson, 2007). In addition, by secreting IFN-γ, cytotoxic $CD8^+$ T cells may also potentiate function of other immune cells, including macrophages and NK cells. Hence, $CD8^+$ T cells are key players during adaptive immune response against intracellular pathogens and tumors (figure 2).

On the other hand, effector $CD4^+$ T cells not only contribute to efficient activation of $CD8^+$ T cells (Bennett et al., 1998; Ridge et al., 1998; Schoenberger et al., 1998) and B cells (Smith et al., 2000), but they also regulate the function of several cells of the innate immune system, such as macrophages and NK cells (figure 2). Depending on the signals that DCs provide to T cells in addition to antigenic pMHC, they can promote the differentiation of distinct $CD4^+$ T cell subsets, including T helper 1 (Th1), Th2 and Th17 (figure 3). A Th1 phenotype on CD4+ T cells is favored by the secretion of IL-12 by DCs. In contrast, a Th2 phenotype is favored in the absence of IL-12 and in the presence of IL-4 during Ag presentation (Watford et al., 2003). Th1 cells secrete predominantly IFN-γ and promote protective cellular immunity against intracellular pathogens and tumor cells (figure 3). In addition, IFN-γ favors secretion of IgG2a and IgG3 by Ag-specific B cells, both isotypes that efficiently couple Ag recognition with complement activation. In contrast, Th2 cells secrete mainly IL-4, which promote secretion of IgE by B cells, an isotype that couples Ag recognition with mast cell degranulation. In addition, Th2 cells contribute to the activation of eosinophils, promoting allergic-type immune responses that can be efficient at eliminating extracellular pathogens, such as helminths and extracellular bacterium (Del Prete, 1998) (figure 3). Acquisition of Th17 phenotype is promoted by IL-23, TGF-β and IL-6, and it is characterized by secretion of IL-17 (figure 3). It is thought that Th17 cells protect against extracellular bacteria, particularly in the gut (Schoenborn & Wilson, 2007), however they have also been extensively associated with autoimmune diseases (McGeachy & Cua, 2008). Another type of CD4+ T cells consists on regulatory T cells (Tregs), which express the transcription factor Foxp3, secrete mainly IL-10 and suppress several of the effector functions of Ag-specific T cells (Nouri-Shirazi et al., 2000) (figure 3). Due to their inhibitory properties, an exacerbated activity of Tregs could impair T cell immunity and promote tumor development or susceptibility to infections. In contrast, a deficient activity of Tregs could lead to the loss of tolerance to self-Ags and thus promote autoimmune responses.

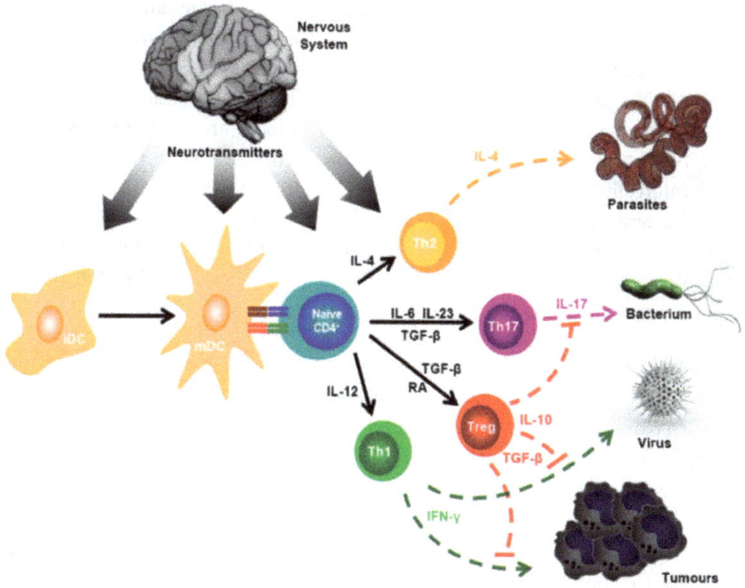

Figure 3. Regulation of immunity by neurotransmitters. DCs contribute to the acquisition of the most appropriate effector phenotype of CD4+ T cells, a process that can be regulated by neurotransmitters at several levels, such as DCs maturation, T cell activation and differentiation.

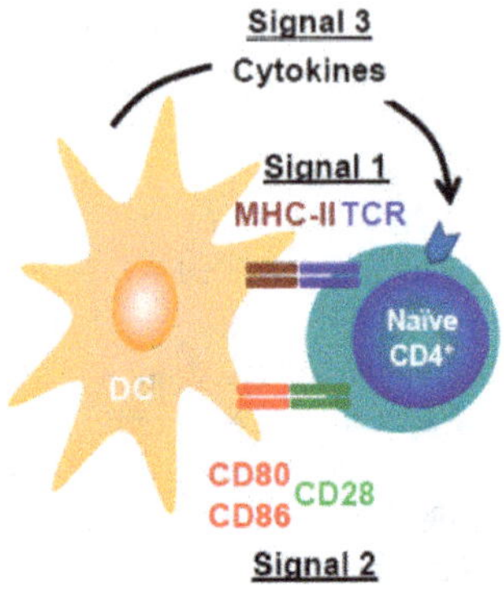

Figure 4. Three kinds of signals are involved in the activation and differentiation of T cells upon their interaction with DCs. The TCR/pMHC interaction determines the specificity of Ag recognition (signal 1). A second complementary signal corresponds to membrane-bound costimulatory molecules, such as CD80 and CD86, which determine the immunogenicity of DCs (signal 2). Finally, soluble molecules secreted by DCs (signal 3) may modulate the acquisition of different functional phenotypes by stimulating their receptors expressed on the $CD4^+$ T cell surface.

DCs are the most potent APCs specialized in the initiation of adaptive immune responses by directing the activation and differentiation of naïve T cells (Lanzavecchia & Sallusto, 2001). These cells can capture both self and foreign Ags in diverse tissues and migrate to secondary lymphoid organs, such as lymph nodes or spleen, to present processed Ags on MHC molecules to T cells. DC-T cell interaction can control and regulate T cell activation, the polarization of the effector phenotype, and the induction of tolerance (Friedl et al., 2005). Important functional components of the DC-T cell synapse are soluble molecules as well as membrane-bound receptors pairs expressed either on the DC or the T cell surface. The most significant intercellular molecular interactions mediating DC-T cell communications are TCR/pMHC interaction, costimulatory molecules with their corresponding receptors and cytokines with their associated receptors. The TCR/pMHC interaction determines the specificity of T cell activation (signal 1, see figure 4). Recognition of pMHC by the TCR is required to trigger all of the later signaling events occurring during T cell function. When Ags are captured by DCs in an inflammatory context generated by cell damage, inflammatory cytokines or pathogen-associated molecular patterns, DCs undergo a maturation process consisting on the expression of elevated levels of costimulatory molecules. CD80 and CD86 (Zheng et al., 2004) are two membrane-bound costimulatory molecules that contribute to the immunogenicity of DCs. Both CD80 and CD86 can bind to CD28 on T cells providing costimulatory signals, which are complementary to TCR-triggered signaling and promote efficient T cell activation (signal 2, see figure 4). The differential expression of these costimulatory surface molecules allows DCs to modulate the nature of T cell responses, promoting either immunity or tolerance, respectively. There is also an important system of soluble proteins and their associated membrane receptors by which DCs regulate the outcome of immune responses. These soluble molecules correspond to a group of regulatory cytokines secreted by DCs, which by stimulating their receptors expressed on the $CD4^+$ T cell surface, modulate the acquisition of different functional phenotypes such as Th1, Th2, Th17 and Tregs, as commented above (see figure 3). Consequently, by a coordinate expression of surface and soluble molecules, DCs can either define the nature of an adaptive immune response against a particular Ag or induce tolerance to self constituents (figure 4). Therefore, DCs are critical and highly relevant at promoting immunity against pathogens and tumors, avoiding the development of autoimmune damage to self healthy tissues.

2.2. Nervous system mediated regulation of immunity

Traditionally, it has been described that the function of immune cells, such as T cells and DCs, is regulated by soluble protein mediators known as cytokines. However, an emerging number of studies have shown that immune system cells can be also regulated by neurotransmitters (Franco et al., 2007). In this regard, it has been described that several receptors for neurotransmitters classically expressed in the nervous system, are also expressed on the surface of immune system cells. For instance, T cells and DCs express some subtypes of glutamate receptors (GluRs), acetylcholine receptors (AChRs), serotonin receptors (5-HTRs), dopamine receptors (DARs), adrenergic receptors, and others (Pacheco et al., 2010). Similar to T cells and DCs, neurotransmitter receptors have been found on the surface of a number of other immune system cells, which implicates that these cells could be regulated by molecules from the nervous system. The identification of these receptors on immune system cells suggests that neurotransmitters play a physiological role in the regulation of the immune response. Thus, deregulation on the activation of these receptors could be involved on the development of autoimmunity or cancer. Moreover, this fact implicates that different physiological or pathological states of the nervous system could be involved in the regulation of immune response. Due to the fact that DCs and T cells express receptors for neurotransmitters present in both compartments, the central nervous system (CNS) and the peripheral nervous system (PNS), immune response can be influenced by these two different parts of the nervous system. In the next sections, some examples of neurotransmitter-mediated regulation of DCs and T cells function (section 2.2.1) as well as the regulation of immunity by the PNS (section 2.2.2) will be analyzed. The interaction of immune cells with the CNS will be analyzed later, in section 2.3, together with neurodegenerative diseases that involve adaptive immune responses inside the CNS.

2.2.1. Neurotransmitter mediated regulation of T cell and DC physiology

By stimulating different neurotransmitter receptors expressed on T cells or DCs, neurotransmitters from diverse sources may strongly regulate the initiation and development of immune responses. Although at low levels, some neurotransmitters can be found in plasma and altered concentrations may reflect different physiological states of the nervous system. Thus, neurotransmitters present in plasma may constitute the primary source of these molecules for immune cells that are found migrating through the blood, such as patrolling naïve or memory T cells or effector T cells migrating toward the place of infection. The existence of innervation of lymphoid tissues by the autonomic nervous system represents an important source of neurotransmitters for immune cells. Innervation of primary lymphoid organs, such as bone marrow and thymus, could regulate the generation and differentiation of new lymphoid and myeloid cells by the nervous system, however, currently, there are not conclusive evidences on this area. On the other hand, the innervation of secondary lymphoid tissues, such as lymph nodes and spleen, by the autonomic nervous system constitutes a very relevant source of neurotransmitters for immune cells (Mignini et al., 2003). Importantly, Ag-presentation and T cell activation occurs in secondary lymphoid tissues, thereby innervation of these organs may represent the main nervous system-mediated regulation of initiation and development of the immune response. Finally, another way by which nervous system may regulate immune cells function through neurotransmitters release is when immune cells, including T cells and DCs, cross the blood-brain-barrier (BBB) and enter into the CNS. Details of how immune cells enter into the CNS are discussed later, in section 2.3. Due to the key role played by T cells and DCs during adaptive immune responses, the contribution of some neurotransmitters in the function of these cells is here analyzed. Specifically, we have focussed the discussion on the involvement of dopamine and acetylcholine in the function of T cells and DCs and in the interaction between these two important immune cells (figure 5).

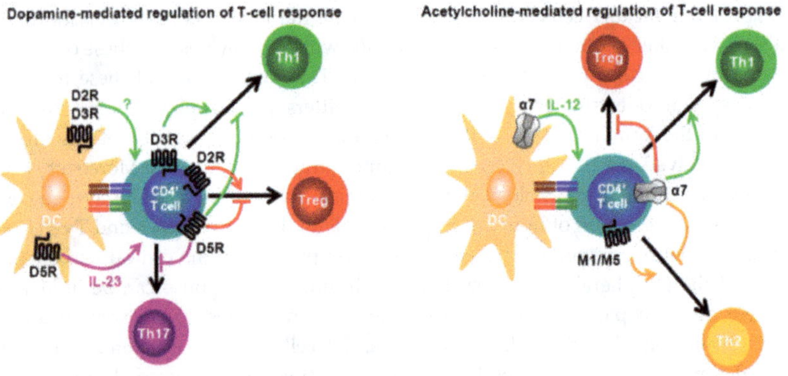

Figure 5. Neurotransmitter-mediated regulation of T cell response. Left panel, stimulation of D5R expressed on DCs promotes secretion of IL-23 favoring Th17 differentiation; while selective stimulation of D2R/D3R on DCs favors polarization of $CD4^+$ T cell responses toward Th1. On the other hand, stimulation of D3R expressed on $CD4^+$ T cells promotes Th1 differentiation, whereas D2R stimulation on these cells would facilitate differentiation to Tregs. Finally, stimulation of D5R expressed on $CD4^+$ T cell inhibits Th1 and Th17 differentiation, and Tregs function. Right panel, stimulation of α7 nAChRs expressed on DCs potentiates secretion of IL-12, improving anti-tumor response *in vivo*. Stimulation of M1/M5 mAChRs expressed on $CD4^+$ T cells potentiates Th2 responses, while stimulation of α7 nAChRs expressed on $CD4^+$ T cells favors Th1 responses and impairs Th2 and Tregs differentiation.

The first example here is dopamine-mediated regulation of immunity. So far, five DARs have been described, all of which are heptaspanning-membrane-receptors that belong to the superfamily of G protein-coupled receptors. All five DARs have been described to be expressed on T cells. In general, the D3R receptor promotes stimulating functions in T cells, while D1R, D2R, D4R and D5R induce inhibitory signals that suppress the T cell function. D3R-stimulation induces secretion of IFN-γ and inhibition of IL-10 and IL-4 synthesis in both $CD4^+$ and $CD8^+$ T cells, which suggests that D3R would favor a Th1-like response. In addition, D3R-triggered signaling not only stimulates chemotaxis by itself, but also potentiates migration of $CD8^+$ T cells induced by some chemokines. On the other hand, stimulation of D1R/D5R impairs proliferation and cytotoxic effector function of $CD4^+$ and $CD8^+$ T cells, respectively, as well as inhibits Tregs function. D2R stimulation promotes enhanced production of IL-10, a cytokine that negatively regulates the function of effector T cells and it could be involved in the polarization toward Tregs. Stimulation of dopamine D4R induces T cell quiescence, a state characterized by decreased cell size and metabolic activity of T cells. In addition to the different effects of DARs stimulation in the T cell physiology, it is important to consider that each DAR display different affinity for dopamine, having higher affinity D3R > D5R > D4R > D2R > D1R (Ki(nM) = 27, 228, 450, 1705, 2340, respectively). Thus, low levels of dopamine (e.g. 50 nM) would stimulate mainly D3R in T cells, favoring Th1-like responses and T-cell migration, whereas moderate dopamine levels (e.g. 300 nM) should stimulate D5R as well, inhibiting T cell function. It is expected that higher dopamine levels, by stimulating multiple DARs, would promote very complex effects in the T cell physiology, probably inhibiting T cell mediated immunity (Pacheco et al., 2009). With regard to the expression of DARs on DCs, D1R and D5R are the DARs expressed at higher levels on the cell surface, whereas D3R and D2R are poorly represented on the DCs surface. It has recently been demonstrated that stimulation of D5R on DCs strongly potentiates production of IL-23, a regulatory cytokine that favors polarization of naïve T cells toward the inflammatory Th17 phenotype. In fact, it was demonstrated that stimulation of D5R on DCs potentiates Th17 responses *in vitro* and *in vivo* (Prado et al., 2012). Other studies have shown that selective stimulation of D2R/D3R or selective inhibition of D1R/D5R on DCs favors polarization of $CD4^+$ T cell responses toward Th1 and impairs the Th17 fate. Thus, depending on the concentration of dopamine, the specific DARs expressed and the kind of immune cell bearing DARs in the place where dopamine is available, this neurotransmitter may induce different effects in the immune response (figure 5, left panel).

The second example is the regulation of T cell responses by the neurotransmitter acetylcholine (Pacheco et al., 2010). This neurotransmitter may promote diverse effects on target cells by stimulating acetylcholine receptors (AChRs) expressed on the surface of those cells. AChRs described are classified in two main groups; namely nicotinic AChRs (nAChRs), which form ion channels and mediate fast excitatory acetylcholine responses, and muscarinic AChRs (mAChRs), which are heptaspanning-membrane-receptors and belong to the superfamily of G protein-coupled receptors. Dopamine, glutamate, serotonin and other neurotransmitters constitute a group of physicochemically and enzymatically stable molecules, which may act on target cells relatively far from where these neurotransmitters were originally released (volume transmission). Thus, substantial amounts of these neurotransmitters can be detected in extracellular fluids including plasma and their role as neurotransmitters is terminated mainly by reuptake into nerve terminals and tissues. In contrast, acetylcholine belongs to the group of labile compounds which, when released, achieve effective concentrations to act over target cells for less than several milliseconds due to its rapid degradation by cholinesterases that are abundant in tissues and plasma (Franco et al., 2007). Thereby, very low acetylcholine levels are detected in plasma and it seems likely that this neurotransmitter primarily acts over immune system by wiring transmission between closely intercellular interactions. Regarding AChRs expression on T cells, it has been described that M3 is the most strongly expressed mAChRs followed by M5 and M4, whereas that M1 and M2 expression is substantially variable among individuals. Muscarinic stimulation on T cells promotes potentiation of T cell activation with strong IL-2 production which is mediated by the M3 subtype. Furthermore, is has been demonstrated that stimulation of the M1 subtype of mAChRs is not required for initial T cell activation, but for subsequent differentiation from naïve $CD8^+$ T cells toward effector cytotoxic $CD8^+$ T cells. In addition, studies suggest that stimulation of M1/M5 mAChRs expressed on $CD4^+$ T cells potentiates Th2 responses *in vivo*. With regard to the function of nAChRs on T cells, it has been demonstrated that nicotine, mainly through $\alpha7$ subunit-bearing nAChRs, triggers intracellular signals. Currently, evidence suggests that stimulation of nAChRs $\alpha7$ on T cells favors Th1 responses, whereas impairs Th2 and Tregs differentiation. Expression of nicotinic as well as muscarinic AChRs has also been described in DCs. However, consequences in DCs physiology upon stimulation of these AChRs have been poorly studied. In this regard, it has been demonstrated that targeting nAChRs $\alpha7$ on DCs *ex vivo* results in increased IL-12 secretion, which in turn, induces enhanced stimulation of cytotoxic $CD8^+$ T cells with a consequent improvement of anti-tumor response *in vivo*. These results suggest that stimulation of nAChRs $\alpha7$ on DCs favors a pro-Th1 phenotype which stimulates polarization of naïve T cells toward Th1. Similar to what happen with dopamine and with other examples of neurotransmitters not mentioned here, depending on the local concentration reached, specific receptors expressed on immune cells, the kind of immune cells located in the microenvironment where acetylcholine is released, this neurotransmitter may exert different and complex regulation of T cell responses (figure 5, right panel).

Finally, as a conclusion in this section, several neurotransmitters may modulate T cell response by stimulating neurotransmitter receptors directly on T cells or by stimulating their receptors on DCs. Consequent neurotransmitter-mediated modulation may favor polarization of the T cell response toward different fates, such as Th1, Th17 or Tregs. Neurotransmitters that induce Th1 responses normally favor anti-viral, anti-tumor and anti-intracellular bacterium response, whereas neurotransmitter contributing to promote strong Th17 and decreased Tregs function may be detrimental in the scenario of autoimmunity. Thus, imbalance on neurotransmitter-mediated regulation of immunity may play an important role in pathologies.

2.2.2. Peripheral nervous system-mediated regulation of immunity

Several studies performed mainly during the last decade have demonstrated a direct regulation of T cell responses by the sympathetic nervous system (SNS). These studies were mainly carried out administering 6-hydroxydopamine, a neurotoxic drug that selectively ablates noradrenergic and dopaminergic neurons, as this molecule is captured specifically through dopamine transporters (DAT) or norepinephrine transporters (NET). Due to the fact that 6-hydroxydopamine cannot cross the BBB, this drug depletes SNS without affecting neurons in the CNS when administered systemically. Norepinephrine is the main neurotransmitter released by the SNS and therefore this is probably the main neurotransmitter responsible for SNS-mediated regulation of immunity, however dopamine has also been involved in the SNS-mediated regulation of T cell response. The evidences point to a dual role of SNS on T cell responses. First, the SNS, probably by releasing norepinephrine, inhibits TGF-β production in the spleen and lymph nodes, thus attenuating generation and function of Tregs. The relevance of this mechanism is evidenced in autoimmunity, as arthritis and experimental autoimmune encephalomyelitis (EAE) are developed with significant less severity in sympathectomized mice than in animals bearing intact SNS (Harle et al., 2008; Bhowmick et al., 2009). Second, by stimulating β2-adrenergic receptors, SNS decreases immunogenicity of APCs in secondary lymphoid organs, attenuating both the Th1 response and the $CD8^+$ T cell mediated cytotoxic activity. The relevance of this mechanism has been demonstrated in two models of infection: influenza and *Listeria monocytogenes*. Efficient clearance of these infectious agents requires potent Th1 and cytotoxic responses. In both cases, sympathectomized mice mounted a stronger antiviral and antibacterial response than those mice with intact SNS (Miura et al., 2001; Grebe et al., 2009). In addition, it is likely that anti-tumor immune response would be strengthened in sympatectomized mice, as this response requires Th1 polarization (see figure 3). The fact that in the first group of studies here described, the SNS potentiates autoimmune responses while in the second group, the SNS attenuates antiviral and antibacterial responses could be explained by a mechanism in which SNS-mediated attenuation of Th1 responses concomitantly favors Th17 responses. However, this possibility or another explanation for these observations is still pending to be clarified. The integrated mechanism mediating SNS-induced regulation of T cell response is schematized in figure 6. Another mechanism that could be relevant in SNS-mediated regulation of immunity involves dopamine. SNS neurons express the enzyme dopamine-β-hydroxylase (DβH), which is necessary to synthesize norepinephrine from dopamine. Dopamine is normally present in low amounts in SNS neurons in normal mice. However, in dopamine-β-hydroxylase deficient mice, all noradrenergic neurons become exclusively dopaminergic. It has been shown that DβH deficient mice develop an attenuated antibacterial response against *Listeria monocytogenes*, when compared with normal mice (Alaniz et al., 1999). Thus, exacerbated dopamine signaling by the SNS results in a decreased Th1 response.

Another PNS-mediated regulation of immunity is the well-documented cholinergic anti-inflammatory pathway, also known as the inflammatory reflex, which results in attenuated activation of splenic macrophages. This pathway does not constitute a nervous system mediated regulation of T cell responses, however, this pathway use T cells as intermediate cells to control innate immunity. Detection of pathogen- or damage- associated molecular patterns by specialized receptors expressed on macrophages, results in a strong production of pro-inflammatory cytokines, such as TNF-α, IL-6, IL-8 and IL-1. When elevated levels of these cytokines are produced, they are transported throughout the blood and detected in the brain as a signal of excessive inflammation. Consequently, the nervous system responds with the inflammatory reflex, which is triggered as action potentials transmitted through the vagus nerve. Outflows from vagus nerve arrive at the coeliac ganglion, which subsequently triggers action potentials in the splenic nerve. Outflows arrive to the spleen, where, by a mechanism dependent on nAChR α7, attenuate the production of inflammatory cytokines by splenic macrophages (Rosas-Ballina & Tracey, 2009). Paradoxically, nerve fibers in spleen, originating in the coeliac ganglion, are noradrenergic, not cholinergic, and utilize norepinephrine as the main neurotransmitter. Recently, it was demonstrated that norepinephrine released from splenic nerve stimulates β1- and β2-adrenergic receptors on memory T cells, which consequently release high amounts of acetylcholine (Rosas-Ballina et al., 2011). Acetylcholine derived from T cells acts subsequently on nAChR α7 expressed on macrophages triggering inhibitory signals that limit inflammatory cytokine release and thus, maintain homeostasis (figure 7). Regarding the sympathetic or parasympathetic origin of this PNS-mediated mechanism, it actually cannot be classified as any of them, as classic parasympathetic signals in the periphery are transduced by the vagus nerve and by mAChRs, not nAChRs. This mechanism is relevant and conceptually very interesting because, first, T cells act as a regulatory bridge between nervous system and innate immunity and, second, T cells work as reservoirs of neurotransmitters, releasing them under specific stimuli, which act subsequently over another immune system cells. This latter concept will be developed in more detail later in this chapter, in section 3.1.

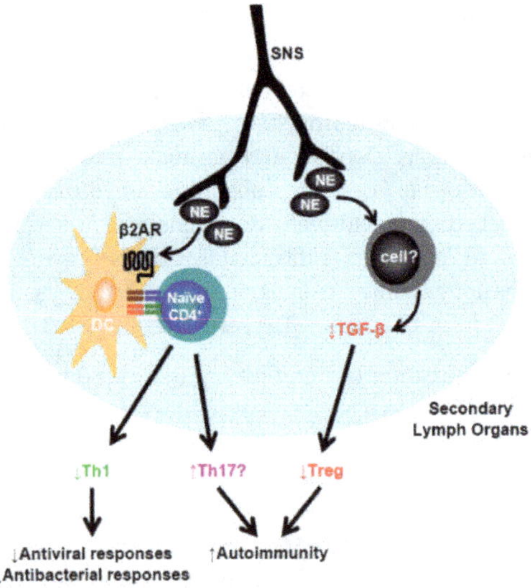

Figure 6. SNS-mediated regulation of T cell response. For explanation see the text. NE, Norepinephrine; β2AR, β2-adrenergic receptors.

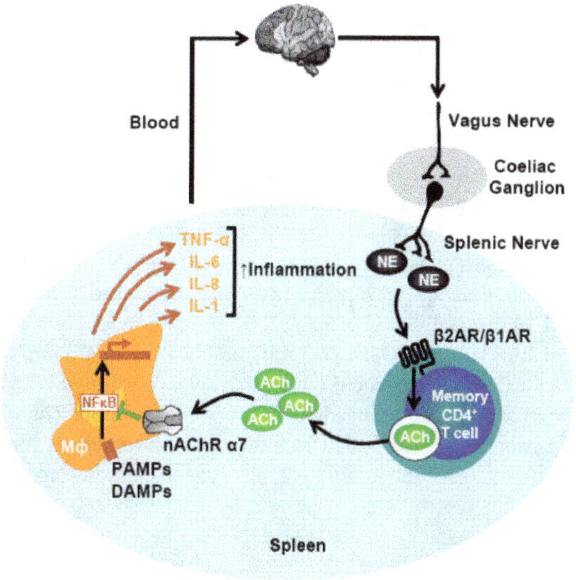

Figure 7. SNS-mediated regulation of innate immunity. Pathogen or damage recognition by spleen macrophages triggers strong production of pro-inflammatory cytokines. Elevated levels of inflammatory cytokines in blood are detected by the nervous system triggering the inflammatory reflex. For explanation see the text. NE, Norepinephrine; DAMPs, Damage-Associated Molecular Patterns; PAMPs, Pathogen-Associated Molecular Patterns.

2.3. Immune system mediated regulation of the nervous system

Entrance of immune system cells in the parenchyma of CNS is restricted, primarily because of the BBB that surrounds parenchymal venules. This barrier is constituted by a first layer of specialized endothelial cells of the blood vessel wall, which are intimately connected by tight junctions. This first layer of cells is surrounded by a basement membrane, and an outer layer constituted by astrocyte feet and microglial cells. Under normal physiological conditions no immune cells are found in the CNS parenchyma. However, some immune cells, including lymphoid and myeloid cells, may infiltrate into the cerebrospinal fluid, which is produced by choroid plexus epithelia and flows throughout the space comprised between the meninges, the subarachnoid space. Importantly, cerebrospinal fluid drains into cervical lymph nodes, enabling peripheral immune cells to survey and respond to CNS Ags under certain pathological conditions. Thus, although the access to the CNS parenchyma of the healthy brain is restricted for immune cells, these cells can enter into the cerebrospinal fluid through choroid plexus. Accordingly, subarachnoid space and choroid plexus of healthy mice are populated by substantial numbers of T cells and heavily populated by myeloid cells. In this regard, DCs that derive from local precursors and exhibit a differentiation and Ag-presenting program similar to spleen DCs, have been identified in the brain of healthy mice (Anandasabapathy et al., 2011). Importantly, cells of the immune system not only have been related with surveillance in the CNS. Under normal conditions, immune mechanisms are activated by environmental/psychological stimuli and they positively regulate the remodeling of neural circuits, promoting memory consolidation, hippocampal long-term potentiation (LTP) and neurogenesis (Yirmiya & Goshen, 2011). According to this, recent studies have demonstrated that spatial memory and learning are $CD4^+$ T cell-dependent in healthy animals. In particular, IL-4-producing T cells accumulate in the subarachnoid space during cognitive tasks. IL-4 produced by these Th2 cells in turn stimulates hippocampal astrocytes to produce and release brain-derived-neurotrophic-factor (BDNF), which acts subsequently on hippocampal neurons (figure 8, left panel), favoring spatial-memory and learning (Derecki et al., 2010). These authors have demonstrated an impaired learning in mice lacking $CD4^+$ T cell or in mice bearing IL-4-deficient CD4+ T-cells. In both cases, normal learning is recovered when mice are reconstituted with normal $CD4^+$ T cell, thus linking immune activity to steady-state cognitive function. The importance of T cells in cognitive function is also supported by more common or natural conditions such as aging, HIV infection and chemotherapy. All of these conditions are associated with decreased T cell function which is accompanied by cognitive impairments (Yirmiya & Goshen, 2011).

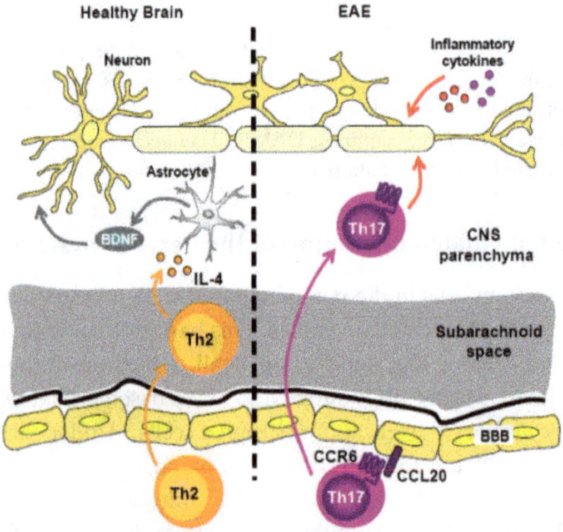

Figure 8. T cell-mediated regulation of the CNS activity. Left panel, in healthy brain, Th2 cells accumulate in the subarachnoid space during cognitive tasks and secrete IL-4, which diffuses into the parenchyma and stimulates astroglia. Consequently, IL-4-stimulated astroglia produces BDNF, which acts on hippocampal neurons favoring acquisition of spatial memory and learning. Right panel, in an autoimmune disease directed against CNS-derived Ags such as EAE, choroid plexus express the chemokine CCL20, which allows CCR6-expressing Th17 cells to cross the BBB and to be recruited into the CNS parenchyma. Subsequently, these Th17 cells mediate the destruction of myelin sheath and axonal damage by the secretion of proinflammatory cytokines and the recruitment of innate immune cells.

On the other hand, the presence of infiltrating immune cells into the CNS parenchyma has been detected in most of the neurodegenerative diseases studied. In a pathological scenario involving the CNS, such as neurodegeneration or imbalance of glial homeostasis, initial neuroinflammatory processes induce brain endothelial cells to express a specialized pattern of adhesion molecules. Adhesion molecules induced by inflammatory processes subsequently allow activated T cells to adhere to vessel walls and to be recruited into the CNS parenchyma. T cells that infiltrate the CNS are previously activated in the periphery, presumably in cervical lymph nodes, in which a sampling of CNS-associated Ags is constantly arriving through the cerebrospinal fluid that drains these lymph nodes. Interestingly, the expression of the chemokine receptor CCR6 by a subset of pathogenic T cells, Th17 cells, and the expression of its corresponding ligand, the chemokine CCL20, by epithelial cells of the choroid plexus have been shown to play an important role in facilitating Th17 cells recruitment to BBB during the development of EAE, the murine model of the autoimmune disease multiple sclerosis. It has been determined that IL-17R and IL-22R are up-regulated on the surface of endothelial cells of BBB in multiple sclerosis lesions. Stimulation of these receptors by IL-17 and IL-22 secreted by Th17 cells induces disruption of tight junctions. Thus, Th17 cells perform a particular and specialized mechanism to cross the BBB (Pacheco et al., 2009). Th17 cells infiltrated into CNS parenchyma interact with microglia, promoting the production of additional inflammatory molecules that up-regulate some key adhesion molecules and chemokines that, in turn, allow subsequent recruitment of another kind of leukocytes through BBB, including Th1 cells and DCs. Once in CNS parenchyma, infiltrating T cells can contribute significantly to regulate the neurodegenerative process by the secretion of different cytokines and the recruitment of innate immune cells. In the case of multiple sclerosis or EAE, CNS-infiltrating T cell acquire aggressive phenotypes and recruit effector cells from the innate immune system which play an important role in mediating demyelination and axonal damage (figure 8, right panel) (Goverman, 2009). Additionally, molecules derived from immune cells can act over glial cells, modulating microglia-mediated neurotoxicity. In this regard, recent studies have shown that peripheral T cells infiltrate into the brain parenchyma at the site of neuronal injury in Parkinson's disease. This T cell-mediated immune response contributes significantly to the destruction of dopaminergic neurons, through a $CD4^+$ T cell-dependent cytotoxic mechanism. These studies support the involvement of pathogenic $CD4^+$ T cell populations, which would induce acquisition of an M1-like pro-inflammatory phenotype by microglia, which is characterized by the secretion of inflammatory factors, such as TNF-α, IL-1β and superoxide. Conversely, in the healthy scenario, other T cell subsets, for instance Tregs and Th2, could contribute to microglial acquisition of an M2-like anti-inflammatory phenotype, which release neurotrophic factors, such as IGF-1 (figure 9), promoting neuronal protection (Appel, 2009; Reynolds et al., 2010). In opposition to the role of CNS-infiltrating T cells in Parkinson's disease and multiple sclerosis, studies carried out in amyotrophic lateral sclerosis have shown that absence of T cells accelerate motoneuron disease while the adoptive transfer of T cells ameliorates disease severity (Lucin & Wyss-Coray, 2009). These results probably could be explained by the contribution of T cells to the acquisition of M2 phenotype by microglia, however further investigation is necessary to confirm this hypothesis. Thus, imbalance on these different kinds of neuro-immune interactions may favor the onset of autoimmune disorders and also may constitute an important component of the pathogenic mechanisms involved in neurodegenerative diseases.

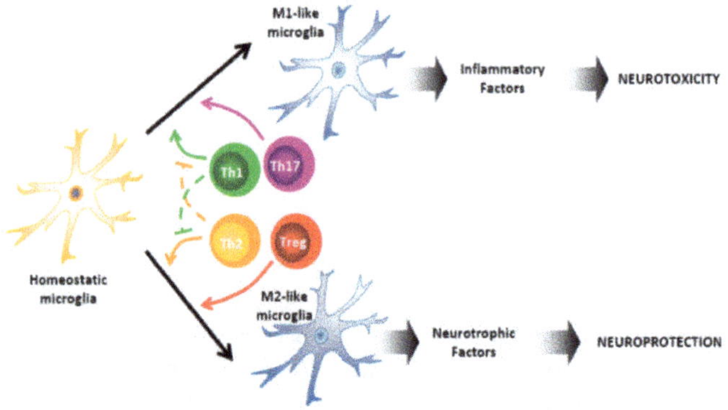

Figure 9. Involvement of $CD4^+$ T cells in the microglia-mediated neurotoxicity or neuroprotection. By recognizing pathogen- or damage- associated molecular patterns, homeostatic microglia undergoes an activation process, which may result in the acquisition of two different phenotypes. The first one is the M1-like phenotype, which produces and releases inflammatory factors, thus promoting neuronal injury. The second one is the M2-like phenotype, which produces neurotrophic factors, sustaining neuroprotection. Different kinds of $CD4^+$ T cell may differentially modulate acquisition of microglial phenotypes. Whereas Th1 and Th17 cells favor acquisition of M1-like phenotype by microglia, Th2 and Tregs contribute to the acquisition of M2-like phenotype. Th1 and Th2 cells may promote antagonistic effects in the acquisition of M2-like and M1-like phenotypes, respectively.

3. The nervous system and immune system use the same mediator molecules

For many years the immune system, including its cells and mediator molecules, was studied and analyzed completely separate from the nervous system and vice versa. Classically, cytokines have been considered the main soluble molecules mediating regulatory communications between immune cells. On the other hand, neurotransmitters have been described as the traditional soluble molecules mediating interactions between neurons or between glial cells and neurons. Currently, it is known that immune system cells can also communicate through neurotransmitters and that nervous system cells can communicate by cytokines as mediators. Evidences from the last decade have shown that immune cells not only express neurotransmitter receptors, but also have the capacity to synthesize and store neurotransmitters in intracellular compartments, which may be released under specific stimuli. Similarly, nervous system cells not only express cytokine receptors, but are also able to produce these molecules. This new knowledge has opened the possibility that immune cells could communicate between themselves through neurotransmitters and that nervous system cells could interact between themselves by cytokines (figure 10). In this chapter, first, the neurotransmitter-mediated communications between immune cells will be discussed (section 3.1); and second, cytokine-mediated interactions between nervous system cells will be also analyzed (section 3.2).

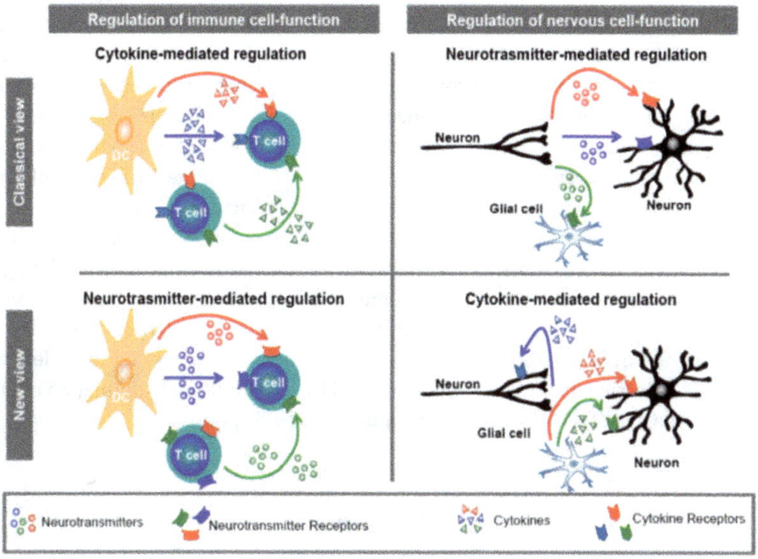

Figure 10. Immune and nervous systems utilize common modulatory molecules. The classical view states that immune system cells and nervous system cells communicate each other by means of cytokines and neurotransmitters, respectively. However, evidence from last decade supports the common usage of these molecules by both systems, resulting in neurotransmitter-mediated communication between immune cells and cytokine-mediated interaction among nervous system cells.

3.1. Neurotransmitters as mediators of immune-mediated immunity

An increasing number of studies have revealed that some cellular components involved in adaptive and innate immune responses such as DCs, T cells, B cells, macrophages and others, are capable of synthesizing and/or capturing classical neurotransmitters, including acetylcholine, dopamine, serotonin and glutamate. Under some stimuli, these cells may release neurotransmitters from vesicular or non-vesicular storages toward extracellular compartment, thus involving autocrine or paracrine communications with different leukocytes. For instance, T cells may release serotonin, norepinephrine, dopamine and acetylcholine, while DCs may release glutamate, dopamine and serotonin (Franco et al., 2007). This fact not only suggests that neurotransmitters could mediate communication between immune cells, but also that this kind of molecules may be involved in bidirectional cross-talk between immune and nervous system (figure 11). Next, some examples of neurotransmitter-mediated communications between immune cells are analyzed and discussed.

DCs capture and synthesize some neurotransmitters, which are stored in intracellular compartments and later used either to regulate their own functions (autocrine process) or to modulate T cells (paracrine process). Recently, it was demonstrated that DCs express the whole machinery to synthesize and store dopamine (Prado et al., 2012). For instance, these cells express tyrosine hydroxylase (TH), which catalyzes the first step necessary for biosynthesis of dopamine. However, these cells do not express DβH, the enzyme necessary to metabolize dopamine and to transform it in epinephrine and norepinephrine. In addition, DCs do not express the plasma membrane dopamine transporter, necessary to take up dopamine from the extracellular compartment. Thereby, DCs synthesize dopamine, but not the other catecholamines, and they can not capture dopamine from the extracellular space. These cells also express enzymes necessary to degrade dopamine in the cytoplasm and vesicular transporters necessary to store dopamine in vesicular compartments. A recent study has indicated that LPS, a pathogen-associated molecular pattern, induces dopamine release from DCs, which subsequently acts in an autocrine manner stimulating D5R expressed on DCs (Prado et al., 2012). D5R stimulation promotes a potent production of IL-23 by DCs, thus conferring these cells with the ability to promote polarization of $CD4^{+}$ naïve T cells toward the inflammatory Th17 phenotype (figure 12; see also figure 5, left panel). Therefore, this autocrine loop mechanism mediated by dopamine and stimulation of D5R on DCs, results to be one of the engines feeding autoimmune responses.

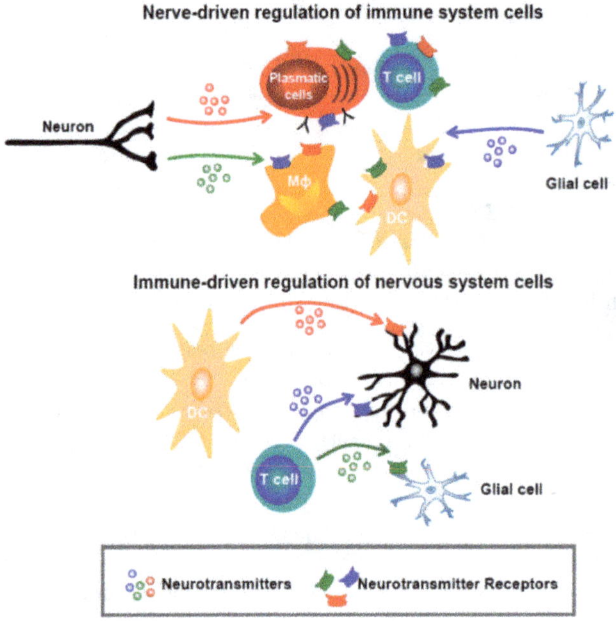

Figure 11. Neurotransmitters mediate bidirectional cross-talk between immune and nervous cells. Neurotransmitter-mediated inter-systemic communication may occur in both directions: the regulation of immune cells by nervous-derived neurotransmitters as well as the modulation of nervous cells by immune-derived neurotransmitters. Similarly, cytokine-mediated inter-systemic communication may occur in both directions, however it has been omitted in the figure for simplification.

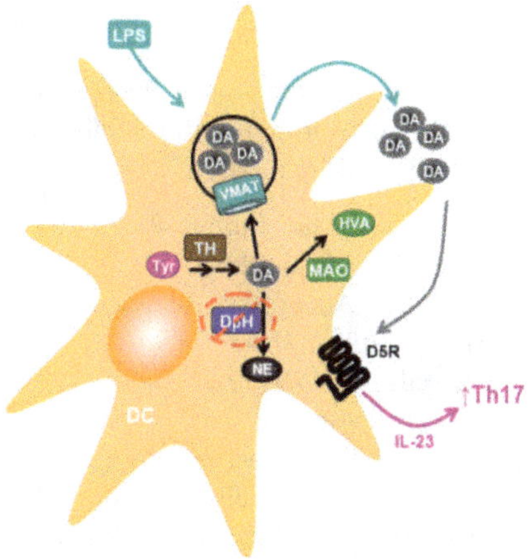

Figure 12. Dopamine-mediated regulation of T cell differentiation by DCs. VMAT, vesicular monoamine transporter; MAO, monoamine oxidase; DA, dopamine; NE, norepinephrine; HVA, homovanillic acid; Tyr, tyrosine. For explanation and other abbreviations see the text.

DCs may also represent a source of acetylcholine and serotonin for T cells (Franco et al., 2007; Pacheco et al., 2010). Acetylcholine released from DCs may constitute an important contribution during first stage of T cell activation. In this regard, it has been described that when undergoing maturation, DCs begin to express choline-acetyl-transferase (ChAT), the enzyme necessary to synthesize acetylcholine (Kawashima et al., 2007). These findings suggest that acetylcholine would be released during Ag presentation to T cells. Acetylcholine released during Ag presentation can, subsequently, exert different effects acting over AChRs expressed on T cells and on DCs (see figure 5, right panel). Regarding DCs as a source of serotonin, these cells do not synthesize this neurotransmitter but express the serotonin transporter (SERT) and therefore can efficiently sequester microenvironmental serotonin. Serotonin is stored in vesicles, which are released via Ca^{2+}-dependent exocytosis during Ag-presentation. SERT expression by DCs is dynamically regulated. Thus, DCs appear to maximize their uptake and storage capacity for serotonin upon maturation or activation, prior to productive encounters with naïve T cells. In contrast, ligation of CD80/CD86 on DCs by the inhibitory ligand expressed on the T cell surface, the Cytotoxic T-Lymphocyte Antigen 4 (CTLA-4), downregulates SERT expression in DCs. Thus, the differential serotonin content of DCs may contribute to modify signaling at the T cell-DC interaction (see below).

Another kind of immune cells that represent a key source of neurotransmitters necessary to regulate immune responses are T cells. Accordingly, it has been described that Tregs constitutively express TH and contain substantial amounts of dopamine and other catecholamines, while effector T cells only contain trace amounts (Cosentino et al., 2007). Moreover, Tregs as well as effector T cells express vesicular monoamine transporters which allow these cells to accumulate catecholamines into vesicular storages. However, only Tregs store and release physiologically relevant amounts of dopamine from vesicular storages. The stimulus responsible for triggering dopamine release from Tregs is still unknown, however it is thought that dopamine from this source would constitute an autocrine loop in which, by stimulating D1R/D5R, this neurotransmitter inhibits the suppresive function of Tregs exerted over effector T cells (Kipnis et al., 2004; Cosentino et al., 2007).

Acetylcholine is also an important mediator produced and released by T cells. Indeed, it seems that this source of acetylcholine is the primary source of this mediator for these cells inside secondary lymphoid organs. All of the components of cholinergic system are expressed in T cells (Pacheco et al., 2010). In this regard, ChAT and acetylcholine stores have been found in memory $CD4^+$ T cells (Rosas-Ballina et al., 2011). In addition to acetylcholine and ChAT, nAChRs, mAChRs, cholinesterase activity and choline transporters have also been found on the surface of these cells. Release of acetylcholine-containing vesicles from memory $CD4^+$ T cells may be triggered by T cell activation or by norepinephrine derived from the PNS in the spleen (see figure 7) (Kawashima & Fujii, 2003; Rosas-Ballina et al., 2011). Acetylcholine release may lead to high local concentrations of this neurotransmitter, which in turn can stimulate AChRs expressed on macrophages and in $CD4^+$ T cells themselves. Stimulation of AChRs expressed on macrophages and on T cells may exert different effects depending on the acetylcholine concentration reached in the microenvironment (see figure 5B, right panel and figure 8).

Another example of neurotransmitter working as mediator of T cell-mediated regulation of immunity is serotonin (Franco et al., 2007; Pacheco et al., 2010). Splenic T cells selectively express tryptophan hydroxylase (TPH), the enzyme that catalyses conversion of L-tryptophan to 5-hydroxytryptophan, the immediate precursor of serotonin. Accordingly, T cells can synthesize serotonin and this capacity is considerably increased following activation. When T cells enter the activation process, serotonin release is induced. Thus, during Ag-presentation by DCs in secondary lymphoid organs, both T cells and DCs (see above) begin to release serotonin in the microenvironment of T cell-DC interaction. Serotonin seems to act early stimulating 5-HT_7 and 5-HT_3 receptors expressed on naïve T cells, potentiating T cell activation and proliferation in these cells. After initial activation, T cells begin to express 5-HT_1 and 5-HT_2 receptors which may promote further serotonin-mediated regulation in the effector function of T cells, such as potentiating polarization of the effector phenotype acquired by T cells.

3.2. Cytokines as mediators of nervous system-mediated nervous function

Cytokines have been long recognized as peripheral mediators of inflammatory reactions, orchestrating complex immune responses against infection and injury. More recently, cytokines have been reported to be synthesized also in the brain by both the glial and the neuronal compartment. Moreover, cytokine receptors have also been found in different regions of the CNS. This section is focused first in the modulation of nervous function by glial cell-derived cytokines and then on the current evidence showing cytokine production by neurons.

One of the normal processes recently found to be regulated by brain-derived cytokines is the LTP in the hippocampus, a mechanism characterized by a sustained enhancement in synaptic transmission following high frequency stimulation, and considered one of the main cellular mechanisms underlying learning and memory. Induction of LTP in hippocampal slice preparations and in freely moving rats caused a clear and persistent increase of IL-1β gene expression on glial cells. Induction of IL-1β production by glial cells requires previous release of neuregulin-1 (Nrg-1) by pre-synaptic neurons, a mediator that binds to its receptor erbB2 on the surface of surrounding glial cells, triggering thus IL-1β production (Ren & Dubner, 2010). IL-1β in turn, stimulates its receptor on the post-synaptic neurons. IL-1-induced signaling on these neurons enhances the capability of responses triggered via ionotropic NMDA glutamate receptor (figure 13). Thus, transient blockade of IL-1 receptors results in the inhibition of LTP maintenance. The key role of IL-1 in the acquisition of LTP is further supported by the complete absence of LTP observed in hippocampal slices obtained from IL-1 receptor knockout mice (Goshen et al., 2007). Moreover, the relevance of IL-1 in the acquisition of memory is supported by studies showing that lack of IL-1 signaling in the brain results in decreased performance on memory and learning tests, suggesting that IL-1-mediated signaling in the hippocampus plays a critical role at establishing LTP and, consequently, learning and memory processes. In addition to its role in the establishment of beneficial processes, it was recently recognized that IL-1β expressed by glial cells may also have a detrimental role in pathology, such as in the genesis and maintenance of persistent pain. Several reports have suggested that IL-1β plays a central role in this process as direct injection of this cytokine into the CNS produces increased sensitivity to pain (hyperalgesia) and enhanced neuronal responses in animals. In line with this, the injection of Freund's adjuvant into the masseter muscle in rats, a model of persistent pain in which occurs muscle inflammation and hyperalgesia, results in increased IL-1β production by activated astroglia in the spinal trigeminal nucleus. On the other hand, evidence shows that IL-1 receptor is also co-expressed with NMDA receptor on post-synaptic neurons involved in pain persistance. Accordingly, it was shown that IL-1 receptor-mediated signaling increases NMDA receptor activity in post-synaptic neurons, increasing synaptic strength and contributing to inflammation-induced pain hypersensitivity. Thus, IL-1β provides a link between astroglial activation and NMDA receptor activation in neurons, a mechanism that results in increased memory acquisition and pain sensitivity.

Another cytokine involved in the regulation of nervous function is IL-6, which is also produced upon induction of LTP both *in vitro* and *in vivo*. However, in contrast to the supportive effect of IL-1β, this cytokine contributes to the extinction of a consolidated LTP. In this regard, it has been shown that IL-6 neutralization after continuous stimulation of pre-synaptic neurons results in a marked prolongation of LTP, an effect that correlates with improved performance of animals in long-term memory tests. In agreement with this, IL-6 knockout mice also present enhanced learning capabilities compared to wild type littermates. This fact indicates that hippocampal expression of IL-6 limits the maintenance of LTP and, therefore, reduces memory and learning. Additional evidence has shown that IL-6 upregulation is tightly localized near the site of stimulation in LTP experiments and is detected only in non-neuronal cells (Besedovsky & del Rey, 2011). This observation has prompted the current model of LTP modulation by cytokines, in which cytokine release by astroglial/microglial cells influence neuron function by acting either in a paracrine or an autocrine manner (figure 13).

Another example of cytokine operating as a modulator between nervous cells is TNFα. In normal conditions, production of TNFα by glial cells enhances synaptic effects by increasing surface expression of ionotropic AMPA glutamate receptors on post-synaptic neurons (figure 13). In this regard, a decreased expression of AMPA receptors on post-synaptic neurons is observed when biological activity of this cytokine is prevented, suggesting that TNFα preserves synaptic strenght at excitatory synapses and that TNFα-mediated signaling may play an important role in synaptic plasticity and in response to neural injury. Indeed, it has been shown that glia-derived TNFα plays a pivotal role in the development of synaptic scaling, one of the two mechanisms involved in the generation of synaptic plasticity, along with LTP, described to date (Stellwagen & Malenka, 2006). In adittion, a recent report has identified neuronal TNF receptor subtype 1, but not 2, as the mediator of TNFα-induced increase in AMPA receptor trafficking and surface expression (He et al., 2012). Regarding the contribution of TNFα in response to neural injury, it has been shown that TNFα levels are increased in a preclinical model of spinal cord injury. In line with these results, it was reported that spinal cord injury induces trafficking of AMPA receptors to the neuronal membrane, a process that enhances excitotoxicity mediated by glutamate (Ferguson et al., 2008). Thus, the current evidence indicates that glia-derived TNFα evokes increased expression of AMPA receptors on post-synaptic neurons, which has been involved not only in beneficial processes such as synaptic plasticity and memory acquisition, but also in injury-triggered neurotoxicity.

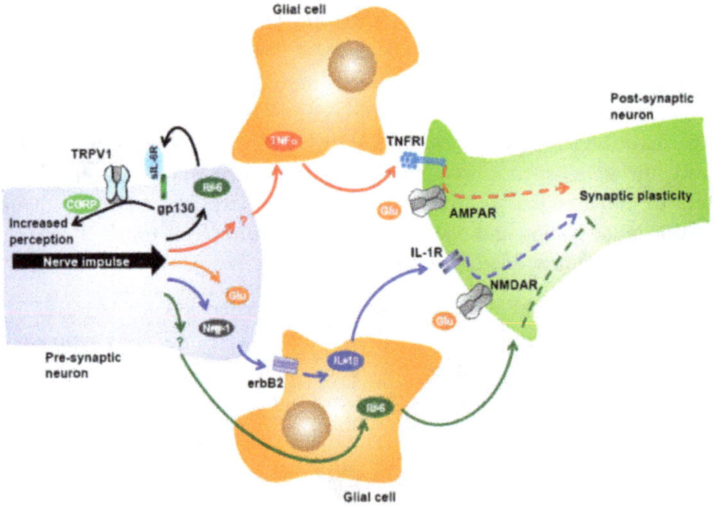

Figure 13. Cytokines as mediators between nervous system cells. Modulation of glutamate (Glu)-mediated signaling in post-synaptic neurons by cytokines produced by glial cells is shown. In addition, regulation of sensitivity of afferent terminals by neuron-derived IL-6 is also shown. For explanation and abbreviations see the text.

In contrast to the numerous studies about the role of cytokines as glia-derived mediators on nervous function, evidence showing modulation of nervous functions by neuron-derived cytokines is very limited. One of the mechanisms involving cytokine release by neurons is IL-6 secretion by primary afferent terminals into the spinal cord in neuropathic pain. It has been shown that arrival of nerve impulses induces the secretion of IL-6, along with several other neuron-derived mediators. In agreement with an autocrine loop, gp130 and gp130-binding subunits of IL-6 receptor have been detected abundantly in afferent neurons under these pathological conditions, nevertheless the binding component for IL-6 itself (glycoprotein 80) has not been detected. Despite stimulation of cultured neurons with IL-6 does not induce any response, addition of a soluble fragment of its receptor results in a rapid sensitization of transient receptor potential cation channel (TRPV1) conductance in response to heat (figure 13). In addition to sensitization of TRPV1 conductance, the addition of a soluble fragment of IL-6 receptor results in the release of calcitonin gene related peptide (CGRP), a mediator involved in pain transmission. In line with a response mediated by a gp130-linked receptor, inhibitors of gp130-mediated signaling prevented the enhancement of TRPV1 sensitivity (Miller et al., 2009). Thus, taken together these results suggest an important role for neuron-derived IL-6 in chronic pain and also indicate that triggering of IL-6 signaling in neurons probably requires the production of soluble IL-6 receptors secreted from other cell types in the vicinity.

4. Conclusion

The interaction between immune system and nervous system involves reciprocal regulation in mammals which can be analyzed in two directions: the immune mediated regulation of nervous system function and the nervous system mediated regulation of immunity. T cells and dendritic cells are two pivotal players in the initiation and development of the immune response and thereby, these two kind of cells become the main convergence points through which immune system regulates nervous system and vice versa. Nervous system mediated interactions occurs not only through classical neurotransmitters produced by nervous cells and acting over immune cells, but also through cytokines derived from cells of the nervous system and exerting their effects on immune system cells. Similarly, immune system mediated regulation of nervous system function occurs by production of both cytokines and neurotransmitters by immune cells which act on nervous system cells. Nervous system mediated regulation of immunity is performed at two levels: 1, by the peripheral nervous system that innervates lymphoid organs in which it releases mediators that act subsequently over immune cells, regulating initiation and polarization of the immune response, and 2, when immune cells cross the blood brain barrier and infiltrate the central nervous system parenchyma, where their effector functions are exposed to regulation mediated by molecules produced by nervous system cells. On the other hand, immune mediated regulation of nervous system functions also occurs in two ways: 1, in the absence of inflammation, immune cells keep patrolling throughout the cerebrospinal fluid where they interact with nervous system cells regulating acquisition of memory, learning and behaviour, and 2, when brain damage or imbalance of glial homeostasis initiate inflammatory processes, immune cells are recruited into the central nervous system parenchyma where they produce molecules which act over glial cells and neurons. Importantly, both nervous and immune systems may produce and also respond to both kinds of mediators: neurotransmitters and cytokines. Studies performed in this area have been carried out mainly in mice, however there are several studies performed in other different mammals, including humans, cats, dogs, and other rodents, confirming results obtained with mice. Thereby, mechanisms involving neuro-immune interactions could be extrapolated to all of the mammals. Imbalance on these different kinds of neuro-immune interaction may favor the onset of autoimmune disorders, contribute to the development of tumors or susceptibility to infectious diseases and also may constitute an important component of pathogenic mechanisms involved in the neurodegenerative diseases such as Parkinson's disease and amyotrophic lateral sclerosis. Due to the similitude of these mechanisms along mammals and taking into account the vast number of murine tools currently developed, including syngenic and congenic strains and also transgenic and knockout mice, studies performed in mouse as model constitute an excellent approach to evaluate neuro-immune mechanisms *in vivo*.

Author details

Rodrigo Pacheco, Francisco Contreras and Carolina Prado
Fundación Ciencia y Vida,
Universidad San Sebastián and Universidad Andrés Bello,
Santiago,
Chile

Acknowledgement

We thank to supporting grants 1095114 from FONDECYT, PFB-16 from CONICYT and 2011-0001-R from Universidad San Sebastián. CP and FC hold CONICYT graduated fellowships.

5. References

Alaniz, R.C., Thomas, S.A., Perez-Melgosa, M., Mueller, K., Farr, A.G., Palmiter, R.D. & Wilson, C.B. (1999). Dopamine beta-hydroxylase deficiency impairs cellular immunity. *Proc Natl Acad Sci U S A,* Vol. 96, No. 5, (Mar 2), pp. (2274-2278)

Amedei, A., Niccolai, E. & D'Elios, M.M. (2011). T cells and adoptive immunotherapy: recent developments and future prospects in gastrointestinal oncology. *Clin Dev Immunol,* Vol. 2011, No. pp. (1-17)

Anandasabapathy, N., Victora, G.D., Meredith, M., Feder, R., Dong, B., Kluger, C., Yao, K., Dustin, M.L., Nussenzweig, M.C., Steinman, R.M. & Liu, K. (2011). Flt3L controls the development of radiosensitive dendritic cells in the meninges and choroid plexus of the steady-state mouse brain. *J Exp Med,* Vol. 208, No. 8, (Aug 1), pp. (1695-1705)

Appel, S.H. (2009). CD4+ T cells mediate cytotoxicity in neurodegenerative diseases. *J Clin Invest,* Vol. 119, No. 1, (Jan), pp. (13-15)

Bennett, S.R., Carbone, F.R., Karamalis, F., Flavell, R.A., Miller, J.F. & Heath, W.R. (1998). Help for cytotoxic-T-cell responses is mediated by CD40 signalling. *Nature,* Vol. 393, No. 6684, (Jun 4), pp. (478-480)

Besedovsky, H.O. & del Rey, A. (2011). Central and peripheral cytokines mediate immune-brain connectivity. *Neurochem Res,* Vol. 36, No. 1, (Jan), pp. (1-6)

Bhowmick, S., Singh, A., Flavell, R.A., Clark, R.B., O'Rourke, J. & Cone, R.E. (2009). The sympathetic nervous system modulates CD4(+)FoxP3(+) regulatory T cells via a TGF-beta-dependent mechanism. *J Leukoc Biol,* Vol. 86, No. 6, (Dec), pp. (1275-1283)

Cosentino, M., Fietta, A.M., Ferrari, M., Rasini, E., Bombelli, R., Carcano, E., Saporiti, F., Meloni, F., Marino, F. & Lecchini, S. (2007). Human CD4+CD25+ regulatory T cells selectively express tyrosine hydroxylase and contain endogenous catecholamines subserving an autocrine/paracrine inhibitory functional loop. *Blood,* Vol. 109, No. 2, (Jan 15), pp. (632-642)

Del Prete, G. (1998). The concept of type-1 and type-2 helper T cells and their cytokines in humans. *Int Rev Immunol,* Vol. 16, No. 3-4, pp. (427-455)

Derecki, N.C., Cardani, A.N., Yang, C.H., Quinnies, K.M., Crihfield, A., Lynch, K.R. & Kipnis, J. (2010). Regulation of learning and memory by meningeal immunity: a key role for IL-4. *J Exp Med,* Vol. 207, No. 5, (May 10), pp. (1067-1080)

Ferguson, A.R., Christensen, R.N., Gensel, J.C., Miller, B.A., Sun, F., Beattie, E.C., Bresnahan, J.C. & Beattie, M.S. (2008). Cell death after spinal cord injury is exacerbated by rapid TNF alpha-induced trafficking of GluR2-lacking AMPARs to the plasma membrane. *J Neurosci,* Vol. 28, No. 44, (Oct 29), pp. (11391-11400)

Franco, R., Pacheco, R., Lluis, C., Ahern, G.P. & O'Connell, P.J. (2007). The emergence of neurotransmitters as immune modulators. *Trends Immunol,* Vol. 28, No. 9, (Sep), pp. (400-407)

Friedl, P., den Boer, A.T. & Gunzer, M. (2005). Tuning immune responses: diversity and adaptation of the immunological synapse. *Nat Rev Immunol,* Vol. 5, No. 7, (Jul), pp. (532-545)

Goshen, I., Kreisel, T., Ounallah-Saad, H., Renbaum, P., Zalzstein, Y., Ben-Hur, T., Levy-Lahad, E. & Yirmiya, R. (2007). A dual role for interleukin-1 in hippocampal-dependent memory processes. *Psychoneuroendocrinology,* Vol. 32, No. 8-10, (Sep-Nov), pp. (1106-1115)

Goverman, J. (2009). Autoimmune T cell responses in the central nervous system. *Nat Rev Immunol,* Vol. 9, No. 6, (Jun), pp. (393-407)

Grebe, K.M., Hickman, H.D., Irvine, K.R., Takeda, K., Bennink, J.R. & Yewdell, J.W. (2009). Sympathetic nervous system control of anti-influenza CD8+ T cell responses. *Proc Natl Acad Sci U S A,* Vol. 106, No. 13, (Mar 31), pp. (5300-5305)

Harle, P., Pongratz, G., Albrecht, J., Tarner, I.H. & Straub, R.H. (2008). An early sympathetic nervous system influence exacerbates collagen-induced arthritis via CD4+CD25+ cells. *Arthritis Rheum*, Vol. 58, No. 8, (Aug), pp. (2347-2355)

He, P., Liu, Q., Wu, J. & Shen, Y. (2012). Genetic deletion of TNF receptor suppresses excitatory synaptic transmission via reducing AMPA receptor synaptic localization in cortical neurons. *Faseb J*, Vol. 26, No. 1, (Jan), pp. (334-345)

Kawashima, K. & Fujii, T. (2003). The lymphocytic cholinergic system and its contribution to the regulation of immune activity. *Life Sci*, Vol. 74, No. 6, (Dec 26), pp. (675-696)

Kawashima, K., Yoshikawa, K., Fujii, Y.X., Moriwaki, Y. & Misawa, H. (2007). Expression and function of genes encoding cholinergic components in murine immune cells. *Life Sci*, Vol. 80, No. 24-25, (May 30), pp. (2314-2319)

Kipnis, J., Cardon, M., Avidan, H., Lewitus, G.M., Mordechay, S., Rolls, A., Shani, Y. & Schwartz, M. (2004). Dopamine, through the extracellular signal-regulated kinase pathway, downregulates CD4+CD25+ regulatory T-cell activity: implications for neurodegeneration. *J Neurosci*, Vol. 24, No. 27, (Jul 7), pp. (6133-6143)

Lanzavecchia, A. & Sallusto, F. (2001). Regulation of T cell immunity by dendritic cells. *Cell*, Vol. 106, No. 3, (Aug 10), pp. (263-266)

Lucin, K.M. & Wyss-Coray, T. (2009). Immune activation in brain aging and neurodegeneration: too much or too little? *Neuron*, Vol. 64, No. 1, (Oct 15), pp. (110-122)

McGeachy, M.J. & Cua, D.J. (2008). Th17 cell differentiation: the long and winding road. *Immunity*, Vol. 28, No. 4, (Apr), pp. (445-453)

Mignini, F., Streccioni, V. & Amenta, F. (2003). Autonomic innervation of immune organs and neuroimmune modulation. *Auton Autacoid Pharmacol*, Vol. 23, No. 1, (Feb), pp. (1-25)

Miller, R.J., Jung, H., Bhangoo, S.K. & White, F.A. (2009). Cytokine and chemokine regulation of sensory neuron function. *Handb Exp Pharmacol*, Vol. No. 194, pp. (417-449)

Miura, T., Kudo, T., Matsuki, A., Sekikawa, K., Tagawa, Y., Iwakura, Y. & Nakane, A. (2001). Effect of 6-hydroxydopamine on host resistance against Listeria monocytogenes infection. *Infect Immun*, Vol. 69, No. 12, (Dec), pp. (7234-7241)

Nouri-Shirazi, M., Banchereau, J., Fay, J. & Palucka, K. (2000). Dendritic cell based tumor vaccines. *Immunol Lett*, Vol. 74, No. 1, (Sep 15), pp. (5-10)

Pacheco, R., Prado, C.E., Barrientos, M.J. & Bernales, S. (2009). Role of dopamine in the physiology of T-cells and dendritic cells. *J Neuroimmunol*, Vol. 216, No. 1-2, (Nov 30), pp. (8-19)

Pacheco, R., Riquelme, E. & Kalergis, A.M. (2010). Emerging evidence for the role of neurotransmitters in the modulation of T cell responses to cognate ligands. *Cent Nerv Syst Agents Med Chem*, Vol. 10, No. 1, (Mar), pp. (65-83)

Park, T.S., Rosenberg, S.A. & Morgan, R.A. (2011). Treating cancer with genetically engineered T cells. *Trends Biotechnol*, Vol. 29, No. 11, (Nov), pp. (550-557)

Parker, D.C. (1993). T cell-dependent B cell activation. *Annu Rev Immunol*, Vol. 11, No. pp. (331-360)

Prado, C., Contreras, F., Gonzalez, H., Diaz, P., Elgueta, D., Barrientos, M., Herrada, A.A., Lladser, A., Bernales, S. & Pacheco, R. (2012). Stimulation of dopamine receptor d5 expressed on dendritic cells potentiates th17-mediated immunity. *J Immunol*, Vol. 188, No. 7, (Apr 1), pp. (3062-3070)

Ren, K. & Dubner, R. (2010). Interactions between the immune and nervous systems in pain. *Nat Med*, Vol. 16, No. 11, (Nov), pp. (1267-1276)

Reynolds, A.D., Stone, D.K., Hutter, J.A., Benner, E.J., Mosley, R.L. & Gendelman, H.E. (2010). Regulatory T cells attenuate th17 cell-mediated nigrostriatal dopaminergic neurodegeneration in a model of Parkinson's disease. *J Immunol*, Vol. 184, No. 5, (Mar 1), pp. (2261-2271)

Ridge, J.P., Di Rosa, F. & Matzinger, P. (1998). A conditioned dendritic cell can be a temporal bridge between a CD4+ T-helper and a T-killer cell. *Nature*, Vol. 393, No. 6684, (Jun 4), pp. (474-478)

Rosas-Ballina, M., Olofsson, P.S., Ochani, M., Valdes-Ferrer, S.I., Levine, Y.A., Reardon, C., Tusche, M.W., Pavlov, V.A., Andersson, U., Chavan, S., Mak, T.W. & Tracey, K.J. (2011). Acetylcholine-synthesizing T cells relay neural signals in a vagus nerve circuit. *Science*, Vol. 334, No. 6052, (Oct 7), pp. (98-101)

Rosas-Ballina, M. & Tracey, K.J. (2009). The neurology of the immune system: neural reflexes regulate immunity. *Neuron*, Vol. 64, No. 1, (Oct 15), pp. (28-32)

Schoenberger, S.P., Toes, R.E., van der Voort, E.I., Offringa, R. & Melief, C.J. (1998). T-cell help for cytotoxic T lymphocytes is mediated by CD40-CD40L interactions. *Nature*, Vol. 393, No. 6684, (Jun 4), pp. (480-483)

Schoenborn, J.R. & Wilson, C.B. (2007). Regulation of interferon-gamma during innate and adaptive immune responses. *Adv Immunol*, Vol. 96, No. pp. (41-101)

Smith, K.M., Pottage, L., Thomas, E.R., Leishman, A.J., Doig, T.N., Xu, D., Liew, F.Y. & Garside, P. (2000). Th1 and Th2 CD4+ T cells provide help for B cell clonal expansion and antibody synthesis in a similar manner in vivo. *J Immunol*, Vol. 165, No. 6, (Sep 15), pp. (3136-3144)

Stellwagen, D. & Malenka, R.C. (2006). Synaptic scaling mediated by glial TNF-alpha. *Nature*, Vol. 440, No. 7087, (Apr 20), pp. (1054-1059)

Watford, W.T., Moriguchi, M., Morinobu, A. & O'Shea, J.J. (2003). The biology of IL-12: coordinating innate and adaptive immune responses. *Cytokine Growth Factor Rev*, Vol. 14, No. 5, (Oct), pp. (361-368)

Wong, P. & Pamer, E.G. (2003). CD8 T cell responses to infectious pathogens. *Annu Rev Immunol*, Vol. 21, No. pp. (29-70)

Yirmiya, R. & Goshen, I. (2011). Immune modulation of learning, memory, neural plasticity and neurogenesis. *Brain Behav Immun*, Vol. 25, No. 2, (Feb), pp. (181-213)

Zheng, Y., Manzotti, C.N., Liu, M., Burke, F., Mead, K.I. & Sansom, D.M. (2004). CD86 and CD80 differentially modulate the suppressive function of human regulatory T cells. *J Immunol*, Vol. 172, No. 5, (Mar 1), pp. (2778-2784)

Chapter 6

Ligand-Induced Cell Adhesion in Synapse Formation

Fernanda Ledda

Additional information is available at the end of the chapter

1. Introduction

Construction of the neural netwoks depends largely on precise contacts between neurons and non-neuronal cells. Numerous studies have described different types of adhesive interactions between cells in the nervous system. These include adhesive contacts between neural cell bodies, axonal attachments to glial cells, axon fasciculation, connection between pre- and postsynaptic specializations as well as to cells outside the nervous system. Although some of the molecules that mediate each of these types of neural adhesive contacts have been characterized, some remain unknown.

Neuronal synapses can be considered as a specialized type of cell-cell interaction that mediates communication between neurons and their target cells. It involves the interaction between two asymmetric partners, the presynaptic specialization, where the synaptic vesicles release neurotransmitters and the postsynaptic density, which contains receptors and adapter scaffold proteins that transduce the neurotransmitter signal [1]. As at other cell-cell junctions, such as epithelial tight junctions or the immune synapse, synaptically-localized neural cell adhesion molecules are not merely static structural components but are often dynamic regulators of synapse function.

In the last years numerous studies provide new insights into the role of adhesion molecules in the formation, maturation, maintenance, function and plasticity of synaptic contacts. Several cell adhesion molecules have been involved in synapse development, including, cadherins, proteocadherins, integrins, NCAM, L1, Fasciclina, Syg, Sidekicks, SynCam, Neurexin-Neuroligin, LRRTM, GDNF/GFRα, Neurexin/Cbl1/GluRδ2, between others [2]. *Trans*-synaptic cell adhesion molecules are particularly attractive candidate mediators of synapse formation because of their potential to bidirectionally coordinate functional and morphological synapse differentiation [3,4].

During neuronal development, specific synaptic circuits are generated by synapse formation between the appropriate pre- and postsynaptic partners and aberrant connectivity can lead to nervous system disorders. The accuracy of synapse formation is fundamental for the normal brain development and depends in part on the controlled spatial and temporal expression of selective adhesion molecules on neuronal surface.

2. Adhesion at the synapses

Many cell adhesion molecules are localized at synaptic sites in neuronal axons and dendrites. These molecules bridge pre- and postsynaptic specializations but do far more than simply provide a mechanical link between cells, they are important elements in the *trans*-cellular communication mediated by synapses. During the last years, some adhesion complexes, which are involved in the formation, maintenance and modulation of synaptic contacts, have emerged. Three particularly interesting molecular systems of pre- and postsynaptic partners that interact *in trans* across the synaptic cleft, have been described: Neurexin-Neuroligin, LRRTM and SynCam.

2.1. Neurexins and neuroligins

Neurexins (Nrxns) and Neuroligins (Nlgns) are the best characterized synaptic cell adhesion system. Neurexins were originally discovered as receptors for α-latrotoxin, a vertebrate-specific toxin present in the black widow spider venom that binds to presynaptic receptors and induces massive neurotransmitter release [5]. There are two types of Nrxns, a longer α-Nrxn (α-Nrxn) and a shorter β-Nrnx (β-Nrxn) isoforms. While α-Nrxn have six extracellular LNS domains (Laminin/ Neurexin/ Sex hormone-binding globulin-domain) with three intercalated EGF-like domains, β-Nrxn only contains a single LNS domain [6,7,8]. Immunofluorescence and subcellular fractionation analysis indicate that Nrxns are located on presynaptic terminals [6,9].

Mammals contain three Nrxn genes (Nrxn1-3), each of which directs the transcription of α- and β-Nrxns from independent promoters [10]. Neurexins are evolutionary conserved and pan-neuronally expressed [10,11]. Homologues of neurexin genes have been described in low vertebrates such as *Danio rerio* [12], and invertebrates such as *Drosophila melanogaster, Caenorhabditis elegans*, honeybees and *Aplysia* [10,13,14,15]. In mammals, alternative splicing of Nrxns can generate thousands of alternatively spliced mRNA transcripts. The ectodomain of α-Nrxns contains five sequences that can be alternatively spliced (S1-5), two of which are also present in β-Nrxns (S4 and S5) [10,11]. Some of these splice sites are localized in the Nrxn binding domain. Interestingly, Nrxn alternative splicing is regionally regulated during development and by neural activity [11,16], and plays an important role in modulating its function at synapses. Indeed, Nrxn distribution to excitatory and inhibitory synapses seems to be regulated by alternative splicing. It has been reported that β-Nrxn without the S4 sequence (-S4) induce differentiation of excitatory synapses while β-Nrxn containing the S4 insert (+S4) promote differentiation of inhibitory synapses [17,18]. Furthermore, α-Nrxn also strongly promotes differentiation of inhibitory synapses [17].

Neuroligins have been identified as endogenous Nrxns ligands [19,20]. As Nrxns, Neuroligins (Nlgns) are type I membrane proteins that consist of an extracellular region, involved in *trans*-synaptic interactions, a single transmembrane sequence and a small cytoplasmic domain that contain a PDZ-domain binding sequence that recruits PSD-95 and other PDZ-domain proteins [21]. The extracellular region contains a domain homologous to acetylcholinesterase, without its enzymatic activity. All Nlgns are enriched in postsynaptic densities. The human genome expresses five Nlgns isoforms (Nlg1-5), and rodent genome contains only 4 isoforms (Nlg1-4). Homologues of Nlgns have been identified in invertebrates including *Drosophila melanogaster* [22,23], honeybees [13,24], *Caenorhabditis elegans* [25] and *Aplysia* [14,15]. In mammals, Nlgns contains two alternative splice sites referred to as SA (in Nlgn1-3) and SB (in Nlgn-1) [20,26]. In contrast to Nrxns, Nlgns are specifically localized to particular synapses. Several studies revealed that Nlgn1 and Nlgn2 are exclusively localized to excitatory and inhibitory synapses respectively. Nlgn3 has been described to be present at both inhibitory and excitatory synapses in hippocampal cells [27,28,29,30,31], while Nlgn4 appears to be localized to inhibitory synapses in some tissues such as, retina, spinal cord, and several lower brain regions; and to excitatory synapses in some tissues such as, hippocampus and cortex [27].

Nrxn-Nlgn complex has been involved in the formation, maturation and function of vertebrate synapses. The evidence indicates that Nrxn-Nlgn bind each other by their extracellular domain to promote adhesion between pre- and postsynaptic specializations, recruiting pre- and postsynaptic molecules to form a functional synapse. Cell based assays of synapse assembly showed that contact of dissociated neurons with Nlgn-expressing fibroblasts can induce the formation of functional presynaptic specializations by recruiting components of the presynaptic machinery in co-cultured neurons [32], while contact of neurons with Nrxn expressing non-neuronal cells can induce postsynaptic differentiation and clustering of postsynaptic receptors in contacting dendrites [31,33].

Recent studies indicate that alternative splicing of *Nrxn* and *Nlgn* mRNA may play an important role modulating the assembly and synapse properties of Nrxn-Nlgn complex [26]. As it has been previously mentioned, Nrxns and Nlgns contain sequences that can be alternatively spliced. Some of these splice sites are localized in the Nrxn-Nlgn binding interface, placing them in a relevant position to modulate Nrxn-Nlgn engagement. It has been proposed that alternative splicing of Nrxns underlies an adhesive code and/or synapse-specific functions [18,34,35]. The evidence indicates that the sequence S4 of Nrxn and insert B of Nlgn have a crucial role in the Nrxn-Nlgn interactions. It has been described that the presence of the insert S4 in β-Nrxn (+S4) strongly reduces binding to Nlgn1 containing the insert B (+B) [26], indicating that Nlgn1(+B), the most common form of Nlgn1, interacts preferentially with β-Nrxn (-S4). Moreover, the presence of insert B inhibits the binding of Nlgn1 to α-Nrxns [36]. On the other hand, the alternative splicing of the segment S4 in Nrxn controls interactions with other Nrxn ligands, such as leucine-rich repeat proteins (LRRTMs) and the Cbln1-GluδR2 complex (See bellow).

The *trans*-synaptic Nrxn-Nlgn complex promotes the assembly and maturation of pre-and postsynaptic machinery. Although the presynaptic signaling events induced by Nrxns are currently not clear, the evidence indicates that neurexins bind directly to the presynaptic scaffolding proteins CASK, MINT1 and to protein 4.1; and could therefore recruit elements of the presynaptic release machinary [37,38]. In agreement with this, it has been shown that RNA interference (RNAi)-mediated supression of the Nlgn-1, -2, -3 in neuronal cultures reduces the number of excitatory and inhibitory synapses [18]. Conversely, Nlgn1 overexpression in neurons has been found to increase the formation of mature presynaptic boutons, enhace the size of the pool of recycling synaptic vesicles and the rate of synaptic vesicle endocytosis [39,40]. Regarding to the postsynaptic consequences of Nrxn-Nlgn interaction, it was described that in excitatory synapses, this adhesion complex is able to capture AMPA receptors through PSD-95 scaffolding proteins [41]. NMDARs are also recruited to Nrxn-Nlgn complex, but this recruitment does not depend on the presence of PSD-95 [42]. On the other hand, studies on Nrxn-Nlgn interaction at inhibitory synapses indicates that the contact between α-Nrxn and Nlgn2 induces clustering of Nlgn2 and recruitment of the post-synaptic scaffolding protein Gephyrin to inhibitory synapses [31]. Complex formation between Nlgn2, Geophyrin and a brain specific GDP/GTP exchange factor, Collybistin, recruits GABA and Glycine receptors to nascent inhibitory synapses. Interestingly, deletion of Nlgn2 in mice perturbs GABAergic and glicinergic synaptic transmission and leads to a loss of postsynaptic specializations specifically at inhibitory synapses [43].

Although the *in vitro* evidence indicates that Nrxn-Nlgn complex induces synapse formation, the analysis of different Nlgn and Nrxn knockout mice are controversial. *In vivo* deletion of all Nrxn or of Nlgn1-3 do not substantially affect synapse formation [30,44], but impair synapse function, suggesting that α-Nrxn and Nlgns are cell adhesion molecules that play an essential role in synapse maturation but are not essential for synapse formation.

2.2. LRRTM

The LRRTM gene family was first described in 2003 [45]. The LRRTM family has four members (LRRTM 1-4) that share similar domain structure with an extracellular domain containing ten extracellular leucine-rich repeats that mediate protein-protein interactions, followed by a single transmembrane domain and a short c-terminal sequence containing a class I PDZ-domain-binding motif. Human and mouse LRRTMs are highly conserved and orthologous genes exist in other vertebrates, but not in invertebrates [45].

In situ hybridization, RT-PCR and immunofluorescence analysis showed that LRRTMs are predominantely expressed in the nervous system and that each LRRTM present a specific and partially overlapping expression pattern [45,46,47]. The four LRRTM are expressed in neurons of the hippocampus, cerebral cortex and in the striatum. LRRTM1 and LRRTM2 are also highly expressed in the thalamus. In contrast, neither LRRTM3 nor LRRTM4 are expressed in these structures. The structural similarities and expression patterns of LRRTMs indicates a possible functional redundancy between them [45].

All four LRRTMs family members are post-synaptic localized and when expressed in non-neuronal cells co-cultured with hippocampal neurons they can induce presynaptic differentiation in contacting axons. LRRTM1 and LRRTM2 selectively promote excitatory, but not inhibitory presynaptic differentiation [46]. In addition, Wit et al (2009) demonstrated that LRRTM2 can interact with the post-synaptic protein PSD-95 and regulate surface expression of AMPA receptors [48].

Independent studies have shown that post-synaptic LRRTM1 and LRRTM2 bind specifically to presynaptic α and β-Nrxn lacking an insert at S4 [49]. Thus, whereas Nlgns bind Nrxn containing or lacking an insert in splice site S4, LRRTMs bind only Nrxns lacking an insert in this splicing site [48,50]. This ability to regulate interaction Nrxn-Nlgn and Nrxn-LRRTM provides an intringuing mechanism for regulating synaptic specificity.

Consistent with the effects of LRRTM on neuronal connectivity, deletion of LRRTM1 in mice revealed altered distribution of the vesicular glutamate transporter vGlut1 *in vivo* [46]. Furthermore, it was demonstrated that lentivirus-mediated knockdown of LRRTM2 *in vivo* decreases the strength of glutamatergic synaptic transmission. Conversely, LRRTM2 overexpression resulted in an increase of excitatory synapses [48,51].

2.3. SynCAM

The SynCAM (Synaptic Cell Adhesion Molecule) family comprises four genes encoding proteins (SynCAM1-4) with three inmuonoglobulin (Ig)-like domains, a single transmembrane region, and a short cytosolic tail with a PDZ type II motif. SynCAM proteins are predominantly expressed in the brain and localize to pre- and postsynaptic sites [52,53].

All SynCAMs are expressed mostly by neurons during the peak period of synaptogenesis around the second postnatal week and remains expressed throughout adulthood in the hippocampus [52].

SynCAM proteins are present at presynaptic and postsynaptic specializations and are involved in homophilic and heterophilic interactions via the extracellular (Ig)-like domains. Interestingly, SynCAM1, 2 and 3, but not SynCAM4, can associate *in trans* by homophilic interactions. However, the evidence indicates stronger heterophilic interactions of SynCAM1/2 and SynCAM3/4 more than homophilic adhesion between each other [54]. It has been shown that heterophilic adhesion complex of SynCAM1/2 drives presynaptic terminal formation in cultured neurons, increasing the number of excitatory synapses [54]. In agreement with a functional role of SynCAM1 in neuronal connectivity, it has been demonstrated that elevated expression of SynCAM in a transgenic model increases functional excitatory synapse number [55]. Conversely, SynCAM1 knockout mice exhibited fewer excitatory synapses. Interestingly, SynCAM1 can alter the plasticity of synapses once they are formed. Thus, SynCAM1 overexpression has been shown to abrogate long term depression (LTD), while it loss increased LTD [55].

3. Control of synapse formation by ligand-induced cell adhesion molecules (LICAM)

During the last years, a novel mechanism of ligand-induced cell adhesion has been described. Unlike other cell adhesion systems, which involve the simple encounter of membrane associated cell adhesion molecules *in trans*, ligand-induced cell adhesion is mediated by membrane-associated proteins but is dependent on the presence of its soluble ligand. This feature may allow a more dynamic response to external stimulus involved in synapse development.

3.1. GDNF and GFRα1

The Glial cell-line Derived Neurotrophic Factor (GDNF) and its glycosylphosphatidylinositol (GPI)-anchor receptor, GFRα1, represent the first example of this new mechanism of cell-adhesion [56]. In this system, GDNF, is able to mediate *trans*-homphilic cell adhesion between cells expressing its receptor, GFRα1. The receptor involved in this process, the GFRα1, can be considered as a ligand-induced cell adhesion molecule (LICAM). The molecular bases underlying *trans*-homphilic interaction mediated by GDNF and GFRα1 is not clear yet. The domains of GFRα1 underlying its LICAM activity have been analyzed using deletion mutants of the receptor. This study revealed that the GFRα1-mediated cell adhesion requires the presence of an intact ligand-binding domain in both interacting partners. In principle, GDNF, as a dimeric protein could promote *trans*-homophilic interactions between receptor-expressing cells. On the other hand, GDNF could act through an allosteric mechanism.

During the last years, numerous studies have shown that GDNF-family ligands contribute to synapse development and maturation [57,58]. The developmental expression pattern of GFRα1 and its ligand, GDNF, during the period of hippocampal synaptogenesis as well as its subcellular localization at pre- and postsynaptic specializations indicated a possible role of GDNF-GFRα1 complex in the formation of neuronal synapses by inducing *trans*-synaptic homophilic cell adhesion. Indeed, microspheres containing GFRα1, mimicking its postsynaptic localization, were able to induce presynaptic differentiation on hippocampal and cortical neurons cultured in the presence of GDNF. This effect was evidenced by recruitment of vesicle-associated synaptic protein, neurotransmitter transporters and activity-dependent vesicle recycling on the hippocampal axons at the sites of contact [56]. Intringuinly, the presynaptic maturation triggered by GDNF and GFRα1 was independent of the canonical receptor Ret and only partially dependent on the neural cell adhesion molecule, NCAM, indicating the existence of an additional signaling molecule involved in this process (Figure 1A)[56,57,59,60]. Whether postsynaptically localized GFRα1 may contribute to postsynaptic maturation remains to be explored. Thus, the ability of GDNF to trigger *trans*-homophilic interactions between GFRα1 molecules represents the first example of regulated cell-cell interactions and a new synaptogenic mechanisms that combines soluble and membrane bound molecules by inducing conformational changes that reorient and expose determinants involved in *trans*-homphilic binding [56,57].

3.2. Cerebelin-GluRδ-neurexin

More recently another example of ligand-induced *trans*-synaptic adhesion interaction has been described. Uemura *et al* (2010) described that the postsynaptic glutamate receptor

(GluR)δ2 interacts with the presynaptic β-Nrnx through the presynaptically secreted glycoprotein, Cerebelin 1 precursor protein (Cbln1) [61].

Based on its amino acid sequence, GluRδ2, is a member of the ionotrophic glutamate receptor family, which plays an essential role in cerebellar Purkinje cells (PC) synapse formation [61,62]. The synaptogenic activity induced by GluRδ2 can be reproduced *in vitro* using primary cultures of cerebellar granule cells (GC) and the extracellular N-terminal domain of GluRδ2 [61] indicating that this domain is critical for its synaptogenic activity. Binding studies demonstrated that postsynaptic GluRδ2 interacts with presynaptic Nrxns, which are known to play a crucial role in presynaptic organization. But this interaction is established through the presynaptically secreted glycoprotein Cbln1. Interestingly, the synaptogenic activity of GluRδ2 is abolished in cerebellar primary cultures from Cbln1 knockout mice and is restored by recombinant Cbln1. In agreement with this, Cbln1-null mice show similar behavioral and physiological phenotypes to those of GluRδ2-null mice confirming that Cbln1 and GluRδ2 are involved in a similar signaling pathway [63]. Direct binding experiments demonstrated that Cbln1 acts as a divalent ligand for postsynaptic GluRδ2 and presynaptic Nrxns, representing a new example of ligand induced *trans*-heterophylic synaptic adhesion. The resulting complex, Nrxn/Cbl1/GluRδ2, mediates synapse formation between cerebellar granule cells and Purkinje cells (Figure 1B). In accordance with this, the amino terminal domain of GluRδ2 and the extracellular domain of β-Nrxn1 suppressed the synaptogenic activity of Cbln1 in cerebellar primary cultures *in vivo* indicating that the interaction of GluRδ2, Cbln1 and Nrxn is essential for cerebellar synapse formation [61].

The evidence indicates that Cbln1 interacts with different subtypes of β-Nrxn and α-Nrxn containing the S4 insert, but not to subtypes lacking the S4 insert, to induce synaptogenesis in cultured cerebellar, hippocampal and cortical neurons. Interestingly, α-Nrxn containing the S4 insert binds to Cbln1 [62] but does not bind to any Nlgs or LRRTMs. Another distinctive feature of the Nrxn/Cbln1/GluRδ2 complex is that it is insensitive to the extracellular $Ca2^{+}$ concentration [62]; while binding of Nrxn to Nlgs and LRRTMs requires extracellular $Ca2^{+}$.

While GluRδ2 is mainly expressed in cerebellar Purkinje cells, GluRδ1 is widely expressed in the developing forebrain including the caudate putamen and hippocampus. In a recent study it has been demonstrated that, in the presence of Cbln1 or Cbln2, GluRδ1 expressed in non-neuronal cells can induce inhibitory presynaptic differentiation on cultured cortical neurons by interacting with Nrxns containing the S4 insert [64].

4. Transient cell-cell interactions in neural development

Trans-synaptic adhesion molecules can affect the function of synapses at multiple levels, from recruiting synaptic proteins during synaptogenesis to regulating synaptic plasticity [1]. Thus, adhesion molecules are involved in dynamic processes such as synapse formation, which involves cell-type specific target recognition and synaptic plasticity, which requires the response to external stimulus or perturbations. A delicate balance between adhesion and de-adhesion cooperates generating robustness and flexibility to ensure normal nervous system development.

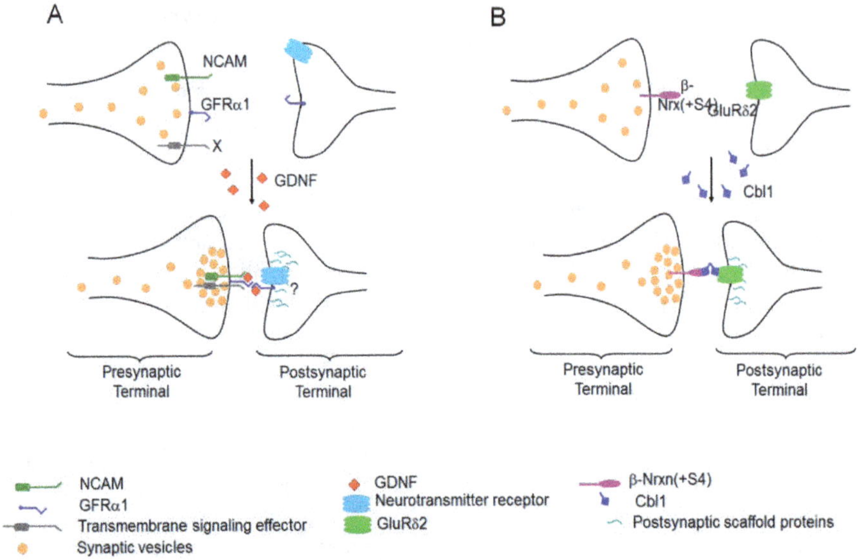

(A) GDNF induce *trans*-homophilic interactions between GFRα1 molecules, which are located at pre- and postsynaptic terminals. In the presence of GDNF, GFRα1, holds the terminals together triggering presynaptic maturation. This is mediated by GFRα1 and partially by the neural cell adhesion, NCAM, suggesting the involvement of an unknown partner (X). The role of GFRα1 in the postsynaptic terminal remains unknown (?).
(B) Cerebelin1 (Cbl1) acts as a bidirectional synaptic organizer by binding presynaptic β-Nrx (+S4) and postsynaptic GluRδ2 inducing pre- and postsynaptic differentiation.

Figure 1. Model describing ligand-induced *trans*-synaptic cell adhesion induced by (A) GDNF and (B) Cerebelin1.

The majority of *trans*-synaptic adhesion mechanisms known to date involve interactions triggered by the encounter between cell adhesion molecules inserted in the pre- and postsynaptic compartments. The activity of these cell-adhesion molecules may be regulated either developmentally or in response to neuronal activity. For instance, activity- dependent alternative splicing of Nrxn transcripts, or post-translational modifications such as glycosilation, polysialilation or palmitoylation drastically modify adhesive properties; binding partners and signaling properties of certain adhesion molecules [26,65,66]. Ligand-induced cell adhesion represents a previously unknown mechanism in which cell adhesion can be regulated not only by the presence of different forms of the associated receptor but also by ligand availability [56]. Furthermore, ligand secretion could also be regulated, for example by neuronal activity. Thus, *trans*-homophilic binding between GFRα1 molecules might be controled by the activity-dependent upregulation of GDNF, which has been shown to be upregulated by seizure activity [67,68]. On the other hand *trans*-heterophilic binding between GluRδ2 and Nrxn might also be regulated by the availability of Cbln1. It has been described that the expression of *Cbln1* mRNA is completely shut down in mature granule cells when the neuronal activity is increased by kainate [69]. Moreover, a recent study revealed that the-ligand binding domain of GluRδ2 can also bind to D-Ser inducing conformational changes of the receptor that might modify the synaptogenic activity of the Nrnx/Cbl1/GluRδ2 complex [70,71] In addition, D-Ser has been shown to be regulated during development and also to be release from astrocytes in an activity-dependent manner [71]. Thus, the activity dependent regulation of each component of the Nrnx/Cbl1/GluRδ2 might increase the plasticity of the system.

Ligand-induced cell adhesion represents a new way to regulate intercellular interactions that may have broad implications not only for the development of the nervous system, but also in other tissues and organs.

5. Role of synaptic cell adhesion systems in nervous system disorders

The ability of *trans*-synaptic cell adhesion molecules to regulate synapse formation, maturation and plasticity supports the idea that deficits in many synaptogenic genes might be associated to neurodevelopmental and/or neuropsychiatric diseases.

Numerous studies indicate a genetic link of mutations in synaptic cell adhesion molecules to autism-spectrum disorders (ASD), in particular to Nlgs and Nrxns [72,73]. Mutations in genes encoding Nrxn1, Nlg3 and Nlg4 have been described to be associated with ASD. These alterations include different type of mutations that have been observed in a small fraction of patients. In particular thirteen different mutations have been described in *Nrxn1* gene: seven point mutations, two distinct translocations and four different deletions [72,74,75,76,77,78,79]; ten different mutations has been found in Nlg4 gene: two frameshifts, five missense mutations and three internal deletions; and a single mutation in *Nlg3* gene (R451C) [72,80,81,82,83]. Moreover, deletions in X-chromosome that includes the Nlg4 locus were detected in patients with autism [79,84,85,86]. Different studies have reported that Nlg4 deletions are also associated with other neurological disturbances including Tourette syndrome, attention deficit hyperactivity disorders, anxiety and depression. In addition, two different deletions of α-Nrxn1 have been observed in families with schizophrenia [58,87,88].

The role of some of these proteins in ASD has been validated in trasgenic animals. Thus, the Nlg3-R451C knockin mouse were reported to show a phenotype that shares some, but not all features with human ASD patients These mice show a modest impairment in social behaviour [89]. Moreover, Nlg-4 knockout mice show deficits in social interactions and communication [90].

In addition, members of SynCAM and LRRTM families have also been associated with nervous system disorders. SynCAM1 has been associated with ASD. Two missense mutations in the SynCAM1 gene of ASD patients and their families have been described. Interestingly, the mutations were located in one domain, which is essential for trans-synaptic interaction [91]. In a recent genetic study, polymorphisms in LRRTM3 were associated with ASD [88]. Moreover, LRRTM1 has been associated with schizophrenia [92].

There is no strong evidence connecting mutations in genes involved in ligand-induced cell adhesion systems with nervous system diseases. So far, only GluRδ1 was found to be associated with schizophrenia [92,93].

Further studies linking mutations in cell adhesion systems with nervous system diseases will contribute to the design of new diagnostic and therapeutic tools for these disorders.

6. Perspectives

It is well established, that cell-adhesion systems are in part responsible for the construction of neural circuits, synapse formation and plasticity. The correct function of the nervous system depends on the establishment of precise synaptic contacts between neurons and its specific targets, and deficits in genes coding for *trans*-adhesion molecules have been associated with learning deficits and cognitive impairments.

During the last years several adhesion molecules have been reported to participate in synapse development, including integrins, cadherins, protocadherins, NCAM, Neurexin-Neuroligin, LRRTM, SynCam, GDNF/GFRα, Nrxn-Cbl1-GluRδ2. The discovery of alternative *trans*-synaptic binding partners, in combination with their differential splice variants and isoforms, gave rise to a much larger spectrum of *trans*-synaptic interactions than was originally though. Furthermore, the mechanism of ligand induced cell adhesion considerably expands the functional repertoire of the ligands and receptors involved in these adhesion complexes and represents a new way to regulate *trans*-synaptic interactions that may have broad implications for the development of the vertebrate nervous system. Regulation of adhesion by soluble ligands allows a dynamic synaptic response to external stimulus during synapse formation and synaptic plasticity.

Based on this, the main challenge will be now to elucidate the complex code by which the *trans*-synaptic cell adhesion systems participate in the different steps of synapse formation, maturation and plasticity and to understand the importance of the different combinations of *trans*-synaptic partners in specific circuits. It will be important to understand how the combination of multiple synaptogenic systems may contribute to synaptic specificity controlling exactly where and when synapses form.

It will be also necessary to address whether individual synapse organizing protein instruct synaptic cell adhesion, or if the *trans*-synaptic interaction results from cooperation among different adhesion molecules. In some cases it is likely that the determination of initial synapse formation is mediated by multiple adhesion systems acting in parallel, as has been evidenced by α-Nrxs and Nlgns knockout mice. The analysis of these animals showed that knockdown of these molecules, does not have consequence in synapse formation, suggesting that other adhesions systems compensate the deficiency in these mice [30,44].

The fact that multiple partners function at the same synapses opens the possibility that they cooperate in the recruitment of the same components to the synapse. Indeed, it has been described that overexpression of Nlgns and LRRTMs in primary hippocampal neurons cooperate in a synergistic manner in glutamate synapse development visualized by an increase in the recruitment of pre-synaptic proteins. Cooperation between different adhesion systems may help to stabilize interactions across the cleft by recruiting the pre- and postsynaptic machinery at multiple points. However the existence of mechanisms that can modulate and modify these interactions should be important, especially for synaptic plasticity.

Further understanding of the molecular pathways and circuit events downstream these cell adhesion organizing systems will be extremely important in light of the role of *trans*-synaptic cell adhesion molecules in neurodevelopmental and cognitive diseases.

Author details

Fernanda Ledda

Laboratory of Molecular and Cellular Neuroscience, Institute of Cellular Biology and Neuroscience (IBCN)-CONICET, Buenos Aires, Argentina,

Laboratory of Molecular and Cellular Neuroscience, Department of Neuroscience, Karolinska Institute, Stockholm, Sweden

7. References

[1] Scheiffele P (2003) Cell-cell signaling during synapse formation in the CNS. Annu Rev Neurosci 26: 485-508.

[2] Yamagata M, Sanes JR, Weiner JA (2003) Synaptic adhesion molecules. Curr Opin Cell Biol 15: 621-632.

[3] Missler M, Sudhof TC, Biederer T (2012)Synaptic cell adhesion. Cold Spring Harb Perspect Biol 4.

[4] Krueger DD, Tuffy LP, Papadopoulos T, Brose N (2012) The role of neurexins and neuroligins in the formation, maturation, and function of vertebrate synapses. Curr Opin Neurobiol.

[5] Ushkaryov YA, Rohou A, Sugita S (2008) alpha-Latrotoxin and its receptors. Handb Exp Pharmacol: 171-206.

[6] Ushkaryov YA, Petrenko AG, Geppert M, Sudhof TC (1992) Neurexins: synaptic cell surface proteins related to the alpha-latrotoxin receptor and laminin. Science 257: 50-56.

[7] Ushkaryov YA, Sudhof TC (1993) Neurexin III alpha: extensive alternative splicing generates membrane-bound and soluble forms. Proc Natl Acad Sci U S A 90: 6410-6414.

[8] Ushkaryov YA, Hata Y, Ichtchenko K, Moomaw C, Afendis S, et al. (1994) Conserved domain structure of beta-neurexins. Unusual cleaved signal sequences in receptor-like neuronal cell-surface proteins. J Biol Chem 269: 11987-11992.

[9] Berninghausen O, Rahman MA, Silva JP, Davletov B, Hopkins C, et al. (2007) Neurexin Ibeta and neuroligin are localized on opposite membranes in mature central synapses. J Neurochem 103: 1855-1863.

[10] Tabuchi K, Sudhof TC (2002) Structure and evolution of neurexin genes: insight into the mechanism of alternative splicing. Genomics 79: 849-859.

[11] Ullrich B, Ushkaryov YA, Sudhof TC (1995) Cartography of neurexins: more than 1000 isoforms generated by alternative splicing and expressed in distinct subsets of neurons. Neuron 14: 497-507.

[12] Rissone A, Monopoli M, Beltrame M, Bussolino F, Cotelli F, et al. (2007) Comparative genome analysis of the neurexin gene family in Danio rerio: insights into their functions and evolution. Mol Biol Evol 24: 236-252.

[13] Biswas S, Russell RJ, Jackson CJ, Vidovic M, Ganeshina O, et al. (2008) Bridging the synaptic gap: neuroligins and neurexin I in Apis mellifera. PLoS One 3: e3542.

[14] Choi YB, Li HL, Kassabov SR, Jin I, Puthanveettil SV, et al. (2011) Neurexin-neuroligin transsynaptic interaction mediates learning-related synaptic remodeling and long-term facilitation in aplysia. Neuron 70: 468-481.

[15] Knight D, Xie W, Boulianne GL (2011) Neurexins and neuroligins: recent insights from invertebrates. Mol Neurobiol 44: 426-440.

[16] Rozic-Kotliroff G, Zisapel N (2007) Ca2+ -dependent splicing of neurexin IIalpha. Biochem Biophys Res Commun 352: 226-230.

[17] Kang Y, Zhang X, Dobie F, Wu H, Craig AM (2008) Induction of GABAergic postsynaptic differentiation by alpha-neurexins. J Biol Chem 283: 2323-2334.

[18] Chih B, Gollan L, Scheiffele P (2006) Alternative splicing controls selective trans-synaptic interactions of the neuroligin-neurexin complex. Neuron 51: 171-178.

[19] Ichtchenko K, Hata Y, Nguyen T, Ullrich B, Missler M, et al. (1995) Neuroligin 1: a splice site-specific ligand for beta-neurexins. Cell 81: 435-443.

[20] Ichtchenko K, Nguyen T, Sudhof TC (1996) Structures, alternative splicing, and neurexin binding of multiple neuroligins. J Biol Chem 271: 2676-2682.

[21] Irie M, Hata Y, Takeuchi M, Ichtchenko K, Toyoda A, et al. (1997) Binding of neuroligins to PSD-95. Science 277: 1511-1515.

[22] Sun M, Xing G, Yuan L, Gan G, Knight D, et al. (2011) Neuroligin 2 is required for synapse development and function at the Drosophila neuromuscular junction. J Neurosci 31: 687-699.

[23] Banovic D, Khorramshahi O, Owald D, Wichmann C, Riedt T, et al. (2010) Drosophila neuroligin 1 promotes growth and postsynaptic differentiation at glutamatergic neuromuscular junctions. Neuron 66: 724-738.

[24] Biswas S, Reinhard J, Oakeshott J, Russell R, Srinivasan MV, et al. (2010) Sensory regulation of neuroligins and neurexin I in the honeybee brain. PLoS One 5: e9133.

[25] Hunter JW, Mullen GP, McManus JR, Heatherly JM, Duke A, et al. (2010) Neuroligin-deficient mutants of C. elegans have sensory processing deficits and are hypersensitive to oxidative stress and mercury toxicity. Dis Model Mech 3: 366-376.

[26] Boucard AA, Chubykin AA, Comoletti D, Taylor P, Sudhof TC (2005) A splice code for trans-synaptic cell adhesion mediated by binding of neuroligin 1 to alpha- and beta-neurexins. Neuron 48: 229-236.

[27] Hoon M, Soykan T, Falkenburger B, Hammer M, Patrizi A, et al. (2011) Neuroligin-4 is localized to glycinergic postsynapses and regulates inhibition in the retina. Proc Natl Acad Sci U S A 108: 3053-3058.

[28] Budreck EC, Scheiffele P (2007) Neuroligin-3 is a neuronal adhesion protein at GABAergic and glutamatergic synapses. Eur J Neurosci 26: 1738-1748.

[29] Song JY, Ichtchenko K, Sudhof TC, Brose N (1999) Neuroligin 1 is a postsynaptic cell-adhesion molecule of excitatory synapses. Proc Natl Acad Sci U S A 96: 1100-1105.

[30] Varoqueaux F, Jamain S, Brose N (2004) Neuroligin 2 is exclusively localized to inhibitory synapses. Eur J Cell Biol 83: 449-456.
[31] Graf ER, Zhang X, Jin SX, Linhoff MW, Craig AM (2004) Neurexins induce differentiation of GABA and glutamate postsynaptic specializations via neuroligins. Cell 119: 1013-1026.
[32] Scheiffele P, Fan J, Choih J, Fetter R, Serafini T (2000) Neuroligin expressed in nonneuronal cells triggers presynaptic development in contacting axons. Cell 101: 657-669.
[33] Nam CI, Chen L (2005) Postsynaptic assembly induced by neurexin-neuroligin interaction and neurotransmitter. Proc Natl Acad Sci U S A 102: 6137-6142.
[34] Craig AM, Kang Y (2007) Neurexin-neuroligin signaling in synapse development. Curr Opin Neurobiol 17: 43-52.
[35] Huang ZJ, Scheiffele P (2008) GABA and neuroligin signaling: linking synaptic activity and adhesion in inhibitory synapse development. Curr Opin Neurobiol 18: 77-83.
[36] Tanaka H, Nogi T, Yasui N, Iwasaki K, Takagi J (2011) Structural basis for variant-specific neuroligin-binding by alpha-neurexin. PLoS One 6: e19411.
[37] Biederer T, Sudhof TC (2001) CASK and protein 4.1 support F-actin nucleation on neurexins. J Biol Chem 276: 47869-47876.
[38] Hata Y, Butz S, Sudhof TC (1996) CASK: a novel dlg/PSD95 homolog with an N-terminal calmodulin-dependent protein kinase domain identified by interaction with neurexins. J Neurosci 16: 2488-2494.
[39] Levinson JN, Li R, Kang R, Moukhles H, El-Husseini A, et al. (2010) Postsynaptic scaffolding molecules modulate the localization of neuroligins. Neuroscience 165: 782-793.
[40] Wittenmayer N, Korber C, Liu H, Kremer T, Varoqueaux F, et al. (2009) Postsynaptic Neuroligin1 regulates presynaptic maturation. Proc Natl Acad Sci U S A 106: 13564-13569.
[41] Mondin M, Labrousse V, Hosy E, Heine M, Tessier B, et al. (2011) Neurexin-neuroligin adhesions capture surface-diffusing AMPA receptors through PSD-95 scaffolds. J Neurosci 31: 13500-13515.
[42] Barrow SL, Constable JR, Clark E, El-Sabeawy F, McAllister AK, et al. (2009) Neuroligin1: a cell adhesion molecule that recruits PSD-95 and NMDA receptors by distinct mechanisms during synaptogenesis. Neural Dev 4: 17.
[43] Poulopoulos A, Aramuni G, Meyer G, Soykan T, Hoon M, et al. (2009) Neuroligin 2 drives postsynaptic assembly at perisomatic inhibitory synapses through gephyrin and collybistin. Neuron 63: 628-642.
[44] Missler M, Zhang W, Rohlmann A, Kattenstroth G, Hammer RE, et al. (2003) Alpha-neurexins couple Ca2+ channels to synaptic vesicle exocytosis. Nature 423: 939-948.
[45] Lauren J, Airaksinen MS, Saarma M, Timmusk T (2003) A novel gene family encoding leucine-rich repeat transmembrane proteins differentially expressed in the nervous system. Genomics 81: 411-421.

[46] Linhoff MW, Lauren J, Cassidy RM, Dobie FA, Takahashi H, et al. (2009) An unbiased expression screen for synaptogenic proteins identifies the LRRTM protein family as synaptic organizers. Neuron 61: 734-749.
[47] Haines BP, Rigby PW (2007) Developmentally regulated expression of the LRRTM gene family during mid-gestation mouse embryogenesis. Gene Expr Patterns 7: 23-29.
[48] de Wit J, Sylwestrak E, O'Sullivan ML, Otto S, Tiglio K, et al. (2009) LRRTM2 interacts with Neurexin1 and regulates excitatory synapse formation. Neuron 64: 799-806.
[49] Siddiqui TJ, Pancaroglu R, Kang Y, Rooyakkers A, Craig AM (2010) LRRTMs and neuroligins bind neurexins with a differential code to cooperate in glutamate synapse development. J Neurosci 30: 7495-7506.
[50] Ko J, Zhang C, Arac D, Boucard AA, Brunger AT, et al. (2009) Neuroligin-1 performs neurexin-dependent and neurexin-independent functions in synapse validation. Embo J 28: 3244-3255.
[51] Ko J, Fuccillo MV, Malenka RC, Sudhof TC (2009) LRRTM2 functions as a neurexin ligand in promoting excitatory synapse formation. Neuron 64: 791-798.
[52] Biederer T, Sara Y, Mozhayeva M, Atasoy D, Liu X, et al. (2002) SynCAM, a synaptic adhesion molecule that drives synapse assembly. Science 297: 1525-1531.
[53] Biederer T, Stagi M (2008) Signaling by synaptogenic molecules. Curr Opin Neurobiol 18: 261-269.
[54] Fogel AI, Akins MR, Krupp AJ, Stagi M, Stein V, et al. (2007) SynCAMs organize synapses through heterophilic adhesion. J Neurosci 27: 12516-12530.
[55] Robbins EM, Krupp AJ, Perez de Arce K, Ghosh AK, Fogel AI, et al. (2010) SynCAM 1 adhesion dynamically regulates synapse number and impacts plasticity and learning. Neuron 68: 894-906.
[56] Ledda F, Paratcha G, Sandoval-Guzman T, Ibanez CF (2007) GDNF and GFRalpha1 promote formation of neuronal synapses by ligand-induced cell adhesion. Nat Neurosci 10: 293-300.
[57] Ledda F (2007) Ligand-induced cell adhesion as a new mechanism to promote synapse formation. Cell Adh Migr 1: 137-139.
[58] Wang CY, Yang F, He XP, Je HS, Zhou JZ, et al. (2002) Regulation of neuromuscular synapse development by glial cell line-derived neurotrophic factor and neurturin. J Biol Chem 277: 10614-10625.
[59] Paratcha G, Ledda F (2008) GDNF and GFRalpha: a versatile molecular complex for developing neurons. Trends Neurosci 31: 384-391.
[60] Paratcha G, Ledda F, Ibanez CF (2003) The neural cell adhesion molecule NCAM is an alternative signaling receptor for GDNF family ligands. Cell 113: 867-879.
[61] Uemura T, Lee SJ, Yasumura M, Takeuchi T, Yoshida T, et al. (2010) Trans-synaptic interaction of GluRdelta2 and Neurexin through Cbln1 mediates synapse formation in the cerebellum. Cell 141: 1068-1079.
[62] Matsuda K, Miura E, Miyazaki T, Kakegawa W, Emi K, et al. (2010) Cbln1 is a ligand for an orphan glutamate receptor delta2, a bidirectional synapse organizer. Science 328: 363-368.

[63] Hirai H, Pang Z, Bao D, Miyazaki T, Li L, et al. (2005) Cbln1 is essential for synaptic integrity and plasticity in the cerebellum. Nat Neurosci 8: 1534-1541.

[64] Yasumura M, Yoshida T, Lee SJ, Uemura T, Joo JY, et al. (2012) Glutamate receptor delta1 induces preferentially inhibitory presynaptic differentiation of cortical neurons by interacting with neurexins through cerebellin precursor protein subtypes. J Neurochem.

[65] Doherty P, Moolenaar CE, Ashton SV, Michalides RJ, Walsh FS (1992) The VASE exon downregulates the neurite growth-promoting activity of NCAM 140. Nature 356: 791-793.

[66] Niethammer P, Delling M, Sytnyk V, Dityatev A, Fukami K, et al. (2002) Cosignaling of NCAM via lipid rafts and the FGF receptor is required for neuritogenesis. J Cell Biol 157: 521-532.

[67] Airaksinen MS, Arumae U, Rauvala H, Saarma M (1999) [Neurotrophic growth factors in the development and plasticity of nervous system]. Duodecim 115: 595-605.

[68] Kokaia Z, Airaksinen MS, Nanobashvili A, Larsson E, Kujamaki E, et al. (1999) GDNF family ligands and receptors are differentially regulated after brain insults in the rat. Eur J Neurosci 11: 1202-1216.

[69] Iijima T, Wu K, Witte H, Hanno-Iijima Y, Glatter T, et al. (2011) SAM68 regulates neuronal activity-dependent alternative splicing of neurexin-1. Cell 147: 1601-1614.

[70] Naur P, Hansen KB, Kristensen AS, Dravid SM, Pickering DS, et al. (2007) Ionotropic glutamate-like receptor delta2 binds D-serine and glycine. Proc Natl Acad Sci U S A 104: 14116-14121.

[71] Henneberger C, Papouin T, Oliet SH, Rusakov DA (2010) Long-term potentiation depends on release of D-serine from astrocytes. Nature 463: 232-236.

[72] Sudhof TC (2008) Neuroligins and neurexins link synaptic function to cognitive disease. Nature 455: 903-911.

[73] Betancur C, Sakurai T, Buxbaum JD (2009) The emerging role of synaptic cell-adhesion pathways in the pathogenesis of autism spectrum disorders. Trends Neurosci 32: 402-412.

[74] Feng J, Schroer R, Yan J, Song W, Yang C, et al. (2006) High frequency of neurexin 1beta signal peptide structural variants in patients with autism. Neurosci Lett 409: 10-13.

[75] Kim HG, Kishikawa S, Higgins AW, Seong IS, Donovan DJ, et al. (2008) Disruption of neurexin 1 associated with autism spectrum disorder. Am J Hum Genet 82: 199-207.

[76] Szatmari P, Paterson AD, Zwaigenbaum L, Roberts W, Brian J, et al. (2007) Mapping autism risk loci using genetic linkage and chromosomal rearrangements. Nat Genet 39: 319-328.

[77] Yan J, Noltner K, Feng J, Li W, Schroer R, et al. (2008) Neurexin 1alpha structural variants associated with autism. Neurosci Lett 438: 368-370.

[78] Zahir FR, Baross A, Delaney AD, Eydoux P, Fernandes ND, et al. (2008) A patient with vertebral, cognitive and behavioural abnormalities and a de novo deletion of NRXN1alpha. J Med Genet 45: 239-243.

[79] Marshall CR, Noor A, Vincent JB, Lionel AC, Feuk L, et al. (2008) Structural variation of chromosomes in autism spectrum disorder. Am J Hum Genet 82: 477-488.

[80] Jamain S, Quach H, Betancur C, Rastam M, Colineaux C, et al. (2003) Mutations of the X-linked genes encoding neuroligins NLGN3 and NLGN4 are associated with autism. Nat Genet 34: 27-29.

[81] Laumonnier F, Bonnet-Brilhault F, Gomot M, Blanc R, David A, et al. (2004) X-linked mental retardation and autism are associated with a mutation in the NLGN4 gene, a member of the neuroligin family. Am J Hum Genet 74: 552-557.

[82] Yan J, Oliveira G, Coutinho A, Yang C, Feng J, et al. (2005) Analysis of the neuroligin 3 and 4 genes in autism and other neuropsychiatric patients. Mol Psychiatry 10: 329-332.

[83] Talebizadeh Z, Lam DY, Theodoro MF, Bittel DC, Lushington GH, et al. (2006) Novel splice isoforms for NLGN3 and NLGN4 with possible implications in autism. J Med Genet 43: e21.

[84] Chocholska S, Rossier E, Barbi G, Kehrer-Sawatzki H (2006) Molecular cytogenetic analysis of a familial interstitial deletion Xp22.2-22.3 with a highly variable phenotype in female carriers. Am J Med Genet A 140: 604-610.

[85] Lawson-Yuen A, Saldivar JS, Sommer S, Picker J (2008) Familial deletion within NLGN4 associated with autism and Tourette syndrome. Eur J Hum Genet 16: 614-618.

[86] Macarov M, Zeigler M, Newman JP, Strich D, Sury V, et al. (2007) Deletions of VCX-A and NLGN4: a variable phenotype including normal intellect. J Intellect Disabil Res 51: 329-333.

[87] Kirov G, Gumus D, Chen W, Norton N, Georgieva L, et al. (2008) Comparative genome hybridization suggests a role for NRXN1 and APBA2 in schizophrenia. Hum Mol Genet 17: 458-465.

[88] Walsh T, McClellan JM, McCarthy SE, Addington AM, Pierce SB, et al. (2008) Rare structural variants disrupt multiple genes in neurodevelopmental pathways in schizophrenia. Science 320: 539-543.

[89] Tabuchi K, Blundell J, Etherton MR, Hammer RE, Liu X, et al. (2007) A neuroligin-3 mutation implicated in autism increases inhibitory synaptic transmission in mice. Science 318: 71-76.

[90] Jamain S, Radyushkin K, Hammerschmidt K, Granon S, Boretius S, et al. (2008) Reduced social interaction and ultrasonic communication in a mouse model of monogenic heritable autism. Proc Natl Acad Sci U S A 105: 1710-1715.

[91] Zhiling Y, Fujita E, Tanabe Y, Yamagata T, Momoi T, et al. (2008) Mutations in the gene encoding CADM1 are associated with autism spectrum disorder. Biochem Biophys Res Commun 377: 926-929.

[92] Francks C, Maegawa S, Lauren J, Abrahams BS, Velayos-Baeza A, et al. (2007) LRRTM1 on chromosome 2p12 is a maternally suppressed gene that is associated paternally with handedness and schizophrenia. Mol Psychiatry 12: 1129-1139, 1057.

[93] Fallin MD, Lasseter VK, Avramopoulos D, Nicodemus KK, Wolyniec PS, et al. (2005) Bipolar I disorder and schizophrenia: a 440-single-nucleotide polymorphism screen of 64 candidate genes among Ashkenazi Jewish case-parent trios. Am J Hum Genet 77: 918-936.

Section 4

Cell Interactions in Normal and Disease Conditions

Chapter 7

Cell-Cell Interactions and Cross Talk Described in Normal and Disease Conditions: Morphological Approach

Nasra Naeim Ayuob and Soad Shaker Ali

Additional information is available at the end of the chapter

1. Introduction

The contact between cells and their microenvironment is fundamental both during development and for the preservation of tissue structure. Picking out the signals coming from the surrounding environment enable cells to react promptly to changes that may occur. Various molecular mechanisms explain the ability of cells to sense the microenvironment could be grouped into two major classes: (1) the transmission of signals in the form of soluble molecules which interact with cellular receptors, such as growth factors, cytokines, hormones, etc., and (2) the interaction of cells with structural components of their environment, namely other cells and the extracellular matrix (ECM) [1].

Cell–cell interactions are central to the function of many organ systems. A common theme for heterotypic cell interactions is the interaction of parenchymal cells with nonparenchymal neighbors with resultant modulation of cell growth, migration, and/or differentiation. Specifically, these interactions are of fundamental importance in physiology [2, 3], pathophysiology [4, 5], cancer [6, 7] developmental biology [8, 9], wound healing [10, 11], and attempts to replace tissue function through 'tissueengineering' [12, 13]. Further understanding of how cell– cell interactions modulate tissue function will allow us to gain fundamental biological insight as well as suggest approaches that will allow the manipulation of tissue function in vitro for therapeutic applications in vitro for therapeutic applications [14].

2. Cell-cell interaction and cross talk phenomena during embryonic period

Cells are the true miracle of evolution. Once the basic building block, the eukaryotic cell, became available, the form of metazoans evolved by changing the arrangement of cells with respect to each other. Cell- cell interaction in embryo was described in literature to have a vital role in cell differentiation and fate of developing cells, a process generally referred to as embryonic induction [15-17].

Jessell and Meltont [18] in their studies on the diffusible Factors in Vertebrate Embryonic Induction reported that one group of cells control the fate of neighboring cells. Inductive interactions involve two primary components. He added that the process involves a signal that is generated by the inducing cell and a receptive system that directly or indirectly controls gene expression in the responding cell. The competence of cells to respond to the ligand also contributes to the extent of induction. A good example of cellular interaction reported in this article is the induction of mesodermal development. In Xenopus blastula, vegetal blastomeric release extracellular signals that induce adjacent animal ectoderm (animal cap cells) to develop into mesodermal tissue such as muscles or mesothelia [19-21].

Inductive interactions involve two primary components: a signal that is generated by the inducing cell and a receptive system that directly or indirectly controls gene expression in the responding cell. The ligands that constitute inductive signals can be anchored to the cell surface or secreted from cells. Thus, the extent of induction can be controlled by regulating ligand production or by limiting its range of action. The competence of cells to respond to the ligand also contributes to the extent of induction. Competence may be controlled by modifying the expression or function of the appropriate receptors, the intracellular signal transduction pathway, or the transcription of target genes (Figure 1). Inductive signals can also control multicellular pattern if the response of similar cells to different concentrations of the same signal results in different cell fates [18].

Schmidt et al. [22] reported that Signals originating from embryonic ectoderm have a role in the development of underlying somites and neural crest which is mediated by Wnt family of secreted signaling molecules that controls a wide range of developmental processes in all metazoans. Neural crest is a population of multipotent progenitor cells that arise from the neural ectoderm in all vertebrate embryos and form a multitude of derivatives including the peripheral sensory neurons, the enteric nervous system, Schwann cells, pigment cells and parts of the craniofacial skeleton. Schmidt et al. [22] reported that neural crest induction requires an ectodermal signal. Signaling molecules of the Wnt, BMP, and FGF families and their downstream effectors have been shown to mediate neural crest induction [23-24].

Dorsolateral bending of the neural plate, an undifferentiated pseudo-stratified epithelium, is essential for neural tube closure which if failed spina bifida results. Ybot-Gonzalez et al. [25] pointed to the cellular interaction between neural crest cells and overlying neuroectoderm via molecular signaling that regulate the formation of dorsolateral hinge points (DLHPs) via antagonism of Bmp signaling that underlies the regulation of DLHP formation during mouse spinal neural tube closure.

3. Cellular interaction in nervous system

Glial cells are widely distributed throughout the nervous system. They have been found to have an impact on chemical synaptic transmission. Interplay among Schwann cells, the nerve and the muscle will provide insights into a better understanding of mechanisms underlying neuromuscular synapse formation and function.

Feng and Ko [26] reported that perisynaptic Schwann cells (PSCs), which are the glia juxtaposed to the nerve terminal at the neuromuscular junction (NMJ) play active and essential roles in synaptic function, maintenance, and development. The authors also mentioned that PSCs can respond to nerve activity by increasing intracellular calcium and are capable of modulating synaptic function in response to pharmacological manipulations. Schwann cell-derived factors can also promote synaptogenesis and enhance synaptic transmission in tissue culture

Feng and Ko [27] had studied the role of glial cells in the formation and maintenance of the neuromuscular junction. The authors reported that during development, PSCs grow beyond nerve terminals and guide nerve terminal extension. Nerve terminals retract or stop extension after PSC ablation by complement-mediated lysis in vivo, suggesting that PSCs can promote synaptic growth and maintenance at developing NMJs.

Schwann cell-conditioned medium (SC-CM), with culture medium consisting of 45% Leibovitz's L-15 medium (Invitrogen), 45% Ringer's solution (in mM: 115 NaCl, 2 CaCl2, 2.5 KCl, and 10 HEPES; pH 7.4), and 10% fetal calf serum (Invitrogen), which may be mediated by transforming growth factor-beta1, can promote synapse formation in Xenopus nerve-muscle culture. In addition, SC-CM contains small molecules (within 500-5000 Da), which can enhance spontaneous synaptic activities acutely and potently at developing frog NMJs. In adult muscles, PSCs can detect evoked synaptic activities and are capable of modulating transmitter release. Nerve terminals retract and synaptic efficacy is reduced at 1 week, but not within the first few hours, after PSC ablation. Thus, PSCs are essential for the long-term, but not short-term, maintenance of synaptic structure and function at the adult NMJ. After nerve injury, adult PSCs sprout extensive processes, which guide regenerating nerve terminals. Schwann cells express agrin and neuregulins, which may help the postsynaptic differentiation and synaptic repair. Furthermore, neuregulin-ErbB signaling pathways play an essential role in synapse-glial interactions at the NMJ. These recent findings suggest that PSCs play multiple roles and actively participate in synaptic development, modulation, maintenance, and repair of the vertebrate NMJ [27].

It was found that PSC interaction with nerve terminals play an important role in re-innervations at frog NMJs: regenerating NTs induce PSCs to sprout, and PSC sprouts, in turn, lead and guide the elaboration of NTs. After nerve injury, PSCs sprout profusely and PSC processes guide regenerating nerve terminals [26, 28].

4. Brain pericytes implication in blood brain barrier and pathological disorders

Brain microvascular pericytes are important constituents of the neurovascular unit. These cells are physically the closest cells to the microvascular endothelial cells in brain capillaries. They significantly contribute to the induction and maintenance of the barrier functions of the blood-brain barriers [29]. The highest pericyte coverage around microvessels is found in the central nervous system (CNS). It is not clear why the CNS needs higher vascular pericyte coverage than other organs, but one of the possibilities is that pericytes contribute to the formation of the blood–brain barrier.

Brain pericytes was suggested in early studies to be as a source of macrophage activity. Results substantiate this functional role via success demonstration of macrophage markers, phagocytosis and antigen presentation. Coupled with current knowledge on the entry of lymphoblasts into brain tissue and perivascular areas as potentially being the primary site of cellular interactions for production of immune responses, this places the pericytes in a position to significantly contribute to central nervous system (CNS) immune mechanisms.

However, it has been shown from some studies with rat bone marrow chimeras that lymphocytes do normally enter CNS tissue [30]. It appears that only immature lymphocytes or lymphoblasts can gain access and they stay there 1 to 2 days [31]; this seems to be a continuous process though, which would mean the constant presence of a lymphoblast population. The immature nature of the cells probably explains why they were not previously detected. These lymphoblasts could then mature, become activated and participate in an immune response.

Other functions of pericytes in brain are controlling of blood flow, regulation of vascular development and immune responses [32].

5. Vascular cell-cell interactions through junctions

The vascular system is considered an excellent example that demonstrates cell adhesion and its regulation. Endothelial cell adhesion plays an essential role in the vascular response to pathological conditions, such as inflammation, ischemia wound healing and, in particular, cancer. Certainly, tumor-associated angiogenesis is key to cancer progression and metastasis, and vascular adhesion molecules are undoubtedly major players in this context [1].

It has become clear that vascular intercellular adhesion exhibits cell type-specific features that account for the specialized roles of the adhesive junctions in the endothelium [1].

In endothelial cells, tight junctions being often intermingled with adherens junctions along the intercellular boundaries rendering this junctional organization not as rigid. Adding together, endothelial cells do not contain desmosomes, although some desmosomal components are found in the complex adherens, a junctional structure specific to certain specialized vascular districts, such as a subset of lymphatic vessels and of veins [33].

The main difference between epithelial and endothelial AJs is that the latter do not contain E-cadherin but an endothelial-specific cadherin, called vascular endothelial (VE) cadherin. The expression of VE-cadherin is essentially restricted to cells of the endothelial lineage and starts very early during the differentiation of endothelial cell precursors [34].

Although VE-cadherin is found in all endothelial cell types, its levels vary in different vascular districts and during angiogenesis, including tumor vascularization. Indeed, the expression of VE-cadherin is enhanced in activated, cancer-associated vessels, suggesting a causal involvement in tumor angiogenesis [35].

Experimental evidence in vivo as well as in endothelial cell cultures pointed to an interplay between VE-cadherin-mediated adhesion and endothelial cell survival (i.e., resistance to programmed cell death or apoptosis). The molecular basis of this cross-talk probably lies in the ability of VE-cadherin to activate the phosphatidyl inositol-3 kinase (PI3K) pathway, an enzymatic cascade that ultimately leads to the inhibition of apoptosis [36].

Vascular Tight Junctions: Junctional adhesion molecules (JAMs) form a group of transmembrane proteins belonging to the immunoglobulin (Ig) superfamily, due to the presence of two Ig domains in their extracellular portion. JAMs appear to be associated with TJs rather than being integral components. As suggested by their name, a prominent feature of JAMs is their ability to promote intercellular adhesion via homophilic binding [37]. However, it is not clear to what extent the pro-adhesive function of JAMs is relevant in vivo.

The JAM family appears to play an important role in the recruitment of various proteins to the TJs. Indeed, JAM-A associates with zonula occludens-1 (ZO-1), cingulin, and occludin, inducing their localization at TJs (Figure 1) [38].

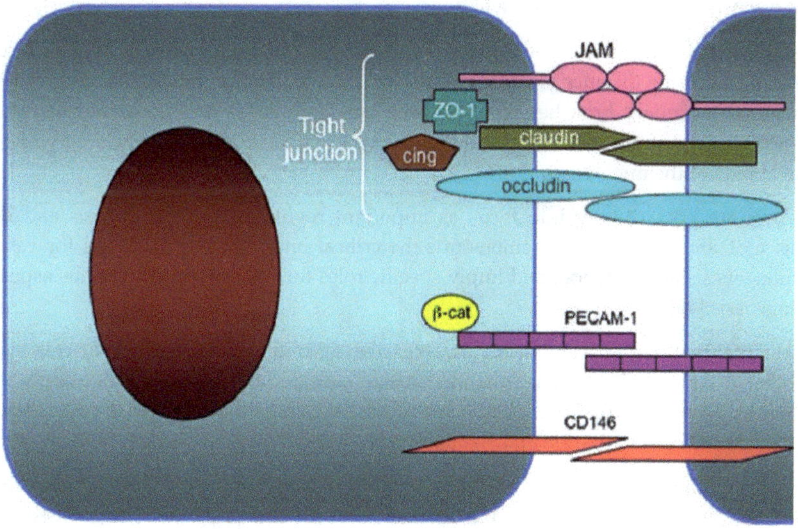

Figure 1. Vascular tight junctions. The molecular organization of the tight junction between endothelial cells is illustrated in a schematic manner, together with the non-junctional adhesion provided by PECAM-1 and CD146. Quoted from Cavallaro [1].

Recent observations have raised the possibility that JAMs are involved in tumor angiogenesis. Indeed, an antibody against JAM-C was reported to interfere with cancer growth by preventing neovascularization [39]. A major function of JAM proteins is their ability to regulate the trafficking of leukocytes and dendritic cells across the endothelium, a process that has crucial implications for the inflammatory response. Interfering with JAM function in vivo, e.g., by using neutralizing antibodies, blocks the transendothelial migration of monocytes and neutrophils in experimental models of inflammation. Vascular JAMs facilitate leukocyte endothelium interactions by heterophilic binding to blood cell integrins [40].

Some adhesion molecules expressed in endothelial cells do not show a specific association to junctional complexes. Platelet-endothelial cell adhesion molecule-1 (PECAM-1, also known as CD31) mediates inter-endothelial adhesion through homophilic binding. In addition, PECAM-1 has been implicated in a broad spectrum of vascular processes, including endothelial cell migration, survival, remodeling and angiogenesis [41].

Endothelial molecules, such as JAMs, PECAM-1 and CD146, that are involved in inflammatory infiltration could also facilitate the trafficking of tumor cells across the vascular wall. Hence, the therapeutic inhibition of adhesion molecules promoting transendothelial migration of inflammatory cells could prove useful also as a strategy to repress the metastatic dissemination of tumor cells [1].

Endothelial/Pericyte Interactions: Pericytes is the term for vascular mural cells embedded within the vascular basement membrane of blood microvessels, where they make specific focal contacts with the endothelium [42].

Morphologically, the pericytes exhibit a small, oval cell body with multiple processes extending for some distance along the vessel axis; these primary processes then give rise to orthogonal secondary branches which encircle the vascular wall. The contour of the cells conforms to that of the adjacent vascular element; also, they are usually enclosed within the basal lamina of the microvasculature [1].

Pericytes are now coming into focus as important regulators of angiogenesis and blood vessel function. Genetic data demonstrate the critical importance of pericytes for vascular morphogenesis and function, and imply specific roles for the cell type in various aspects of angiogenesis [43].

Development of a vascular system involves the assembly of two principal cell types - endothelial cells and vascular smooth muscle cells/pericytes (vSMC/PC) - into many different types of blood vessels. Senger and Davis [44] stated that Pericyte coverage leads to vessel remodeling, maturation and stabilization.

Pericyte-endothelial interaction mediated by cytokines: Insight into the molecular mechanisms of endothelial–pericyte interactions has accelerated during the past 1 to 2 years. Discovery of cytokine regulation confirmed the molecular cell talk between the two types of cells. Intercellular communication between endothelial and mural cells are mediated by many cytokines such as transforming growth factor β, angiopoietins, platelet-derived growth factor, spingosine-1-phosphate, and Notch ligands and their respective receptors [42]. Bergers and

Song [45] reported that Endothelial-derived PDGF-BB and HB-EGF coordinately regulate pericyte recruitment during vasculogenic tube assembly and stabilization (Figure 2).

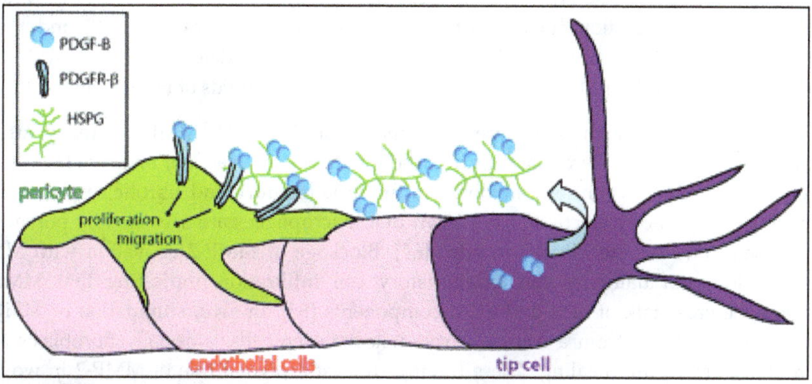

Figure 2. PDGF-B/PDGFR-β signaling is necessary for pericyte recruitment during angiogenesis. PDGF-B is synthesized and secreted by the migratory tip cells at the leading edge of angiogenic sprouts. Binding of PDGF-B to HSPG is important for localization of PDGF-B to the vicinity to the developing vessel. Pericytes, which express PDGFR-β, are dependent on of endothelium-derived PDGF-B for proliferation and migration [42].

Pericytes and pathological disorders: Pericyte as a multipotent progenitor cell of pathophysiological importance is gaining increasing attention. Bergers and Song [45] reported that when vessels lose pericytes, they become hemorrhagic and hyperdilated, which leads to conditions such as edema, diabetic retinopathy, and even embryonic lethality. Motegi, et al [46] studied the role of Pericyte-Derived MFG-E8 Regulates Pathologic Angiogenesis. Recent interest in pericytes also stems from their potential involvement in diseases [43-47] such as diabetic microangiopathy [48, 49] tissue fibrosis [50] cancer [51] atherosclerosis [52] and Alzheimer''s disease [53, 54].

6. Mesh-induced foreign-body reaction in hernea

The fact that tissue cells respond to biomaterial implantation is illustrated by granuloma formation and cell infiltration surrounding mesh materials over time. Common cellular components of such a reaction are infiltrating macrophages. These cells have the propensity to synthesize a plethora of pro inflammatory cytokines [transforming growth factor-β (TGF-β), platelet-derived growth factor (PDGF), vascular endothelial growth factor (VEGF)) and are regarded as key players directing the extent of fibrosis with influence on the phenotypic behavior of surrounding fibroblasts [55, 56].

Residential fibroblasts independently contribute to the regulation of tissue remodeling and wound healing. They occur as activated myofibroblasts encapsulating the mesh filaments and are constitutionally involved in extracellular matrix (ECM) remodeling by synthesizing type-I and type-III collagen. Furthermore, fibroblasts are the source of enzymes involved in matrix degradation such as matrix metalloproteinases (MMPs) that may affect the ongoing foreign-body reaction. MMPs are the most abundant proteases in wound healing [57] and MMP-2 (72-kDa collagenase, gelatinase A) enzymatic activities are up-regulated in diseases associated with inflammatory reaction such as arthritis [58], cancer [59], atheroma [60] and tissue ulceration [10]. A pivotal role for MMP-2 in hernia disease was determined by a study that detected elevated levels of MMP-2 enzymatic activity in wound fluids of hernia patients [61].

Beyond their capability to hydrolyze components of the ECM, MMP-2 directly affects cellular phenotypes, proliferation rates and the inflammatory reaction, and several studies indicate that MMP-2 is centrally involved in the inflammatory and fibrotic response [62]. Regarding foreign-body reaction, it is known that macrophages are activated by polymeric nanoparticles and secrete MMP-2 in vitro [64]. Blockage of MMP-2 activation with MMP inhibitor Ilomastat dampens the inflammatory cell infiltration, indicating that MMP-2 mediates the cross-talk of cells and ECM components [65]. In vivo, stimulation of MMP-2 expression may result from a complex cross-talk between cells, especially fibroblasts and macrophages in wound healing. These findings hint at a pivotal role of MMP-2 in wound healing and foreign body reaction and suggest the investigation of the molecular mechanisms that govern MMP-2 gene transcription after biomaterial implantation [66].

Transgenic as well as knockout models were established to elucidate the underlying gene regulation required for wound healing [64]. Meshes interfere with MMP-2 gene regulation due to soluble factors, ECM modification or cell cross-talk. In MMP-2/Lac Z transgenic mice the impact of mesh implantation on MMP-2 gene expression can be evaluated and compared to MMP-2 enzymatic activity, protein synthesis and expression/binding of transcription factors [66].

7. Dendritic cell–NK cell cross-talk

The interaction of NK cells with the professional antigen-presenting cells of the immune system, the dendritic cells (DC), in regulating both innate resistance and adaptive immunity. DC is antigen-presenting cells, cornerstones between pathogen entry and lymph nodes that quickly respond to foreign antigens. Located in peripheral organs, skin, and mucosal surfaces, DC sample the environment for self and foreign material [67].

The molecular mechanisms involved in NK cell triggering by human DC start to be unraveled. Mature DC or immature DC in the presence of maturation stimuli, such as LPS or *Mycobacterium tuberculosis* or IFN, are able to activate NK cells [68, 69]. The crucial role of IL-12 in IFN-secretion by + human NK cells stimulated by monocyte- or CD34 –derived DC and LPS or by peripheral blood m DC in response to TLR3 or TLR8 legends has been formally demonstrated. Other cytokines, such as IL-18, and/or cellular contacts are also involved [70, 72]. However, NK cell activation by DC also requires direct cell-to-cell contacts and depends on the adhesion molecule LFA-1 [73].

The formation of DC/NK cell conjugates was found to depend on cytoskeleton remodeling and lipid raft mobilization in DC. BM-DC derived from mice with loss of function of the Wis kott Aldrich syndrome protein, a major cytoskeletal regulator expressed in hematopoietic cells; fail to promote NK cell lytic activity and IFN-secretion [71]. Moreover, disruption of the DC cytoskeleton with pharmacological agents abolished the DC-mediated NK cell activation. Therefore, the cross-talk between LPS-activated DC and NK cells is dictated by functional synapses [71].

8. Cellular interaction in normal and fibrosed heart muscle fibers

Cells in the heart interact through both paracrine and autocrine pathways and by direct contact with the formation of gap and adherens junctions and desmosomes. In adherens junctions cadherins on one cell bind to cadherins on another cell in contact and link intracellularly to the actin cytoskeleton via catenins. Desmosomes link to intermediate filaments. Adherens junctions and desmosomes mechanically connect cardiomyocytes and so distribute contractile force within the myocardium. Gap junctions provide intercellular channels for ionic communication that allows the rapid and coordinated spread of excitation throughout the heart [74].

The conversion of fibroblasts to myofibroblasts is central to the development of cardiac fibrosis in response to hypertension [75] in ventricular hypertrophy, hypertension, or infarction; the number of fibroblasts in the heart has been shown to increase [75]. A scar tissue formed after myocardial infarction also is stiffened by both reparative fibrosis and actively contracting cardiac fibroblasts. These myofibroblasts influence myocardial function by increased collagen secretion and contractility causing a stiffening of the heart muscle that can lead to diastolic dysfunction and heart failure [76] and also by possibly interfering with the electrical connectivity of the cardiomyocytes [75]. In this example Genin et al described how tissue constructs serve as model systems in which to study how fibroblasts and cardiomyocytes interact to control contractile force and tissue stiffness. They dissect here the electrical and mechanical cell-cell phenomena that might underlie the above observations. [74]

From the mechanical perspective, a possible explanation is that the eventual domination of the construct by the proliferative myofibroblasts stiffens the tissue constructs in a way that retains the ability to produce a steady baseline force, perhaps exerted mainly by myofibroblasts, while losing the ability to generate myocyte-dependent twitch force. The stiffening of fibrotic myocardium can result from secretion of excessive ECM from the myofibroblasts [77], and likely from increased ECM remodeling and increased myofibroblast contractility as well. The stiffening of the extracellular environment by myofibroblasts and associated rise in baseline force may overwhelm the actomyosin contractile mechanism in the cardiac myofibroblasts, constraining it to a low number of cross-bridge connections by limiting the motion of the contractile apparatus.

In addition to these mechanical effects, myofibroblasts can impair both heart and EHT contractile function by distorting excitatory conduction. Under normal conditions the numerous fibroblasts in the heart maintain the ECM that provides the underlying structure for a continuous network of cardiomyocytes in electrical contact via gap junctions. The spread of electrical excitation in this network is organized to stimulate an orderly contraction first of the atria and then the ventricles to promote optimal pumping efficiency of the heart. Evidently, the presence of the fibroblasts in normal heart muscle does not perturb this orderly impulse conduction. Myofibroblasts can distort the propagation of the excitatory wave both by disrupting the normal interactions (gap junction formation) among the cardiomyocytes and by forming gap junctions and therefore electrical contact with the cardiomyocytes [74].

Myofibroblasts can be coupled electrotonically to cardiomyocytes in vitro via gap junctions mediated by the connexins Cx43 and Cx45 [78, 79]. In these experiments strands of cardiomyocytes were coated with cardiac fibroblasts that had converted to the myofibroblast phenotype [74].

This coupling suggests that myofibroblasts might not only provide a barrier to the electrical interaction of cardiomyocytes but might also provide a conductive link between them. This was demonstrated in vitro by connecting two strands of neonatal rat ventricular cardiomyocytes by a band of cardiac myofibroblasts [80].

These experiments demonstrate that myofibroblasts both distributed throughout the heart muscle and in border zones of healing infarcts can play a complex role in impulse propagation, imposing steric blockage, providing alternative but slower conduction pathways and predisposing the tissue to arrhythmia [77]. Finally, a model system similar to that described above has demonstrated that contact with myofibroblasts can cause spontaneous activation of cardiomyocytes that could be analogous to ectopic activity in the heart [78].

9. Cell- cell interaction in lung parenchyma cells

Cellular interactions in lung parenchyma were reviewed by Fehrenbach, 2001 [81]. The main functional units of lung parenchyma are the alveoli; an air filled sacs. Alveoli are lined by two types of cells; alveolar type I (type I pneumocyte) which are of considerable little thickness allowing gaseous exchange between alveolar air and pulmonary capillaries forming what is known as blood-air barrier Figure (3, 4).

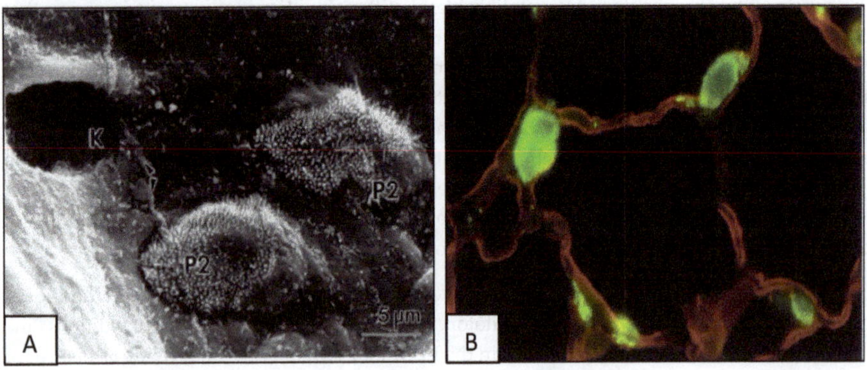

Figure 3. A: showing scanning microscopy of two great alveolar cells (Type II pneumocytes) B: alveolar cells (Type II pneumocytes) stained by immunofluroscence for surfactant protein D (green). Quated from Fehrenbach [81].

The second type of cells is alveolar type II or great Type II pneumocyte which is usually cuboidal in shape and rich in organelles involved in secretion of phospholipid material known as surfactant [81].

10. TypeII-type I- alveolar cell interaction

Cell- cell interaction is well represented in lung parenchyma between type II alveolar cells and other resident cells of the lung . Direct laterl cell-cell contact with type I alveolar cell is maintained by cell junction complex that induce gap junction [82]. This contact allow mechanical stimulation of type I cell to modulate exocytosis rate of surfactant secteted by type II via transimmision of calcium ion oscillation throuhgh gap junction. On the other hand, direct inhubitory interaction between the two cells have been postulated by Mason and McCormack [83] to supress Type II proliferation. This explain proliferative activity of the later in case of injury of type I pneumocytes. Type II pneumocyte also was proposed by Kapanci et al. [84] as stem cell of the adult that differentiate to Type I pneumocyte and subsequent repair and re-establishing of air – blood barrier [85, 86].

11. TypeII-typeII interaction

Khalil et al. [87] reported that inhbition of alveolar type II cells can be mediated via paracrine action via Type II cell–derived transforming growth factor (TGF) as in case of bleomycin-induced experimental lung fibrosis.

Removal of cells dying by apoptosis is essential to normal development, maintenance of tissue homeostasis, and resolution of inflammation. Surfactant protein A (SP-A) and surfactant protein D (SP-D) are high abundance pulmonary collectins implicated in apoptotic cell clearance in vitro. Other collectins, such as mannose-binding lectin and the collectin-like C1q, have been shown to bind to apoptotic cells and drive ingestion through interaction with calreticulin and CD91 on the phagocyte in vitro. However, only C1q has been shown to enhance apoptotic cell uptake in vivo. Similar to C1q and mannose-binding lectin, SP-A and SP-D bound to apoptotic cells in a localized, patchy pattern and drove apoptotic cell ingestion by phagocytes through a mechanism dependent on calreticulin and CD91. These results suggest that the entire collectin family of innate immune proteins (including C1q) works through a common receptor complex to enhance removal of apoptotic cells, and that collectins are integral, organ-specific components of the clearance machinery [88].

12. Alveolar-endothelial cell interaction

Embryonic studies showe that pulmonary endothelial cells exhbit inductive activity on foetal lung alveolar epithelium [89]. This effect was well studied in tissue culture. A paracrine mechanism of action was suggested to be exerted on alveolar cells via endothelin cytokine produced by capillary endothelium. on the other hand, celluar interaction is well represented by the fact that alveolar type II cells cells may act as transducers of an inflammatory signalfrom the alveolus to the capillaryendothelium

13. Cell- cell interaction between type II alveolar cells and mobile interstial pumonary cells

Lung stroma or interalveolar connective tissue is rich in mobile ceels needed for pulmonary defense mechaniusm, those cells include alveolar macrophage dervide form circylating monocytes.

14. Type II alveolar- alveolar macrophage - interaction

This type of interaction was represented by recopricoal effect on proliferation of both cells via cytokines production such as hepatocyte growth factor [90] andheparin-binding epidermal growth factor-like protein secreted by macrophage and RANTES and MCP-1 produced by alveolar type II cells [91]. Furthermore, Stamme et al. [92] reported that SP-A released from AE2 cellsmay modulate macrophage functions such as, oxygen radical release [93], and nitric oxide production [92].

15. Type II alveolar- leucocyte interaction

Cytokines produced by type II alveolar cells were reported to influence differentiation of leucocytes (neutrophils-basophiles ands eosinophils) and have arole in lung parenchyma inflammatory reactions. On the other hand, alveolartype II cells can exert inhibitory effect on lymphocytes. Direct interaction of pneumocytes with migrating monocytes wasreported to be mediated by b2-integrins CD11b/CD18and b1-integrins as well as by CD47 [94]. The concept of considering typeIIalveolar cells as the "defender of the alveolus" by Fehrenbach, 2001 implies thatsevere damage or loss of AE2 cells results in a considerablevulnerability of the alveolus such as lung fibrosis [95]. In spite of all knowledge reported by Fehrenbach [82], he added that more studied of mystery of cell–cell interactionsof AE2 cells still remains to be expanded.

16. Type II - Fibroblasts interaction

Findingd reported by Fehrenbach [82] and Shannon & Deterding 1997 [96] showed that reciprocal cell–cell relationship between type II alveolar cells and fibroblast control the modelling of alveoles during lung morphogenesis as well as duringremodelling associated with alveolar repair following lung injury [97, 98]. Both direct and indirectcell–cell interactions have been reported.

Alvealar E2 cells have beenreported to secrete a factor that eitheir inhibits or stimulate fibroblast proliferation [99, 100]. In contrast, however, an increase in fibroblastproliferation was seen if both cell populations grown in coculturewere able to establish direct cell–cell contacts [100]. Transmission electron microscopy has demonstrated structural interaction between fibroblast processes and type II alveolar cell membranes [99]. Immunoelectronmicroscopy indicated that CD44v6 islocalised at the tips of these foot processes [97]. TheCD44 molecules constitute a family of integral membrane glycoproteins that act as receptors of hyaluronan and osteopontin,for example, and are well established as beinginvolved in epithelial cell migration and differentiation [101]

17. Cell- cell interaction in mammalian testicular tissue

The mammalian testis represented a mixed gland where the exocrine; the seminiferous tubules is responsible for male gamete formation in the process named spermatogenesis. While interstitial (Leydig) cells form the endocrine part that is involved in testosterone production.

Seminiferous tubule presents a highly complex cellular interacting system. Well documented interactions and communications take place between Sertoli and germ cells at different stages of their development [102]. At morphological level, Sertoli cell in testis of many species is associated with ~30–50 germ cells at each stage of the spermatogenic cycle in the epithelium. In his interesting article concerning Cell Junction Dynamics in the Testis Yan Cheng and Mruk [103] mentioned that germ cells largely rely on Sertoli cells for structural and nutritional support [104]. Blood-testis barrier (BTB) formed by tight Junction between the lateral of Sertoli cells [104-108]. Serves for isolation of developing haploid germ cells from body immune response. In addition, cell-cell communications via paracrine factors and signaling molecules were also observed. Sertoli cells in this way can provide developing germ cells with the needed nutrients and biological factors [109-110]. Germ cell-Sertoli cell interactions were studied early by Zabludoff et al [111] who found that regulation of CP-2 (a novel Sertoli cell product) synthesis and secretion by the Sertoli cell is dependent on paracrine signals or direct cell contact with the germ cells.

The findings of Sharpe et al [113] showed that the functions of all of the cell types in the testis are interwoven in a highly organized manner. the authors emphasize that in normal adult rat testis there is a complex interaction between the Leydig cells, the Sertoli (and/or peritubular) cells, the germ cells, and the vasculature, and that testosterone, but not other Leydig cell products, plays a central role in many of these interactions, they added that The Leydig cells drive spermatogenesis via the secretion of testosterone which acts on the Sertoli and/or peritubular cells to create an environment which enables normal progression of germ cells through stage VII of the spermatogenic cycle. In addition, testosterone is involved in the control of the vasculature, and hence the formation of testicular interstitial fluid, presumably again via effects on the Sertoli and/or peritubular cells.

18. Cellular interaction in mammalian skin

Skin is the largest body organ which serves a number of important functions for the welfare of the organism. It has unique structure being derived from two different embryonic sources, namely the ectoderm which give rise the outer epithelial component, the epidermis and epidermal derivatives and the mesoderm which are the source of dermal connective tissue elements [114].

To be an effective body barrier cellular interaction either via structural or cytokine contact between both epidermal and dermal components of skin. Epithelial-mesenchymal interactions control epidermal growth and differentiation. It was found that reciprocal stimulatory effects between keratinocytes and dermal fibroblasts and micro vascular endothelial cells via induction of paracrine growth factor gene expression. The superficial epidermal layer of skin consists of a variety of cells, namely, keratinocytes, the dominant cell type, melanocytes, the coloring cells of skin, Langerhan cells responsible for epidermal defense and Merkel sensory cell [115].

Cell talk is will represented in the process of skin pigmentation (Melanogenesis). Melanin is synthesized and packaged in organelles containing melanogenic enzymes (tyrosinase gene family of proteins), the melanosomes, which are trans located down to the tips of the melanocyte dendrites and then transferred to the neighboring keratinocytes, where they form melanin cap over the nuclei to protect DNA from UV damage [116-117]. Many researches confirmed the presence of cellular interaction between melanocyte and keratinocytes for regulation of skin pigmentation [118-120] which in case of failure result in pathological pigmentation.

Solar Lentigo; macular brown pigmentation appearing after chronic sun exposure is considered as a component of photoaging. Lesional keratinocytes express enhanced levels of endothelin-1 (ET-1) [121] and (stem cell Factor) SCF [122] that stimulate melanocyte proliferation and melanin formation.

Ephelides (Freckles) which are small, discrete brown macules usually <0.5 cm in diameter appear on exposed areas among children and young adults, especially in fair-haired and fair-skinned individuals [123]. Histologically, the melanocytes are normal or reduced in number when compared with adjacent normal skin, but melanin production is increased owing to UV stimulation. Large numbers of mature melanosomes are evident in dendritic melanocytes [124].

Post-inflammatory Hyperpigmentation following inflammation occurring mainly in Fitzpatrick skin types IV-VI [123]. Post-inflammatory hyperpigmentation (PIH) represents a pathophysiological response to cutaneous inflammation, such as acne, atopic dermatitis, discoid lupus erythemasosus, erythema dyschromicum perstans, fixed drug eruption, generalized drug eruption, idiopathic eruptive macular pigmentation, impetigo, insect bites, irritant and allergic contact and photocontact dermatitis, lichen planus, lichen simplex chronicus, morphea, pityriasis rosea, polymorphous light eruption, psoriasis, burn, abrasive and postsurgical trauma, and viral exanthem [125]. Melanocytes can either be stimulated by the inflammatory process to become hyper-functional, thus secreting more melanin, or the number of melanocytes can increase. Epidermal hyperpigmentation (such as that associated with acne) occurs when increased melanin is transferred to keratinocytes, whereas dermal pigmentation (e.g., associated with lichen planus and cutaneous lupus erythematosus) occurs when the basement membrane is disrupted causing melanin to fall into the dermis and resides within melanophages [125].

19. Keratinocyte-Langerhans cell interaction

Skin is an immunological organ consisting of epidermal cells, i.e. keratinocytes and Langerhans cells (LCs, antigen-presenting dendritic cells), and both innate and acquired immune systems operate upon exposure of the skin to various external microbes or their elements [126].

Langerhans cells are dendritic cells (antigen-presenting immune cells) of the skin and mucosa, and contain large granules called Birbeck granules. They are present in all layers of the epidermis, but are most prominent in the stratum spinosum [127]. These Birbeck granules, are rod-like membrane-bound structures with regular cross-striations, one end of which frequently distends in a vesicle so that they resemble a tennis racket or Ping-Pong paddles (15 to 50nm in length and 4 nm thick). These granules form as a result of clathrin-assisted endocytosis; however, their function is not known [127, 128].

Langerhans cells derive from the cellular differentiation of monocytes with the marker "Gr-1" (also known as "Ly-6G/Ly-6C") [129]. In skin infections, the local Langerhans cells take up and process microbial antigens to become fully functional antigen-presenting cells. Presenting the processed antigens to T cells resulting in T-cell differentiation and activation has an important role in innate cutaneous immunity [130]. Toll-like receptors (TLR) are involved to enhance the ability of LCs to present a specific antigen to T cells [127]. Toll-like receptors (TLRs) are a class of conserved receptors that recognize pathogen-associated molecular patterns (PAMPs) present in microbes [131]. Hari et al. also reported that these receptors are expressed on several skin cells including keratinocytes, melanocytes, and Langerhans cells [131].

20. Cell-cell interaction in dermal resident cells

Dermis is the deep connective tissue component of skin; it is the habitat of many cells either fixed (permanent) or transient visitors. Fibroblasts represent the dominant population of cells with only few hematopoietic cells residing in these tissues [132]. These cells are responsible for collagen production and serve to maintain the extracellular matrix and stromal connective tissues. They do this by secreting compounds that serve as precursors to components of the extracellular matrix, which go on to form collagens, glycosaminoglycan's, glycoproteins, and reticular and elastic fibers. Fibroblasts has well known role in wound healing processes [133] as well as abnormal scar formation [134]. Adipocytes aggregate under the dermal tissue to offer a secondary protective barrier to body organs.

Dermal cell interaction was described in normal as well as in pathological conditions of human skin. A complex pattern of cell talk or cellular interaction was observed by Ali et al. [134] in case of keloid scars. Fibroblast-lymphocytes, mast cells and macrophages contact were described by authors in such lesions (Figure 4, 5). Walsh et al. [135] reported the role of human mast cells as "gatekeepers" of the dermal microvasculature and indicate that mast cell products other than vasoactive amines influence endothelium in a proinflammatory fashion. Mast cells of considerable number, size and degranulation were found by Ali et al. [134] to dominate in some lesions of keloid scar. This suggested the interaction between those two cells and their implication in scar formation. A paracrine loop between adipocytes and macrophages via free fatty acid and free fatty acids and tumor necrosis factor alpha signaling was reported by Suganami et al [136] and Andrade et al. to aggravate inflammatory changes [137].

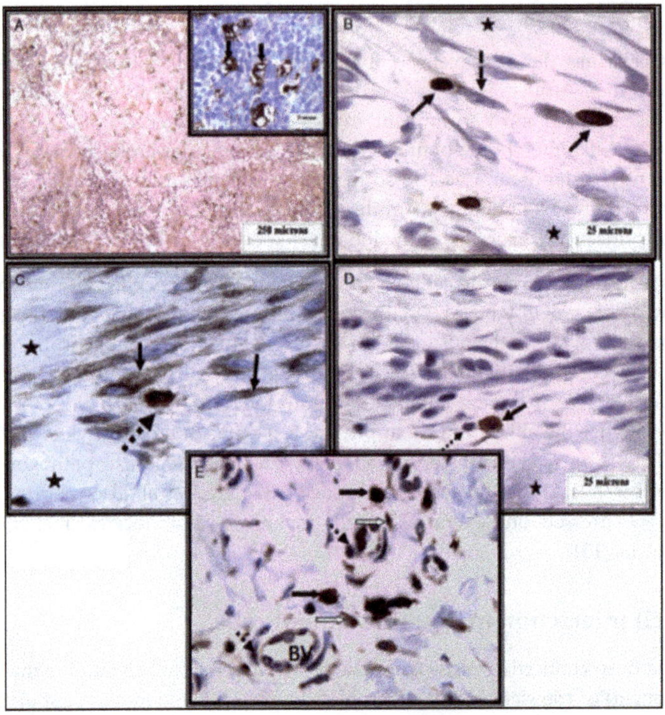

Figure 4. Cell interaction between different cell population in keloid dermal connective tissue, Quoted from Ali et al. [134].

The extracellular matrix is important because its composition determines the physical properties and integrity of dermal connective tissue [128]. Fibroblast growth factor (FGF) signaling is involved in a wide range of important organically activities with differential effects in several cell types. Ali et al. [134] found an interesting association of fibroblasts with lymphocytes, mast cells and macrophages known to be increased in dermal tissue in case of inflammatory processes characterized skin injury. A sort of cell interaction seemed to occur between fibroblasts and these immune cells was suggested. This interaction was termed cell talk by Lim et al [138].

Mast cells were the third type of immune cells that were interestingly found in large numbers of keloid scars examined in this study. This explained the finding that most patients with abnormal scars complained of itching as a symptom and erythema as a sign [139-140]. Both resolved by corticosteroid treatment. In this study, mast cells were found in close contact with fibroblasts. Mast cell activation is a characteristic feature of chronic inflammation, a condition that may lead to fibrosis as a result of increased collagen synthesis by a fibroblast [140-144].

Fibroblasts were also found to produce a mast cell growth factor that supposedly regulates mast cell survival, differentiation, and granule synthesis, whereas, mast cells were shown to affect the biochemical properties of fibroblasts, which can lead to fibrosis. Fibroblasts can modulate the functions of both mast cells and eosinophils, which also increase the amount of melanocytes and melanin pigment [141-143].

Obesity is associated with decreased dermal elasticity. This denotes the presence of an adipocyte–fibroblast interaction. Azure and Amano [145] found that enlarged adipocytes have negative regulation of dermal fibroblasts through release of free fatty acids.

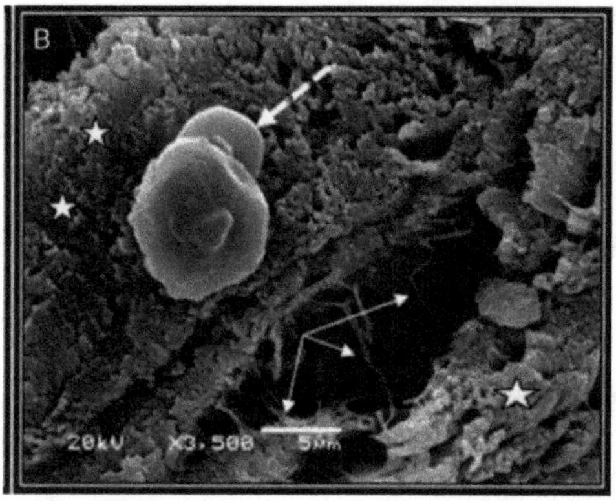

Figure 5. Mast cell-lymphocyte interaction by scanning microscopy, Quoted from Ali et al. [134].

Eosinophil infiltration into the inner dermal compartment is a predominant pathological feature of atopic dermatitis [146]. The interaction between eosinophils and fibroblasts under IL-31 and IL-33 stimulation differentially activated extracellular signal-regulated kinase, c-Jun N-terminal kinase, p38 mitogen-activated protein kinase, nuclear factor-κB and phosphatidylinositol 3-kinase–Akt pathways.

Eosinophil infiltration into the inner dermal fibroblast layer causing inflammation in atopic dermatitis has been well established]. Investigation of the interaction between eosinophils and fibroblasts may therefore help to elucidate the mechanism of initiating local inflammatory response in atopic dermatitis [147].

Author details

Nasra Naeim Ayuob
Department of Anatomy, Faculty of Medicine, King Abdulaziz University, Jeddah, Saudi Arabia
Department of Histology and Cytology, Mansoura University, Egypt

Soad Shaker Ali
Department of Anatomy, Faculty of Medicine, King Abdulaziz University, Jeddah, Saudi Arabia
Department of Histology and Cytology, Assuit University, Egypt

Acknowledgement

The authors would like to thanks their families who provided support to allow them working in this chapter. Special thank for Mohamed Basem Eldeek; Dr. Nasra son and Dr. Basem Eldeek, Associate Professor of Public Health, Faculty of Medicine, King Abdulaziz University, Dr. Nasra Husband for supporting her during working in this chapter.

21. References

[1] Cavallaro U (2008) Adhesion Molecules in the Vascular Cell Cross. In Marme D, Fusenig N, editors. Tumor Angiogenesis, Part 2, Germen: Springer, 289-308, DOI: 10.1007/978-3-540-33177-3_16.

[2] van Breemen, C., Skarsgard, P., Laher, I., McManus, B., Wang, X. (1997) Endothelium-smooth muscle interactions in blood vessels. *Clin. Exp. Pharmacol. Physiol.* 24,989-992[Medline]

[3] Clark, B. R., Keating, A. (1995) Biology of bone marrow stroma. *Ann. N.Y. Acad. Sci.* 770, 70-78[Medline]

[4] Davies, P. F. (1986) Biology of disease: vascular cell interactions with special reference to the pathogenesis of atherosclerosis. *Lab. Invest.*55,5-24[Medline]

[5] Grinnell, A. D. (1995) Dynamics of nerve-muscle interaction in developing and mature neuromuscular junctions. *Physiol. Rev.*75,789-834[Abstract/Free Full Text]

[6] Camps, J. L., Chang, S. M, Hsu, T. C., Freeman, M. R., Hong, S. J., Zhau, H. E, von Eschenbach, A. C., Chung, L. W. (1990) Fibroblast-mediated acceleration of human epithelial tumor growth *in vivo*. *Proc. Natl. Acad. Sci. USA* 87,75 79[Abstract/Free Full Text]

[7] Hornby, A. E., Cullen, K. J. (1995) Goldberg, I.D. Rosen, E. M. eds.*Epithelial–Mesenchymal Interactions in Cancer* ,249-272

[8] Aufderheide, E., Chiquet-Ehrismann, R., Ekblom, P. (1987) Epithelial-mesenchymal interactions in the developing kidney lead to expression of terascin in the mesenchyme. *J. Cell Biol.* 105,599-608[Abstract/Free Full Text]

[9] Taderera, J. V. (1967) Control of lung differentiation *in vitro*. *Dev. Biol.*16,489-512[Medline]

[10] Grinnell, F. (1992) Wound repair, keratinocyte activation and integrin modulation. *J. Cell Sci.* 101,1-5[Free Full Text]

[11] Brown, L. F., Yeo, K. T., Berse, B., Yeo, T. K., Senger, D. R., Dvorak, H. F., Van De Water, L. (1992) Expression of vascular permeability factor (vascular endothelial growth factor) by epidermal keratinocytes during wound healing. *J. Exp. Med.* 176,1375-1379[Abstract/Free Full Text]

[12] Morgan, J. R., Yarmush, J. R. (1997) Bioengineered skin substitutes.*Sci. Med.* 4,6-16

[13] L'Heureux, N., Paquet, S., Labbe, R., Germain, L., Auger, F.A. (1998) A completely biological tissue-engineered human blood vessel. *FASEB J*12,1331-1340[Abstract/Free Full Text].

[14] Bhatia, S. N., Balis, U. J., Yarmush, M. L., Toner, M. Effect of cell–cell interactions in preservation of cellular phenotype: cocultivation of hepatocytes and nonparenchymal cells. (The FASEB Journal. 1999;13:1883-1900.)

[15] Spemann, H. (1938). Embryonic Development and Induction (New Haven, Connecticut: Yale University Press).

[16] Jacobson, A. G. (1966) Inductive processes in embryonic development. Science 752,25-34.

[17] Gurdon JB. (1987). Embryonic induction-molecular prospects. Development 99, 285-306.

[18] Jessell' TM. and Meltont DA. Diffusible Factors in Vertebrate Embryonic Induction Review. Cell, Vol. 66, 257-270, January 24, 1992, Copyright 0 1992 by Cell Press

[19] Slack, J. M., Darlington, 8. G., Heath, J. K., and Godsave, S. F. (1987). Mesoderm induction in early Xenopus embryos by heparin-binding growth factors. Nature 326, 197-200.

[20] Kimelman, D., Abraham, J. A., Haaparanta, T., Palisi, T. M., and Kirschner, M. W. (1988). The presence of fibroblast growth factor in the frog egg: its role as a natural mesoderm inducer. Science 242, 1053-1056.

[21] Paterno, G. D., Gillespie, L. L., Dixon, M. S., Slack, J. M., and Heath, J. K. (1989). Mesoderm inducing properties of INT-2 and kFGF: two oncogene-encoded growth factors related to FGF. Development 706, 79-63.

[22] Schmidt C, McGonnell I, Allen S, Patel K (2008): The role of Wntsignalling in the development of somites and neural crest. AdvAnatEmbryol Cell Biol. 2008;195:1-64.

[23] Barembaum, M. and Bronner-Fraser, M. (2005). Early steps in neural crest Barembaum and Bronner-Fraser ,2005 and Raible ,2006specification. Semin. Cell Dev. Biol. 16, 642-646.

[24] Raible, D. W. and Ragland, J. W. (2005). Reiterated Wnt and BMP signals in neural crest development. Semin. Cell Dev. Biol. 16, 673-682.

[25] Ybot-Gonzalez P, Gaston-Massuet C, Girdler G, Klingensmith J, Arkell R, Greene N D.E. and Copp. A J. Neural plate morphogenesis during mouse neurulation is regulated by antagonism of Bmp signaling. Development 134, 3203-3211 (2007)

[26] Feng Z, Ko CP. Neuronal glia interactions at the vertebrate neuromuscular junction. CurrOpinPharmacol. 2007 Jun;7 (3):316-24. Epub 2007 Mar 30.

[27] Feng Z, Ko CP.The role of glial cells in the formation and maintenance of the neuromuscular junction. Ann N Y Acad Sci. 2008;1132:19-28.

[28] Koirala S, Qiang H, Ko CP. Reciprocal interactions between perisynaptic Schwann cells and regenerating nerve terminals at the frog neuromuscular junction. J Neurobiol. 2000 Sep 5;44(3):343-60.

[29] Kovac A, Erickson MA, Banks WA. Brain microvascular pericytes are immunoactive in culture: cytokine, chemokine, nitric oxide, and LRP-1 expression in response to lipopolysaccharide. J Neuroinflammation. 2011 Oct 13;8:139.

[30] Hickey WF, Hsu BL, Kimura H. T-lymphocyte entry into the central nervous system J. Neurosci. Res., 28 (1991), pp. 254–260

[31] Hickey WF, Vass K, Lassmann H. Bone marrow-derived elements in the central nervous system: an immunohistochemical and ultrastructural survey of rat chimeras J. Neuropathol. Exp. Neurol., 51 (1992), pp. 246–256

[32] Rucker HK, Wynder HJ, Thomas WE. Cellular mechanisms of CNS pericytes. Brain Res Bull. 2000 Mar 15; 51(5):363-9.

[33] Hammerling B, Grund C, Boda-Heggemann J, Moll R, Franke W (2006) The complexus adhaerens of mammalian lymphatic endothelia revisited: a junction even more complex than hitherto thought. Cell Tissue Res 324:5567

[34] Gory S, Vernet M, Laurent M, Dejana E, Dalmon J, Huber P (1999) The vascular endothelial-cadherin promoter directs endothelial-speci? c expression in transgenic mice. Blood 93:184192

[35] Prandini MH, Dreher I, Bouillot S, Benkerri S, Moll T, Huber P (2005) The human VE-cadherin promoter is subjected to organ-speci? c regulation and is activated in tumour angiogenesis. Oncogene 24:29923001

[36] Carmeliet P, Lampugnani MG, Moons L, Breviario F, Compernolle V, Bono F, Balconi G, Spagnuolo R, Oostuyse B, Dewerchin M, Zanetti A, Angellilo A, Mattot V, Nuyens D, Lutgens E, Clotman F, de Ruiter MC, Gittenberger-de Groot A, Poelmann R, Lupu F, Herbert JM, Collen D, Dejana E (1999) Targeted deficiency or cytosolic truncation of the VE-cadherin gene in mice impairs VEGF-mediated endothelial survival and angiogenesis. Cell 98:147157

[37] Bazzoni G, Martinez-Estrada OM, Mueller F, Nelboeck P, Schmid G, Bartfai T, Dejana E, Brockhaus M (2000a) Homophilic interaction of junctional adhesion molecule. J Biol Chem 275:3097030976

[38] Bazzoni G, Martinez-Estrada OM, Orsenigo F, Cordenonsi M, Citi S, Dejana E (2000b) Interaction of junctional adhesion molecule with the tight junction components ZO-1, cingulin, and occludin. J Biol Chem 275:2052020526

[39] Lamagna C, Hodivala-Dilke KM, Imhof BA, Aurrand-Lions M (2005) Antibody against junctional adhesion molecule-C inhibits angiogenesis and tumor growth. Cancer Res 65:57035710

[40] Ebnet K, Suzuki A, Ohno S, Vestweber D (2004) Junctional adhesion molecules (JAMs): more molecules with dual functions? J Cell Sci 117:1929

[41] Jackson DE (2003) The unfolding tale of PECAM-1. FEBS Letters 540:7–14 40. Armulik A, Abramsson A, Betsholtz C. Review Endothelial/Pericyte Interactions. Circulation Research. 2005; 97: 512

[42] Gerhardt H and Betsholtz C(2003) Review Endothelial-pericyte interactions in angiogenesis. CELL AND TISSUE RESEARCH, 314 (1) 15-23,

[43] Senger DR and Davis GE. Angiogenesis Cold Spring Harb Perspect Biol August 2011;3:a005090 First published onlineAugust 1, 2011.

[44] Bergers G and Song S. The role of pericytes in blood-vessel formation and maintenance1. NeuroOncol. 2005 October; 7(4): 452–464.

[45] Motegi SI, Leitner WW, Lu M, Tada Y, Sárdy M, Wu C, Chavakis T, Udey MC. Pericyte-Derived MFG-E8 Regulates Pathologic Angiogenesis. Arteriosclerosis thrombosis and vascular biology (2011), 31(9): 2024-2034

[46] Allt G, Lawrenson JG. Pericytes: cell biology and pathology. Cells Tissues Organs. 2001;169(1):1-11.

[47] Yamagishi SI, Hsu CC, Taniguchi M, Harada SI, Yamamoto Y, Ohsawa KS, Kobayashi KI, Yamamoto H. Receptor-Mediated Toxicity to Pericytes of Advanced Glycosylation

End Products: A Possible Mechanism of Pericyte Loss in DiabeticMicroangiopathy. Biochemical and Biophysical Research Communications. Volume 213, Issue 2, 15 August 1995, Pages 681–687
[48] Motiejūnaitėa R, Kazlauskas A. Review Pericytes and ocular diseases. Experimental Eye Research, 86 (2), February 2008, Pages 171–177
[49] Kalluri R and Neilson EG. Epithelial-mesenchymal transition and its implications for fibrosis. J Clin Invest. 2003;112(12):1776–1784. doi:10.1172/JCI20530.
[50] Kalluri R. 2003. Angiogenesis: Basement membranes: structure, assembly and role in tumour angiogenesis. Nature Reviews Cancer 3, 422-433
[51] Juchem G, Weiss DR, Gansera B, Kemkes BM, Mueller-Hoecker J, Nees S. Pericytes in the macrovascular intima: possible physiological and pathogenetic impact. Am J Physiol Heart Circ Physiol. 2010 Mar;298(3):H754-70. Epub 2009 Dec 18.
[52] Stewart PA, Hayakawa K, Akers MA, Vinters HV. A morphometric study of the blood-brain barrier in Alzheimer's disease. Lab Invest. 1992 Dec;67(6):734-42.
[53] Farkas E, De Jong GI, de Vos RA, Jansen Steur EN, Luiten PG. Pathological features of cerebral cortical capillaries are doubled in Alzheimer's disease and Parkinson's disease. ActaNeuropathol. 2000 Oct;100(4):395-402.
[54] Chapman HA: Disorders of lung matrix remodeling. J Clin Invest 2004, 113: 148-157.
[55] Li Y, Yang J, Dai C, Wu C, Liu Y: Role for integrin-linked kinase in mediating tubular epithelial to mesenchymal transition and renal interstitial fibrogenesis. J Clin Invest 2003, 112. 503-516.
[56] Woessner JF Jr: MMPs and TIMPs an historical perspective Mol Biotechnol 2002, 22: 3349
[57] Ishikawa T, Nishigaki F, Miyata S, et al.: Prevention of progressive joint destruction in collagen-induced arthritis in rats by a novel matrix metalloproteinase inhibitor, FR255031. Br J Pharmacol 2005, 144: 133143
[58] Mandal M, Mandal A, Das S, Chakraborti T, Chakraborti S Clinical implications of matrix metalloproteinases. Molecular and Cellular Biochemistry 2003, 252: 305-329.
[59] Wu M, Li YG: The expression of CD40-CD40L and activities o matrix metalloproteinases in atherosclerotic rats. Mol Cel Biochem 2006, 282: 141-146.
[60] Agren MS: Gelatinase activity during wound healing. Br J Dermatol 1994, 131: 634-640.
[61] Turck J, Pollock AS, Lee LK, Marti HP, Lovett DH: Matrix metalloproteinase 2 (gelatinase A) regulates glomerular mesangia cell proliferation and differentiation. J Biol Chem 1996, 271. 15074-15083.
[62] Marti HP, Lee L, Kashgarian M, Lovett DH: Transforming growth factor-beta 1 stimulates glomerular mesangial cel synthesis of the 72-kd type IV collagenase. Am J Pathol 1994. 144: 82-94.
[63] Arbeit JM, Hirose R: Murine mentors: transgenic and knockout models of surgical disease. Ann Surg 1999, 229: 21-40
[64] Harendza S, Pollock AS, Mertens PR, Lovett DH: Tissue-specific enhancer-promoter interactions regulate high level constitutive expression of matrix metalloproteinase 2 by glomerular mesangial cells. J Biol Chem 1995, 270: 18786-18796
[65] Lynen-Jansen P., Klinge, D.H.U. Lovett, Mertens P.R. Biomaterials: Disturbing Factors in Cell Cross-Talk and Gene Regulation. In Schumpelick V. and Fitzgibbons RJ. Recurrent Hernia. Springer, Medizin Verlag Heidelberg. 2007, II, 63-67.

[66] Steinman RM. (2003) Someinterfaces of dendritic cell biology.Apmis 111:675–697.
[67] Fernandez NC et al. (2002) Dendritic cells (DC) promote natural killer (NK) cell functions: dynamics of the human DC/NK cell cross talk. Eur Cytokine Netw 13:17–27.
[68] Yu Y et al. (2001) Enhancement of human cord blood CD34+ cell-derived NK cell cytotoxicity by dendritic cells. J Immunol 166:1590–1600.
[69] Gerosa F et al. (2005) The reciprocal interaction of NK cells with plasmacytoid or myeloid dendritic cells profoundly affects innate resistance functions. J Immunol 174:727–734
[70] Borg C et al. (2004) NK cell activation by dendritic cells (DCs) requires theformation of a synapse leading to IL-12 polarization in DCs. Blood 104:3267–3275
[71] Gerosa F et al. (2002) Reciprocal activating interaction between natural killer cells and dendritic cells. J Exp Med 195:327–333
[72] Poggi A et al. (2002) NK cell activation by dendritic cells is dependent on LFA-1-mediated induction of calcium-calmodulin kinase II: inhibition by HIV-1 Tat C-terminal domain. J Immunol 168:95–101
[73] Genin G M., Abney T M., Wakatsuki T, and Elson E L. Cell-Cell Interactions and the Mechanics of Cells and Tissues Observed in Bioartificial Tissue Constructs, chapter 5 in Section II: Cooperative Cell Behavior and Mechanobiology. In A. Wagoner Johnson and Brendan A.C. Harley (eds.), Mechanobiology of Cell-Cell and Cell-Matrix Interactions, DOI 10.1007/978-1-4419-8083-0_5,# Springer Science+Business Media, LLC 2011.
[74] Camelliti, P., Borg, T. K., and Kohl, P., 2005, "Structural and Functional Characterisation of Cardiac Fibroblasts," Cardiovasc Res, 65, pp. 40–51.
[75] Kass, D. A., Bronzwaer, J. G., and Paulus, W. J., 2004, "What Mechanisms Underlie Diastolic Dysfunction in Heart Failure?," Circ Res, 94, pp. 1533–42.
[76] Rohr, S., 2009, "Myofibroblasts in Diseased Hearts: New Players in Cardiac Arrhythmias?," Heart Rhythm, 6, pp. 848–56.
[77] Miragoli, M., Gaudesius, G., and Rohr, S., 2006, "Electrotonic Modulation of Cardiac Impulse Conduction by Myofibroblasts," Circ Res, 98, pp. 801–10.
[78] Kohl, P., Camelliti, P., Burton, F. L., and Smith, G. L., 2005, "Electrical Coupling of Fibroblasts and Myocytes: Relevance for Cardiac Propagation," J Electrocardiol, 38, pp. 45–50.
[79] Gaudesius, G., Miragoli, M., Thomas, S. P., and Rohr, S., 2003, "Coupling of Cardiac Electrical Activity over Extended Distances by Fibroblasts of Cardiac Origin," Circ Res, 93, pp. 421–8.
[80] Fehrenbach H, 2001: Review alveolar epithelial type II cell :defender of alveolar revisted Respir Res 2001,2:33-46
[81] Kasper M, Traub O, Reimann T, Bjermer L, Grossmann H, Müller M, Wenzel KW: Upregulation of gap junction protein connexin43 in alveolar epithelial cells of rats with radiation-induced pulmonary fibrosis. Histochem Cell Biol 1996, 106: 419–424.
[82] Mason RJ, Leslie CC, McCormick-Shannon K, Deterding RR, Nakamura T, Rubin JS, Shannon JM: Hepatocyte growth factor is a growth factor for rat alveolar type II cells. Am J Respir Cell Mol Biol 1994, 11:561–567.

[83] Kapanci Y, Weibel ER, Kaplan HP, Robinson FR: Pathogenesis and reversibility of the pulmonary lesions of oxygen toxicity in monkeys. II. Ultrastructural and morphometric studies. Lab Invest 1969, 20:101–117.
[84] Witschi H: Proliferation of type II alveolar cells: a review of common responses in toxic lung injury. Toxicology 1976, 5:267–277.
[85] Bitterman PB, Polunovsky VA, Ingbar DH: Repair after acute lung injury. Chest 1994, 105:118S–121S.
[86] Khalil N, O'Connor RN, Flanders KC, Shing W, Whitman CI: Regulation of type II alveolar epithelial cell proliferation by TGF-beta during bleomycin-induced lung injury in rats. Am J Physiol 1994, 267:L498–507.
[87] Vandivier RW, Ogden CA, Fadok VA, Hoffmann PR, Brown KK, Botto M, Walport MJ, Fisher JH, Henson PM and Greene KE. Role of Surfactant Proteins A, D, and C1q in the Clearance of Apoptotic Cells In Vivo and In Vitro: Calreticulin and CD91 as a Common Collectin Receptor Complex. The Journal of Immunology,2002, 169: 3978 –3986.
[88] Smith SK, Giannopoulos G: Influence of pulmonary endothelial cells on fetal lung development. Pediatr Pulmonol 1985, 1: S53–S59.
[89] Mason RJ, McCormack FX: Alveolar type II cell reactions in pathologic states. In Lung surfactant: Basic research in the pathogenesis of lung disorders. Edited by Müller B, von Wichert P. Basel; Karger, 1994:194–204.
[90] Worgall S, Singh R, Leopold PL, Kaner RJ, Hackett NR, Topf N, Moore MA, Crystal RG: Selective expansion of alveolar macrophages in vivo by adenovirus-mediated transfer of the murine granulocyte-macrophage colony-stimulating factor cDNA. Blood 1999, 93:655–666.
[91] Stamme C, Walsh E, Wright JR: Surfactant protein A differentially regulates IFN- - and LPS-induced nitrite production by rat alveolar macrophages. Am J Respir Cell Mol Biol 2000, 23: 772–779.
[92] Weissbach S, Neuendank A, Pettersson M, Schaberg T, Pison U: Surfactant protein A modulates release of reactive oxygen species from alveolar macrophages. Am J Physiol 1994, 267: L660–666.
[93] Rosseau S, Selhorst J, Wiechmann K, Leissner K, Maus U, Mayer K, Grimminger F, Seeger W, Lohmeyer J: Monocyte migration through the alveolar epithelial barrier: adhesion molecule mechanisms and impact of chemokines. J Immunol 2000, 164: 427–435.
[94] Kuwano K, Hagimoto N, Kawasaki M, Yatomi T, Nakamura N, Nagata S, Suda T, Kunitake R, Maeyama T, Miyazaki H, Hara N: Essential roles of the Fas-Fas ligand pathway in the development of pulmonary fibrosis. J Clin Invest 1999, 104:13–19.
[95] Shannon JM, Deterding RR: Epithelial-mesenchymal interactions in lung development. In Lung growth and development. Edited by McDonald JA. New York; Marcel Dekker, Inc., 1997: 81–118.
[96] Kasper M, Haroske G: Alterations in the alveolar epithelium after injury leading to pulmonary fibrosis. Histol Histopathol 1996, 11:463–483.
[97] O'Reilly MA, Stripp BR, Pryhuber GS: Epithelial-mesenchymal interactions in the alteration of gene expression and morphology following lung injury. Microsc Res Tech 1997, 38: 473–479.

[98] Adamson IY, Hedgecock C, Bowden DH: Epithelial cell–fibroblast interactions in lung injury and repair. Am J Pathol 1990, 137:385–392.

[99] de Lara LV, Becerril C, Montano M, Ramos C, Maldonado V, Melendez J, Phelps DS, Pardo A, Selman M: Surfactant components modulate fibroblast apoptosis and type I collagen and collagenase-1 expression. Am J Physiol 2000, 279:L950–798.

[100] Bajorath J: Molecular organization, structural features, and ligand binding characteristics of CD44, a highly variable cell surface glycoprotein with multiple functions. Proteins 2000, 39:103–111.

[101] William W. Wright1, , Sonya D. Zabludoff1, , Tarja-Leena Penttilä, Martti Parvinen. Germ cell-sertoli cell interactions: Regulation by germ cells of the stage-specific expression of CP-2/cathepsin LmRNA by sertoli cells. Developmental GeneticsVolume 16, Issue 2, pages 104–113.

[102] Yan Cheng C.and Dolores D. Mruk .2002. Cell Junction Dynamics in the Testis: Sertoli-Germ Cell Interactions and Male Contraceptive Development . *Physiol Rev January 10, vol. 82 no. 4* 825-874

[103] Courot M, Hochereau-De Reviers Mt, And Ortavant R. The Testis, edited by Johnson AD, Gomes WR, and Vandemark NL. New York: Academic, 1970, vol. 1, p. 339–432.

[104] Jegou B. The Sertoli-germ cell communication network in mammals. Int Rev Cytol 147: 25–96, 1993.

[105] Dym M And Cavicchia Jc. Further observations on the blood-testis barrier in monkeys. Biol Reprod 17: 390–403, 1977.

[106] Dym M And Cavicchia Jc. Junctional morphology of the testis. Biol Reprod 18: 1–15, 1978.

[107] Dym M And Fawcett Dw. The blood-testis barrier in the rat and the physiological compartmentation of the seminiferous epithelium. Biol Reprod 3: 308–326, 1970.

[108] Byers S, Jegou B, Maccalman C, And Blaschuk O. Sertoli cell adhesion molecules and the collective organization of the testis. In: The Sertoli Cell, edited by Russell LD and Griswold MD. Clearwater, FL: Cache River, 1993a, p. 461–476.

[109] Byers S, Pelletier Rm, And Suarez-Quian C. Sertoli cell junctions and the seminiferous epithelium barrier. In: The Sertoli Cell, edited by Russell LD and Griswold MD. Clearwater, FL: Cache River, 1993b, p. 431–446.

[110] Enders Gc. Sertoli-Sertoli and Sertoli-germ cell communications. In: The Sertoli Cell, edited by Russell LD and Griswold MD. Clearwater, FL: Cache River, 1993, p. 448–460.

[111] Zabludoff Sd, Karzai Aw, And Wright Ww. Germ cell-Sertoli cell interactions: the effect of testicular maturation on the synthesis of cyclic protein-2 by rat Sertoli cells. Biol Reprod 43: 25–33, 1990.

[112] Sharpe RM. Regulation of spermatogenesis. In: The Physiology of Reproduction (2nd ed.), edited by Knobil E and Neill JD. New York: Raven, 1994, p. 1363–1433.

[113] Cui D. 2011. Atlas of Histology with Functional and Clinical Correlations. Lippincott William & Wilkin, Philadelphia, 400.

[114] Smola H, G Thiekötter, and NE Fusenig (1993) Mutual I nduction of growth factor gene expression by epidermal-dermal cell interaction Journal of cell biology vol. 122 no. 2 417-429

[115] Boissy, RE: 2003 .Melanosome transfer to and translocation in the keratinocytes. *Exp Dermatol 2003 12: 5–12,*

[116] Imokawa G. 2004. Autocrine and Paracrine Regulation of Melanocytes in Human Skin and in Pigmentary Disorders. Pigment Cell Research 17(2): 96–110.

[117] Duval, C, Regnier, M, Schmidt, R: 2001 Distinct melanogenic response of human melanocytes in mono-culture, in co-culture with keratinocytes and in reconstructed epidermis, to UV exposure. Pigment Cell Res 14: 348–355,

[118] Cardinali G, Simona Ceccarelli†,‡, Daniela Kovacs*, Nicaela Aspite*, Lavinia Vittoria Lotti†, Maria Rosaria Torrisi*,†,‡ and Mauro Picardo 2005 . Keratinocyte Growth Factor Promotes Melanosome Transfer to Keratinocytes. Journal of Investigative Dermatology 125, 1190–1199;

[119] Costin GE, Hearing VJ. 2007 Human skin pigmentation: melanocytes modulate skin color in response to stress. Faseb J;2:976-94.

[120] Kadono S, Manaka I, Kawashima M et al (2001) The role of the epidermal endothelin cascade in the hyperpigmentation mechanism of lentigo senilis. J Invest Dermatol 116: 571–577

[121] Hattori H, Kawashima M, Ichikawa Y et al (2004) The epidermal stem cell factor is over-expressed in lentigo seniles: implication for the mechanism of hyperpigmentation. J Invest Dermatol 122:1256–1265

[122] Yoko F. 2010. Disorders of Pigmentation. In: Thomas Krieg, David R. Bickers and Yoshiki Miyachi, Therapy of Skin Diseases, A Worldwide Perspective on Therapeutic Approaches and Their Molecular Basis, Springer-Verlag Berlin Heidelberg 2010. Pp. 525 - 537

[123] Rhodes AR, Albert LS, Barnhill RL et al (1991) Sun-induced freckles in children and young adults. A correlation of clinical and histopathologic features. Cancer 67: 1990–2001.

[124] Lynde CB, Kragt JN, Lynde CW (2006) Topical treatments for melasma and postinfl ammatory hyperpigmentation. Skin Ther Lett 11:1–6

[125] Sugita K, K Kabashima, K Atarashi, T Shimauchi, M Kobayashi, and Y Tokura 2007. Innate immunity mediated by epidermal keratinocytes promotes acquired immunity involving Langerhans cells and T cells in the skin. Clin Exp Immunol. 2007 January; 147(1): 176–183.

[126] Barbara Y; Heath, John W. (2000). Wheater's Functional Histology, 5th Ed.. Churchill Livingstone. p. 162.

[127] Gartner LP, Hiatt JI. 2009 Color Atlas of Histology, 5th Edition, lipincott Williams &Wikins, Baltimore, Maryland, USA, pp.232.

[128] Ginhoux F, Tacke F, Angeli V, Bogunovic M, Loubeau M, Dai X, Stanley E, Randolph G, Merad M (2006). "Langerhans cells arise from monocytes in vivo". Nat Immunol 7 (3): 265–73.

[129] Loser K, Beissert S. 2007 Dendritic cells and T cells in the regulation of cutaneous immunity. Adv Dermatol. 23:307-33.

[130] Hari A, Flach TL, Shi Y, Mydlarski PR. 2010 Toll-like receptors: role in dermatological disease. Mediators Inflamm. 2010; 2010:437246.

[131] Mine S, Fortunel NO, Pageon H, et al. 2008 Aging alters functional lyhuman dermal papillary fibroblasts but not reticular fibroblasts:a new view of skin morphogenesis and aging. PLos ONE. 3:e4066.
[132] Darby IA, Tim D Hewitson(2007). Fibroblast differentiation in wound healing and fibrosis. International Review Of Cytology Volume: 257, Issue: 07, Pages: 143-179
[133] Ali SS., Ayuob NN. and Hajrah N H. (2011). Cell Talk: A Phenomenon Observed in Keloid Scar by Immunohistochemical Study. Applied Immunohistochemistry and Molecular Morphology, 19 (2), 153–159.
[134] Walsh L J, G Trinchieri, H A Waldorf, D Whitaker, and G F Murphy(1991) Human dermal mast cells contain and release tumor necrosis factor alpha, which induces endothelial leukocyte adhesion molecule PNAS 15, vol. 88 no. 10 4220-4224
[135] Suganami T, Nishida J, Ogawa Y. A paracrine loop between adipocytes and macrophages aggravates inflammatory changes: role of free fatty acids and tumor necrosis factor alpha. Arterioscler Thromb Vasc Biol. 2005 Oct;25(10):2062-8. Epub 2005 Aug 25.
[136] Andrade Z.A., J. de-Oliveira-Filho and A.L.M. Fernandes (1998) Interrelationship between adipocytes and fibroblasts during acute damage to the subcutaneous adipose tissue of rats: an ultrastructural study. Braz J Med Biol Res, May, Volume 31(5) 659-664.
[137] Lim IJ, Phan TT, Bay BH, et al. 2002 Fibroblasts cocultured with keloid keratinocytes: normal fibroblasts secrete collagen in a keloid likemanner. Am J Physiol Cell Physiol. 283:C212–C222.
[138] Al-Attar A, Mess S, Thomassen JM, et al. Keloid pathogenesis and treatment. Plast Reconstr Surg. 2006;117:286–300.
[139] Liu W, Wu X, Gao Z, et al. Remodeling of keloid tissue into normal-looking skin. J Plast Reconstr Aesthet Surg. 2008;61: 1553–1554.
[140] Ribatti D, Vacca A, Marzullo A, et al. Angiogenesis and mast cell density with tryptase increase simultaneously with pathological progression in B-cell non-Hodgkin's lymphoma. Int J Canc. 2000; 85: 171–175.
[141] Noli C, Miolo A. The mast cell in wound healing. Veterin Dermatol.2001;12:303–313.
[142] Tomita M, Matsuzaki Y, Edagawa M, et al. Association of mast cells with tumor angiogenesis in esophageal squamous cell carcinoma. Dis Esophag. 2001;14:135–138.
[143] Abel M, Vliagoftis H. Mast cell-fibroblast interactions induce matrix metalloproteinase-9 release from fibroblasts: role for IgE-mediated mast cell activation. J Immunol. 2008;180:3543–3550.
[144] Ezure T, Amano S. (2011). Negative regulation of dermal fibroblasts by enlarged adipocytes through release of free fatty acids. J Invest Dermatol. 131(10):2004-9.
[145] Wong CK, Leung KML, Qiu1 HN, Chow JYS, Choi AOK, Lam CWK. 2012 Activation of Eosinophils Interacting with Dermal Fibroblasts by Pruritogenic Cytokine IL-31 and Alarmin IL-33: Implications in Atopic Dermatitis. PLoS ONE 7(1): e29815.
[146] Simon D, Braathen LR, Simon HU (2004) Eosinophils and atopic dermatitis. Allergy 59: 561–570.

Chapter 8

A Systems Approach to Understanding Bone Cell Interactions in Health and Disease

P. Pivonka, P.R. Buenzli and C.R. Dunstan

Additional information is available at the end of the chapter

1. Introduction

Bone is an important organ performing three essential physiological functions: mechanical support, mineral homeostasis (such as calcium and phosphate) and support of haematopoiesis. In fact, bone diseases in the elderly are associated with high morbidity and increased mortality. Osteoporosis and related skeletal complications are amongst the most important diseases impacting both the quality of life of our aging population and contributing costs to our health care system.

These different physiological functions of bone involve complex regulatory mechanisms at different spatial and temporal scales. For example, calcium and phosphate homeostasis is controlled on the whole organism scale and involves several organs such as parathyroid glands, gut, kidney and bone (Figure 1). On the other hand, bone structure and its mechanical properties are controlled on the tissue scale by the processes of bone modelling and bone remodelling. Bone modelling enables bones to adapt their shape during growth and in response to the prevailing mechanical loads. Bone remodelling removes micro-damage accumulating in the bone matrix during repeated mechanical loading. In this book chapter, we will focus on the implications of bone cell interactions in the control of bone remodelling and in the development of its disorders.

The renewal of bone during bone remodelling is achieved by a sequence of bone resorption and bone formation (Figure 2). This process both enables the repair of microcracks in the bone matrix and the balance between resorption and formation can be modulated for mineral homeostasis. The main cell types participating in bone remodelling are osteoblasts (bone forming cells), osteoclasts (bone resorbing cells) and osteocytes (load-sensing cells). During remodelling, these cell types are spatially and temporally organised in functional structures called basic multicellular units (BMUs). The remodelling sequence operated by a BMU follows well-defined phases. It starts with an activation phase consisting of recruitment

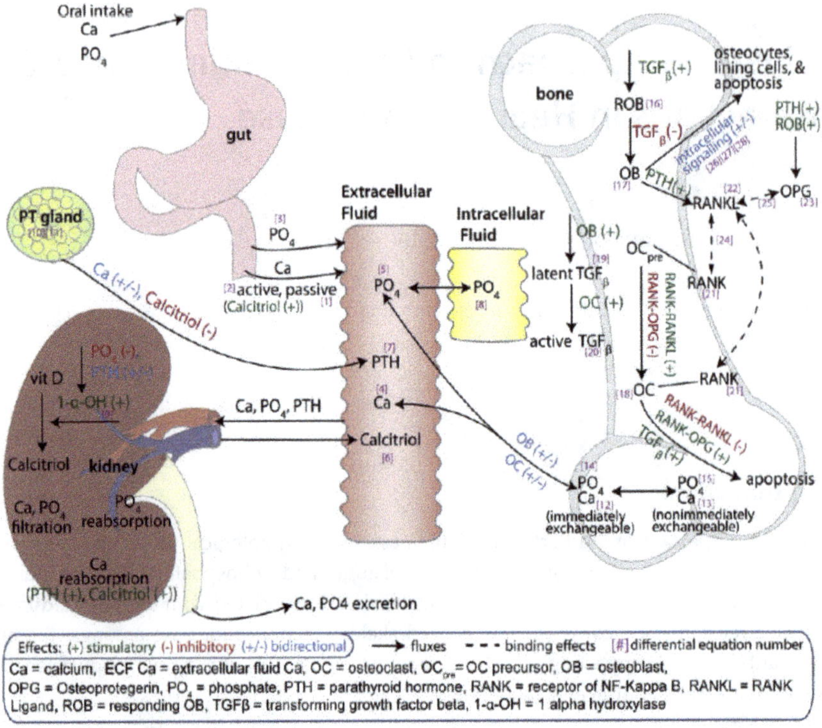

Figure 1. Schematic representation of whole body calcium regulation and local bone tissue regulation via bone remodelling (figure from Peterson and Riggs 2010, reproduced with permission).

of precursor cells. This is followed by a resorption phase characterised by active osteoclasts (OC_{a}s) removing bone matrix and a formation phase where active osteoblasts (OB_{a}s) lay down osteoid which then becomes mineralised to form new bone matrix. After completion of the local renewal of the bone matrix, bone cells become quiescent (resting phase) until a new remodelling event is initiated.

Over the past decades, a large number of regulatory factors produced by hormonal glands (such as parathyroid- and thyroid glands), lymphocytes, bone cells involved in sensing the mechanical environment (such as osteocytes and lining cells), and even tumour cells, have been shown to influence the bone remodelling sequence. These regulatory factors are essential components of the cell-cell signalling network between the bone-resorbing and bone-forming cells, and will be briefly reviewed in Section 2. Many bone disorders such as osteoporosis, Paget's disease and cancer-related bone diseases are related to disruptions in the biochemical or cellular components of this signalling network. These disruptions lead to imbalances between bone resorption and bone formation in the BMU remodelling sequence, and/or to changes in bone turnover as expressed by the activation frequency of BMUs (Riggs and Parfitt 2005).

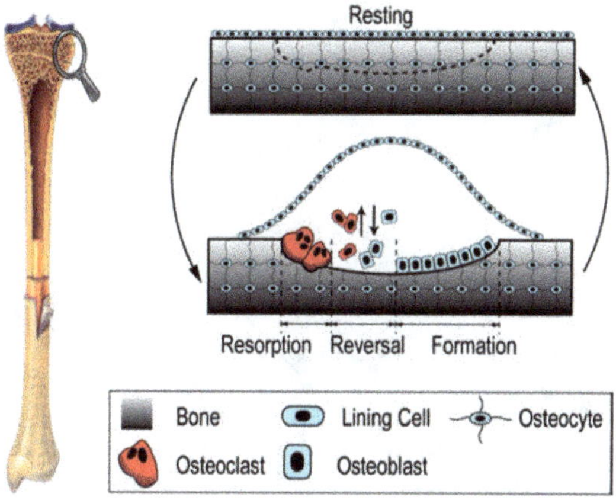

Figure 2. Bone remodelling sequence executed by different bone cell types within Basic Multicellular Units (BMUs).

Bone diseases are often multifactorial and can exhibit high inter-individual variability. It is of paramount importance to characterise and treat these diseases in a patient-specific way. Different patients exhibit different temporal evolutions of bone mass and bone cell populations. These differences can be used to define a patient-specific "disease signature", for example by the quantitative characterisation of both bone resorption rate and bone formation rate (Figure 3). The quantitative definition of such a signature is of prime importance for mathematical modelling. Mathematical modelling in biology in the next decades is expected to develop into clinical tools to help predict the evolution of a disease in an individual and to find its optimum treatment regimes. In this contribution, a mathematical model of bone cell interactions during bone remodelling is presented. This model is applied to simulate a catabolic bone disease (e.g. osteoporosis) and to investigate various treatment strategies.

A major challenge in bone biology is to understand the spatio-temporal mechanisms of action of cell-cell signalling and the interdependence between the bone structure and the bone cells actively modifying this structure at different time and length scales. Cell-cell and cell-structural interactions are complex, forming various feedback loops that are essential for bone homeostasis. Disruption of cell-cell and cell-structure feedback loops can lead to bone fractures (Chavassieux, Seeman et al. 2007). In osteoporosis for example, an important signalling pathway (i.e., the RANK-RANKL-OPG pathway – see Section 2) is strongly activated, leading to an excess of resorption over formation during remodelling (Vega, Maalouf et al. 2007). Growth factors released from the bone matrix during bone resorption further stimulate the resorptive (catabolic) pathway in a positive feedback loop, which increases remodelling activity and bone loss. To maintain bone homeostasis, formative (anabolic) pathways need to be stimulated to counterbalance the bone catabolic responses, and so to maintain skeletal integrity.

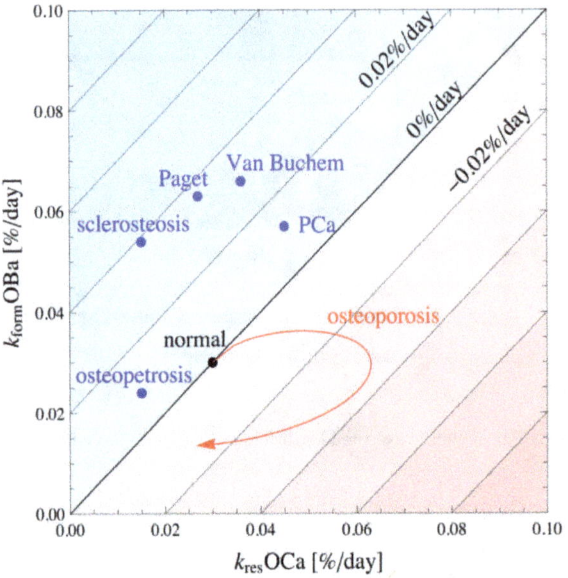

Figure 3. Rate of bone formation [%/day] versus rate of bone resorption [%/day] diagram: utilized to represent bone disease signature; red area indicates catabolic regimes while blue area indicates anabolic regimes; points on diagonal indicate bone homeostasis and rate of bone turnover.

A systems approach to bone remodelling can help advance our current understanding of the complex system formed by the bone cells and their signalling network (Cassman, Arkin et al. 2005; Smallwood 2011). Many cellular and bio-molecular behaviours have been discovered by purely experimental research. This has led to a 'profusion' of biological data, summarised in detailed "component network models". However, bone remodelling is a highly dynamic system with interdependent regulatory controls operating on multiple components. These interconnections form feed-forward and feedback controls and may lead to system behaviours that cannot be easily predicted from the behaviour of the individual components (e.g. non-linear bifurcations, instabilities, snap-through behaviours) (Strogatz 2001). Understanding how this biological system functions as a whole requires a quantitative analysis of its dynamic network behaviour by mathematical and computional modelling.

Computational approaches provide a powerful integrative methodology to investigate the complex system behaviour of bone remodelling, enabling quantitative systematic testing of various experimental and theoretical hypotheses *in-silico* (Pivonka and Komarova 2010;

Webster and Müller 2011). As such, computational modelling can be considered an additional methodology to improve our understanding of bone biology in the same way as *in-vitro* cell culture experiments and *in-vivo* animal models are experimental methodologies. In biology, *in-vitro* and *in-vivo* models are used as controlled simplifications of the studied processes. Most experimental systems investigate physiological extremes such as maximal inhibition of specific pathways, complete ablation of a gene, or multi-fold gene over-expression. The relevance of the data obtained in these experimental manipulations for human physiology relies on the assumptions that a) the modeled processes are fundamentally similar to those occurring in humans; b) the role of the quantities of interest that are studied or manipulated is more important than the role of the quantities that are kept constant or that are not measured; and c) the effects of extreme changes reflect normal physiology.

The main drawbacks and limitations of mathematical modelling are due to the complexity of the biological systems under investigation, which requires sophisticated models. Various classes of mathematical models can be used depending on the length and time scales to investigate. As a general rule, an increase in model complexity is associated with an increase in the number of model parameters. This raises the important question of calibrating and validating a mathematical model (Babuska and Oden 2004). Many parameters introduced in mathematical models have not yet been measured or are out of experimental reach using currently available methods. For these reasons, these mathematical models are often qualitative. The development of more quantitative models will rely on the collaborative efforts of experimental biologists and applied mathematicians/bioengineers.

This book chapter is organized as follows: In Section 2, a brief overview of bone biology is presented, with emphasis on the biochemical regulation of bone remodelling, bone diseases and currently available therapeutic interventions. In Section 3, we first present a general approach to modelling a system of cell interactions mediated by signalling molecules based on receptor-ligand binding reactions. This approach is then used to present a comprehensive bone cell population model for bone remodelling. In Section 4, we apply this model to simulate a catabolic bone disease (osteoporosis) and investigate different in-silico treatment strategies. In particular, we apply the receptor-ligand binding reaction scheme to include the effect of the drug denosumab, a monoclonal antibody to RANKL. Finally, we provide a summary and conclusions in Section 5.

2. Bone biology background

Bone has multiple functions within the body with the most important being to provide support to muscles, ligaments, tendons and joints to enable movement. For this function the bone needs to maintain structure, strength, and rigidity without compromising lightness. Secondly bone provides a readily accessible store of calcium to support calcium homeostasis. Other roles include providing a protected environment for bone marrow and support for a haematopoietic stem cell niche (Calvi, Adams et al. 2003), and acting as an endocrine organ regulating energy metabolism, perhaps through osteocalcin (Clemens and Karsenty 2011).

To achieve the above functions a number of cellular processes are required. For the structural demands, there is a need for a control system for the detection of bone strain and microdamage, and signalling systems for an appropriate reparative response, which would involve the directed induction of bone resorption to remove damaged bone and new bone formation to replace it (Kidd, Stephens et al. 2010). To maintain the structural integrity of bone, signalling processes are required for the coordination of bone resorption and bone formation. In fact, on the inner endosteal surfaces of bone and within Haversian systems within cortical bone, there is strong spatial and temporal coupling of bone resorption and bone formation. Bone resorption is initially induced and bone formation follows such that, by the end of this remodeling cycle, an approximate balance is achieved between the amounts of bone removed and replaced (Jaworski 1984). In contrast on periosteal bone surfaces where the main function is modelling bone, bone resorption tends to be associated with growth related shaping of the bone, and intermittent cycles of bone formation contribute to a gradual expansion in total bone diameter without the requirement for prior bone resorption.

Bone contributes also to calcium and phosphate homeostasis as an accessible internal store for these essential minerals. To fulfill the body's calcium homeostatic demands for a tightly controlled blood level of calcium ions, three organs respond to systemic hormones to provide complementary mechanisms for increasing or decreasing blood calcium levels. The loss of calcium from the kidney in urine can be altered by changing the rate of calcium reabsorption from the renal tubules. The proportion of calcium and phosphate absorbed from food can be changed, and the calcium and phosphate flux in and out of bone can be altered. Parathyroid hormone (PTH) is secreted by the parathyroid gland when blood calcium levels are low and acts on the kidney to increase calcium reabsorption from the distal tubules to decrease loss of calcium in the urine. Simultaneously it decreases phosphate reabsorption resulting in increased urinary phosphate loss. PTH also acts on bone to increase bone resorption by osteoclasts and may also induce a calcium and phosphate flux from labile calcium phosphate within bone and in particular that deposited around osteocytes (Parfitt 1976; Talmage and Mobley 2008). PTH also acts to induce activation of vitamin D in the kidney to produce 1,25 dihydroxyvitamin D which acts primarily as a hormone to increase calcium and phosphate absorption from the gut, and can increase bone resorption by osteoclasts to increase calcium and phosphate release from bone. Calcitonin is produced by the thyroid gland in response to elevated blood calcium levels and acts directly on osteoclasts to inhibit their activity thus reducing calcium and phosphate release from bone by osteoclasts though this effect is reversible (Chambers 1982). It also may be important in protecting the skeleton during prolonged bone stress due to pregnancy and lactation (Hurwitz 1996; Woodrow, Sharpe et al. 2006).

As conflicting structural and homeostatic signals may be concurrently present, there is a requirement for the ability to integrate multiple signals such that responses are tailored to the particular local bone environment; for example less structurally important bone should be targeted for calcium mobilization during times of high calcium demand as observed as preservation of trabecular rods seen in the maternal skeleton during lactation (Liu, Ardeshirpour et al. 2012).

To achieve these diverse requirements there are three highly specialized bone cells present in bone:

1. Osteocytes reside within the bone matrix and interact with bone and communicate to adjacent osteocytes via a dense and interconnecting canaliculae network containing osteocyte cell processes. Osteocytes orchestrate the detection and cellular responses to microdamage, fracture and changing bone strain and are the bone cells present in by far the largest numbers , comprising 90% of all bone cells (Lanyon 1993).
2. Osteoblasts are cells of mesenchymal origin which line the surface of bone and are capable of bone formation by progressive bone matrix deposition and subsequent mineralization. Cells of this lineage have receptors for PTH and 1,25 dihydroxyvitamin D and a number of local regulatory factors (Rodan and Martin 1981). After completion of their role in bone formation, the final fate for these cells is either cell death or further differentiation to osteocytes, which occurs concurrently with encasement within bone matrix.
3. Osteoclasts are large multinucleated cells of haematopoietic origin able to resorb bone by adhering to bone surfaces, secreting acid to demineralise the bone and proteolytic enzymes to break down the collagenous bone matrix. However, they are typically unable to respond directly to pro-resorptive hormones and require the presence of osteoblast lineage cells to locally regulate their differentiation and activity.

Communication between osteocytes, osteoblasts and osteoclasts enables a spatially and temporally coordinated and directed response through the integration of multiple catabolic and anabolic signals leading to skeletal responses to both physiological demands and pathological challenges.

2.1. Control of bone resorption

Cells of the osteoblast lineage maintain a pool of osteoclast precursors through the constitutive secretion of M-CSF which is a differentiation and survival factor for early osteoclast precursors which express the m-CSF receptor c-fms (Wiktor-Jedrzejczak, Bartocci et al. 1990). This pool of cells is then available for recruitment to the osteoclast population when needed to provide capability for either bone modeling (sculpting of bone shape), or remodeling (renewal of bone) or for initiating bone resorption to release calcium.

2.1.1. RANK-RANKL-OPG pathway

The regulation of bone resorption is through integration of multiple pro- and anti-resorptive stimuli with convergence of output into a dominant mediating pathway. Cells of the osteoblast lineage integrate multiple pro-and anti-resorptive signals, including hormonal, mechanical and pathological signals, with output through the changing balance of their expression of a cytokine, receptor activator of NFkappaB ligand (RANKL) (Yasuda et al 1998, Lacey et al 1998), and its inhibitor, osteoprotegerin (OPG) (Simonet, Lacey et al. 1997; Tsuda, Goto et al. 1997). RANKL is expressed as either a membrane bound cytokine or released in a soluble form by cells of the osteoblast lineage. Osteocytes, osteoblasts and osteoblast precursors can express RANKL and there is at present some controversy regarding which predominates. Selective knockdown of RANKL in osteocytes produces an apparently moderately severe osteopetrosis in mice indicating a significant role for osteocytes as a source for RANKL (Nakashima, Hayashi et al. 2011). RANKL binds to its receptor RANK (Nakagawa, Kinosaki et al. 1998) on the surface of osteoclast precursors inducing a number of intracellular signalling pathways to drive differentiation to an osteoclast phenotype, activating osteoclastic bone resorption and increasing osteoclast survival. OPG is a decoy receptor for RANKL, and if secreted in excess by cells of the osteoblast lineage, will bind RANKL and prevent its association with RANK, thereby inhibiting osteoclast differentiation and activity and promoting osteoclast apoptosis. Thus excess expression of RANKL relative to OPG will promote bone resorption and an excess of OPG relative to RANKL will inhibit bone resorption (http://www.rankligand.com/).

Typically, pro-resorptive hormones such as PTH and calcitriol will increase RANKL expression and decrease OPG expression within bone leading to increased bone resorption, while bone protective agents such as estradiol and testosterone tend to increase the expression of OPG relative to RANKL thus reducing bone resorption (Horwood, Elliott et al. 1998; Michael, Härkönen et al. 2005). Further, PTHrP, IL-1 and TNFα are examples of local regulatory factors which increase RANKL expression by cells of the osteoblast lineage during pathological bone loss related to cancer (PTHrP) or inflammation (IL-1 and TNFα). Local regulatory factors such as Mechanical signals, in contrast to hormones, are locally expressed and alter RANKL and OPG expression in a spatially restricted manner. Reduced bone strain or the presence of microfractures or fatigue damage, increase expression of RANKL relative to OPG, while increased bone strain tends to decrease the expression of RANKL, relative to OPG expression, by osteocytes (Kulkarni, Bakker et al. 2010). The pro-resorptive hormones and other local regulatory factors increase RANKL by binding to their own specific receptors in osteoblasts, activating the specific signalling pathways associated with these receptors, leading, through transcriptional activation of the RANKL promoter, to increased RANKL expression.

While RANKL/RANK signalling is necessary and dominant in the regulation of bone resorption, other modulating signalling molecules can signal directly to osteoclasts to magnify or diminish their responses to RANKL. Examples are inflammatory cytokines which can magnify osteoclast responses (Lam, Takeshita et al. 2000) and the systemic hormone calcitonin which is a potent but reversible direct inhibitor of osteoclast activity via activation of the calcitonin receptor on mature osteoclasts (Chambers 1982). Other sources of RANKL are also possible in bone but these tend to contribute to the regulation of bone resorption during disease. Activated t-cells in particular secrete RANKL and can induce bone resorption during infection or chronic inflammation (Teng, Nguyen et al. 2000).

2.1.2. Current therapeutic interventions of bone resorption – anti-resorptive drugs

There are a number of treatments currently clinically available and in development for inhibiting bone resorption. These target the osteoclast with different treatments inhibiting osteoclast differentiation, osteoclast function, osteoclast survival or a combination of these.

The bisphosphonate family of drugs have been available for many years and act by binding tightly to bone mineral, being internalized by osteoclasts by endocytosis during bone resorption where they act to inactivate osteoclast function and to induce osteoclast apoptosis. Bisphosphonate treated bone is characterized by reduced numbers of osteoclasts, most of which are withdrawn from the bone surface, and many of which are undergoing apoptosis (programmed cell death). The later nitrogen containing bisphosphonates such as zoledronate are highly potent and long acting inhibitors of osteoclast function, with dosing frequency of up to one year being effective (Kavanagh, Guo et al. 2006).

Calcitonin has also been available for many years and acts by reversibly inactivating osteoclasts, however it lacks the sustained activity to be effective where the pathological stimuli for bone resorption are present such as when metastatic cancer foci are present in bone, though it is used as a treatment for postmenopausal osteoporosis or when rapid suppression of bone resorption is required (Mazzuoli, Passeri et al. 1986).

A recently approved treatment for osteoporosis and metastatic bone disease is the RANKL/RANK signalling inhibitor denosumab, which is a monoclonal antibody to RANKL which mimics the action of OPG by binding RANKL to prevent its association with RANK. This protein acts to prevent osteoclast differentiation as RANKL signalling is an essential requirement for osteoclast formation. It also acts to reduce osteoclast survival as RANKL is an important survival factor for osteoclasts. Histologically bone treated with denosumab is characterized by a marked reduction in osteoclast numbers (McClung, Lewiecki et al. 2006).

There are a number of treatments in development which target osteoclast function. Inhibitors of the protease cathepsin K inhibit the ability of osteoclasts to break down bone collagen and thus inhibit their function. Bone of animals treated with cathepsin K inhibitors are characterized by the presence of many osteoclasts, and shallow resorption lacunae lined by a layer of demineralised matrix reflecting the ability of osteoclasts to demineralise bone but not to break down the matrix (Eastell, Nagase et al. 2011).

Systemic or local factor	Effect on RANKL expression	Effect on OPG expression	Effect on bone resorption
PTH	↑↑	↓↓	↑↑↑
$1,25(OH)_2D_3$	↑↑	↓↓	↑↑↑
Glucocorticoids	↑	↓↓↓	↑↑
Estradiol		↑	↓
Testosterone		↑	↓
Interleukin-1β	↑↑	↓↓	↑↑↑
TNFα	↑↑	↓↓	↑↑↑
Prostaglandins	↑↑	↓↓	↑↑↑

Table 1. Regulation of the relative expression of RANKL and OPG by a selection of systemic and local factors (from (Horwood, Elliott et al. 1998), (Tsukii, Shima et al. 1998), (Lee and Lorenzo 1999), (Hofbauer, Dunstan et al. 1998), (Brändström, Jonsson et al. 1998), (Hofbauer, Gori et al. 1999), (Hofbauer, Khosla et al. 1999), (Hofbauer, Lacey et al. 1999), (Vidal, Sjögren et al. 1998), (Michael, Härkönen et al. 2005))

2.2. Control of bone formation

In contrast to the regulation of osteoclasts, regulation of osteoblasts and bone formation is more complex with evidence of extensive signal duplication and redundancy. To control bone formation there are a large number of growth factors with a range of actions on cell behaviour. There are several classes of growth factors with a range of potent actions on osteoblasts and their precursors, and each class typically having multiple members and receptors, and a significant number of signalling inhibitors/enhancers. The regulatory functions of these can be divided into the following:

1. Inducing proliferation of the osteoblast precursor population.
2. Directing lineage commitment of pluripotent mesenchymal stem cells into the osteoblast lineage.
3. Inducing differentiation of osteoblast precursors into mature osteoblasts.
4. Regulation of osteoblast activity including both collagen deposition and matrix mineralization.
5. Directing further differentiation of osteoblasts to osteocytes with concurrent encasement in the bone matrix.

Regulation of bone formation occurs through the sequential action of growth factors favouring these functions in a signalling cascade characterized by feed forward and feed back paracrine and autocrine signalling.

2.2.1. Proliferative agents

Proliferative agents act as priming agents to increase the population of precursors available differentiate into functioning osteoblasts. Important proliferative growth factors include transforming growth factor beta (TGF-ß) (Dallas, Rosser et al. 2002), fibroblast growth factors (FGF), particularly FGF-1 and FGF-2 (Dunstan, Boyce et al. 1999), and platelet derived growth factors (PDGF) particularly PDGF-BB (Caplan and Correa 2011). These factors induce proliferation of osteoblast precursors in bone but tend to inhibit osteoblast differentiation. Thus their role is important in the early stages of osteoblast generation when the critical requirement is to generate a sufficiently large pool of osteoblast precursors from the relatively small number of mesenchymal stem cells present. Levels of these agents are thus increased transiently in the bone environment during bone resorption and after fracture, but progressive differentiation of precursors through to fully mature osteoblasts requires reduction in their levels.

The source of these factors in bone varies depending on the site and the nature of the stimulus for bone formation. Bone injury leads to infiltrates in the bone of inflammatory cells which are the source of PDGF-BB, which in turn is essential to expand the mesenchymal precursor pool for effective fracture repair. In contrast, in endosteal bone in which bone formation is required to balance bone removed by osteoclasts recruited for bone remodeling, a primary source of proliferative signals is the bone matrix itself. Osteoblasts, when forming bone, produce growth factors that are sequestered within the bone matrix. In the case of TGF-β this is as an inactive pro-form of the protein. However osteoclasts activate and release TGF-β during bone resorption, enabling it to be available to induce the proliferation of osteoblast precursors (Dallas, Rosser et al. 2002).

FGF's which are highly proliferative, and IGF's and BMP's which have limited proliferative activity but are stimulators of osteoblast differentiation are also sequestered in bone with potential for release during bone resorption.

2.2.2. *Lineage commitment factors*

The *wnt* family of growth factors and their respective receptors, co-factors and inhibitors provide one source of regulation of lineage commitment of mesenchymal stem cells and the population of osteoblast precursors expanded by the proliferative agents above. There is evidence that mature osteoblasts can produce and secrete Wnt's and these can act on early mesenchymal precursors to induce commitment to differentiation in the osteoblast lineage. Determining the details of wnt action has been complicated by the presence of 19 wnt ligands, 10 wnt receptors not to mention multiple wnt receptor co-factors and inhibitors (Kawano and Kypta 2003). BMP's which are produced by osteoblast lineage cells are sequestered in bone matrix may also contribute to lineage commitment for osteoblast precursors (Abe, Yamamoto et al. 2000), though this is difficult to separate from their differentiation activity. During the inflammation such as occurs during fracture repair, transient exposure to TNF-α may influence commitment of mesenchymal stem cell to the osteoblast lineage (Lu, Wang et al. 2012).

2.2.3. *Differentiation factors*

The BMP's, particularly BMP2, BMP4 and BMP7, are particularly potent differentiation agents with only limited proliferative activity. IGF1 acts to increase osteoblast activity and to increase survival and is also found in bone matrix. The activity of IGF's is modulated by various binding proteins while that of BMP's can be modulated by specific inhibitors such as noggin, chordin and gremlin. They are sourced in bone from osteoblastic cells as autocrine and paracrine factors, and are released from bone matrix during bone resorption. They have an important role directing the differentiation of osteoblast precursors into mature bone forming osteoblasts (Gazzerro and Canalis 2006). This involves migration of precursors to the bone surface, polarization of the cell, activation of expression of collagen type 1 and other bone matrix proteins and the development of extensive intracellular machinery for protein manufacture and secretion. Parathyroid hormone related protein (PTHrP) is a paracrine factor produced predominantly by early osteoblast precursors which acts to induce osteoblast differentiation and survival and also acts as a survival factor for osteocytes (Martin 2005; Michael, Härkönen et al. 2005). PTHrP can also mimic PTH actions to increase RANKL expression and thus also increase bone resorption. When PTHrP is produced in excessive amounts by cancer cells, it can cause profound bone loss (Suva, Winslow et al. 1987). Regulation of further differentiation of osteoblasts to osteocytes is not currently well understood, but presumably is also closely regulated by paracrine or autocrine differentiation signals.

Growth factor family	Effect on osteoblast lineage commitment	Effect on proliferation	Effect on differentiation	Sources	Signalling inhibitors/ modulators
Wnts	↑↑	↑	↑	Osteoblasts lineage cells	DKK-1, DKK2, sFRP, sclerostin etc
FGF's	↑	↑↑↑	↓↓	Bone matrix, Osteoblast lineage cells	
PDGF's	?	↑↑↑	↓↓	Inflammatory cells	
TGF-β	↑	↑↑↑	↓↓	Bone matrix	
IGF's	↑	↑↑	↑↑	Osteoblasts, Bone matrix	IGF binding proteins
BMP's	↑	↑	↑↑↑	Osteoblasts, Bone matrix	Noggin, chordin, gremlin
Prostaglandins	?	↑↑	↑↑	Osteocytes	
PTHrP			↑	Osteoblast precursors	

Table 2. Families of growth factors active in osteoblast regulation (from (Li, Hassan et al. 2008), (Rodda and McMahon 2006), (Zhou, Mak et al. 2008), (Abe, Yamamoto et al. 2000), (Tanaka, Ogasa et al. 1999), (Kawaguchi, Pilbeam et al. 1995), (Dallas, Rosser et al. 2002; Martin 2005; Miao, He et al. 2005))

Regulation of the differentiation of periosteal osteoblasts is likely to differ with regards to growth factors involved compared to those for osteoblasts in Haversian systems or on endosteal surfaces as the formation of periosteal osteoblasts is not preceded by osteoclastic bone resorption and thus growth factors stored in the bone matrix are not released to prime the osteoblast precursor population.

2.2.4. Current therapeutic agents to stimulate bone formation – anabolic drugs

There are a number of treatments that have been found to promote bone formation but it has been found essential that treatments preserve bone material properties and structure in order for improvement in bone strength. Fluoride is a potent anabolic factor used clinically for many years treating osteoporosis but was found to cause the formation of bone in which bone matrix was disorganized and poorly mineralized, and trabeculae were lacking in normal architecture. Its use significantly increased bone mineral density but was actually associated with increased fracture rates (Gutteridge, Price et al. 1990; Gutteridge, Stewart et al. 2002). Currently the most effective anabolic factor is parathyroid hormone (teriparatide) which fulfils the requirement for a factor which, at the doses used, induces the formation of new bone which retains both its normal lamellar structure and trabecular architecture and is treatment is associated with decreased risk of fracture (Paschalis, Glass et al. 2005). Interestingly, PTH is catabolic if continually present, but anabolic when given intermittently (by daily injection) (Onyia, Helvering et al. 2005). Strontium has been claimed to be a modestly anabolic factor when given orally as strontium ranelate and is approved for the treatment of osteoporosis due to it fracture prevention effects (Reginster 2002). However the mechanism of this action is not well determined.

Anabolic treatments in development include a sclerostin neutralizing antibody which in early trials promoted bone formation and increased bone mineral density (Padhi et al 2011). Presumably this antibody treatment acts by reducing the osteocyte mediated down-regulation of bone formation through the expression of the Wnt signalling inhibitor sclerostin. Interestingly one of the actions of intermittent PTH is to reduce osteocyte production of sclerostin and so PTH and the sclerostin neutralizing antibody may work in part by similar stimulatory actions on Wnt signalling (Keller and Kneissel 2005).

3. Mathematical model of bone cell interactions

The above review of some of the most important signalling pathways involved in the coupling of bone resorption and bone formation in bone remodelling indicates the complexity of the communication network between bone cells. While rapid advancement of experimental testing techniques generates large amounts of information on bone remodelling, a challenging problem is to integrate this experimental data into an understanding of system behaviour and to connect observations made at different time and length scales. In the following, an overview of mathematical models employed to integrate this different experimental information into a systems understanding of bone regulation is provided. The complexity of biological systems such as bone is characterized by the respective spatial and temporal scales involved. Due to the large variation of spatio-temporal scales involved in bone remodelling only subsets of these scales are generally investigated depending on the biological questions addressed. Mathematical modelling has been applied at the following scales:

On the *subcellular scale* one may be interested in signal transduction mechanisms such the NF-κB signalling pathway that becomes activated once RANKL binds to its receptor RANK expressed on osteoclastic cells and so triggers gene expression and consequently osteoclastic activity and cell survival. NF-κB signalling plays a pivotal role in the pathogenesis of osteolytic bone disorders. The NF-κB signalling pathway is regulated to maintain bone homeostasis by cytokines such as RANKL, TNF-α and IL-1. NF-κB activation involves stimulus-induced degradation of its inhibitor IkB, which allows for its translocation to the nucleus. Several mathematical models have been developed aimed at understanding how NF-κB translocation and IkB association/dissociation rate constants keep the majority of NF-κB in an inactive state in resting cells (Kearns, Basak et al. 2006).

At the *cell and single BMU scale* one is concerned with modelling different cell behaviour such as resorption and formation properties of bone cells and respective features of the BMU such as resorption speed, number, type and distribution of cells in the cutting and closing cone, blood vessel, etc.. A number of mathematical models have been developed to investigate single BMU behavior using either discrete (agent-based) approaches (Cacciagrano, Corradini et al. 2010; Buenzli, Jeon et al. 2012) or continuous approaches based on partial differential equations (Ryser, Komarova et al. 2010; Buenzli, Pivonka et al. 2011).

On an even larger scale in the following referred to as *tissue scale* one may want to characterize how bone mass changes with time during osteoporosis and other bone diseases. Such data can readily be obtained using micro-computed tomography (micro-CT) which relates measured quantities such as bone volume, bone surface area, trabecular numbers and spacing to a certain representative volume of bone tissue (such as 2x2x2 mm^3). Such as a volume generally contains several BMUs and through serial sectioning histology the number of bone cells in that volume can be estimated. In mathematical terms the size of the chosen volume is crucial for defining average quantities (both mechanical and biological). A volume size where a continuous mathematical description is possible is generally denoted as representative volume element (RVE). The characterisation of an RVE requires a separation of scales between microscopic constitutive elements or components and the scale of interest at which properties are represent as continuous quantities. For example estimating the local mechanical stiffness of bone requires that the RVE contains sufficiently many pores and bone matrix, yet is small enough to represent spatial inhomogeneities (such as due to local bone morphology) and to be considered infinitesimal at this scale. In this context, different mathematical models of bone cell dynamics have been proposed by Komarova et al. (Komarova, Smith et al. 2003), Lemaire et al. (Lemaire, Tobin et al. 2004) and Pivonka et al (Pivonka, Zimak et al. 2008).

On the *whole organ scale*, i.e., the skeletal scale one may be interested in the action of muscles on bones during locomotion. These types of models are commonly referred to as musculoskeletal models and allow analyzing the forces produced in bones, cartilage and ligaments due to contraction and extension of muscles together with establishing in-vivo loading conditions (Shelburne and Pandy 1997; Shim, Hunter et al. 2011).

On the largest scale, i.e., the *whole organism scale* one may be concerned with understanding mineral homeostasis. As has been pointed out in Section 2 bone plays an important role as a reservoir of calcium and phosphate. Several mathematical models have been proposed to describe calcium and phosphorous homeostasis considering different hormonal regulators such as PTH and calcitonin (Raposo, Sobrinho et al. 2002; Peterson and Riggs 2010).

The following applications are concerned with temporal aspects of bone cell interactions using a representative volume element containing several BMUs. Using this scale of representation we investigate changes in bone matrix (volume fraction) and bone cell numbers over time both under pathological conditions and using different therapeutic interventions.

3.1. Modelling cell-cell interactions and cell responses

Receptor-ligand binding, cell signalling and cell response

Bone balance and bone turnover during remodelling are critically dependent on the coordination between osteoclasts and osteoblasts. This coordination is carried out by cell–cell interactions mediated by several signalling molecules summarized in Section 2. In this section, we present a generic approach to modelling such cell–cell communication. This approach is based on fundamental chemical reaction principles of material balance and mass action kinetics, and can be applied to other biological systems (Lauffenburger and Linderman 1996; Lemaire, Tobin et al. 2004; Pivonka, Zimak et al. 2008).

Extracellular ligands (e.g. systemic hormones, autocrine and paracrine factors, growth factors etc.) modulate cell behaviour by binding to specific receptors on cells and activating intracellular signalling pathways. Intracellular signalling mechanisms can be complex and may include a cascade of interconnected downstream pathways involving protein trafficking, translocation to the nucleus, gene transcription factors, protein synthesis, etc. These intracellular mechanisms lead to an overall cell response, such as cell differentiation, proliferation, apoptosis or the expression of signalling molecules or receptors (see Figure 4). Both extracellular and intracellular signalling mechanisms are governed by the chemical reaction principles of material balance and mass action kinetics (Lauffenburger and Linderman 1996). To focus on the modelling of cell-cell interactions mediated by extracellular signalling molecules, the approach followed in this book chapter will treat extracellular and intracellular signalling differently. Extracellular signalling will be modelled explicitly by considering receptor–ligand binding reactions governed by mass action kinetics. By contrast, intracellular signalling will be modelled phenomenologically by assuming that it leads to an overall relation between the "input signal" perceived by a cell via extracellular receptor-ligand binding and the cell response (Figure 4).

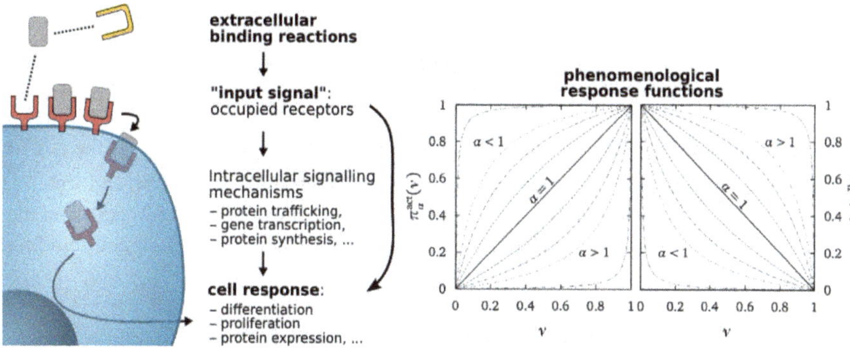

Figure 4. Modelling of cell responses using receptor ligand binding reactions; phenomenological response functions can either promote (π^{act}) or repress (π^{rep}) a certain cell response; relationship between input signal and output signal can be linear (α=1) or non-linear (α≠1).

Extracellular signalling: competitive binding of receptors and ligands

Receptors enable a cell to sense its local biochemical environment. If a ligand binds to a cell surface receptor and particularly if it becomes endocytosed (e.g. activating intracellular signalling mechanisms), that ligand becomes unavailable to signal to the other cells. We refer to this effect as "competitive binding". Competitive binding influences the amount of ligand binding to a single cell in presence of other cells, and so influences the "input signal" that a cell may perceive from its micro-environment. An advantage of considering explicit extracellular receptor–ligand binding reactions in a mathematical model is to account for such competitive binding in a systematic way. Competitive binding may occur either amongst identical receptors (e.g. RANKL binding RANK on several osteoclasts), or amongst different kinds of receptors binding the same ligand (e.g. RANKL binding either RANK or OPG). In the following, we first consider a simple receptor–ligand chemical reaction to illustrate the concept of competitive binding amongst identical receptors. We will then present a more general situation applicable to the RANK–RANKL–OPG signalling system, where there may be competitive binding between several types of RANKL (e.g. expressed on osteoblasts, on osteocytes, or in soluble form) and several types of receptors, (e.g. RANK, OPG and denosumab).

We first consider a single ligand L, assumed to be produced at rate P_L and degraded at rate D_L. This ligand may bind to a single receptor R, produced at rate P_R and degraded at rate D_R (production and degradation may include the generation and apoptosis of cells expressing the ligand or receptor). After binding, the receptor and ligand form a bound complex, which we denote by $\widehat{RL}$. We consider that this complex may either unbind (reverse reaction) or may be endocytosed, in which case both the receptor and the ligand become unavailable for further extracellular reactions. This case effectively corresponds to a disappearance of the bound complex from the extracellular environment. The chemical reaction flows corresponding to this receptor-ligand binding situation can thus be summarised as:

$$\overset{P_R\downarrow}{\underset{D_R\downarrow}{R}} + \overset{P_L\downarrow}{\underset{D_L\downarrow}{L}} \underset{k_{RL}^r}{\overset{k_{RL}^f}{\rightleftharpoons}} \widehat{RL} \xrightarrow[\text{(endocytosis)}]{D_{RL}} \varnothing \tag{1}$$

where $k_{RL}^f, k_{RL}^r > 0$ are the forward and reverse binding reaction rates and D_{RL} is the rate at which the bound complex is degraded from the extracellular environment by endocytosis. Assuming first order reaction rates (i.e. rates proportional to the amount of reactant) and the law of mass action, the rate equations corresponding to this receptor–ligand binding situation are:

$$\begin{aligned}
\partial_t \widehat{RL} &= k_{RL}^f R \cdot L - k_{RL}^r \widehat{RL} - D_{RL} \widehat{RL} \\
\partial_t R &= P_R - D_R R + k_{RL}^r \widehat{RL} - k_{RL}^f R \cdot L \\
\partial_t L &= P_L - D_L L + k_{RL}^r \widehat{RL} - k_{RL}^f R \cdot L
\end{aligned} \tag{2}$$

where the symbols R, L and $\widehat{RL}$ denote the free (unbound) receptor, free ligand and bound complex concentrations (number of molecules per unit volume, e.g. mol/L). Receptor–ligand binding reactions occur on a fast time scale compared to characteristic times of cell behaviours. Local concentrations of receptors and ligands quickly converge to a quasi-steady state (Lemaire, Tobin et al. 2004; Pivonka, Zimak et al. 2008; Buenzli, Pivonka et al. 2012). This steady-stated is determined by setting time derivatives to zero in the system of ordinary differential equations (ODEs) Eqs. (2), which leads to the following system of algebraic equations:

$$R^{st} = \frac{P_R}{D_R + \frac{D_{RL}}{k_{RL}} L^{st}}, \quad L^{st} = \frac{P_L}{D_L + \frac{D_{RL}}{k_{RL}} R^{st}}, \quad \widehat{RL}^{st} = \frac{1}{k_{RL}} R^{st} L^{st} \tag{3}$$

where

$$k_{RL} = \frac{k_{RL}^r + D_{RL}}{k_{RL}^f} \tag{4}$$

is a parameter specific to the binding reactions (1). Equation (3) indicates that competitive binding effects only occur if there is a degradation of the bound complex, i.e. if $D_{RL} > 0$. Indeed, if $D_{RL} = 0$ (no degradation of the bound complex by endocytosis), then the steady-state concentration of free receptors is independent of the amount of free ligands and the steady-state concentration of free ligands is independent of the amount of free receptors. In fact, without bound complex endocytosis, the assumed production and degradation of the receptors and ligands act as "equilibrating reservoirs". This also explains why the steady-state concentrations do not depend on the initial amount of reactants.

The above description can be generalised to a situation in which a number of different kinds of receptors $R = R_1, R_2, \ldots$ may bind a number of different kinds of ligands $L = L_1, L_2, \ldots$. The binding reactions (1) hold for all R and L and lead to the following system of algebraic equations for the steady-state concentrations of receptors, ligands and bound complexes (dropping the steady-state superscript "st" from the notation):

$$R = \frac{P_R}{D_R + \sum_L \frac{D_{RL}}{k_{RL}} L}, \quad L = \frac{P_L}{D_L + \sum_R \frac{D_{RL}}{k_{RL}} R}, \quad \widehat{RL} = \frac{1}{k_{RL}} R \cdot L \tag{5}$$

for all $R = R_1, R_2, \ldots$ and $L = L_1, L_2, \ldots$. As before, competitive binding effects may occur provided the bound complexes are degraded (i.e. $D_{RL} > 0$). Indeed, in that case the existence of several receptors to bind the same ligand lowers the free ligand concentration and the existence of several ligands to bind the same receptor lowers the free receptor concentration. We note here that the production and degradation rates of the ligands and receptors may depend themselves on the concentrations of ligands and receptors in the above formulas, to account for example for saturation or self-limiting mechanisms whereby production of a receptor or ligand stops if they are already available in sufficient amounts (see below the production rates of RANKL and OPG). A closed form or analytical solution of the system of equations (5) is not always possible and a numerical solution is often needed. For example, substituting the first equation into the second in (5) leads to a system of equations involving only the ligand concentrations L. If there is only one ligand type, this reduces to a single polynomial equation in L whose order is equal to the number of receptors binding that ligand plus one.

Intracellular signalling and cell response

The raw "input signal" communicated to a cell by extracellular signalling depends on the number of receptors on the cell that are bound to a ligand (occupied receptors). It is convenient to take as input signal the fraction ν_{RL} of occupied receptors on the cell, equal to

$$\nu_{RL} = \frac{\widehat{RL}}{R_{tot}} = \frac{\widehat{RL}}{R + \widehat{RL}} = \frac{L}{k_{RL} + L}, \tag{6}$$

where R_{tot} is the concentration of both free and bound receptors. The fraction of occupied receptors on a cell only depends on the concentration of extracellular free ligand L and a single binding parameter k_{RL}. In absence of bound complex degradation (i.e. if D_{RL}=0), k_{RL} is the so-called dissociation binding constant (Lauffenburger and Linderman 1996).

Cells may behave in several different ways in response to extracellular signalling molecules. A signalling molecule may either "activate" or "repress" a certain cell response, such as differentiation, proliferation, apoptosis, or the expression of other signalling molecules and receptors (see e.g. Tables 1 and 2). In a mathematical model, these cell responses are associated with certain model parameters. An effective way of modelling the response of a cell to a signalling molecule is to specify a phenomenological relation between the strength of the input signal to that cell (its fraction of occupied receptors) and the model parameter representing the cell response. This relation integrates potentially complex intracellular signalling mechanisms that are not explicitly modelled, and so is of phenomenological nature. In fact, such a relation is prone to experimental determination, and has been measured experimentally in other contexts.[1] In the following, "activation" or "repression" of a cell response by a ligand will always be represented by modulating the model parameter associated with that cell response with a function of the fraction of occupied receptors. To this effect, we introduce two classes of functions of ν_{RL}: activator functions $\pi_\alpha^{\text{act}}(\nu_{RL})$ and repressor functions $\pi_\alpha^{\text{rep}}(\nu_{RL})$, where α is a shape parameter (see Figure 4). These activator and repressor functions are defined as follows:

$$\pi_{\text{act}}^{\alpha}(\nu) = \frac{\nu}{\alpha + \nu(1-\alpha)}, \qquad \pi_{\text{rep}}^{\alpha}(\nu) = 1 - \pi_{\text{act}}^{\alpha}(\nu) = \frac{\alpha(1-\nu)}{\nu + \alpha(1-\nu)}, \qquad \alpha > 0. \tag{7}$$

---

[1] For example, in the context of human fibroblasts stimulated by epidermal growth factor (EGF), the mitogenic response of the fibroblasts has been shown to depend linearly on the fraction of occupied EGF receptors Lauffenburger, D. A. and J. J. Linderman (1996). Receptors: Models for Binding, Trafficking, and Signaling Oxford, Oxford University Press., Fig. 6-7, p.249.

For example, if D is a model parameter representing a rate of cell differentiation, its "activation" by a signalling molecule L will be represented by $D\pi_{\alpha}^{\mathrm{act}}(\nu_{RL})$, whereas its "repression" by L will be represented by $D\pi_{\alpha}^{\mathrm{rep}}(\nu_{RL})$. Since receptor occupancy only depends on the free ligand concentration, one can also directly write the overall relation between free ligand concentration and the strength of the cell response from Eqs. (6) and (7):

$$\pi_{\mathrm{act}}^{L} \equiv \pi_{\mathrm{act}}^{\alpha}\left(\nu_{RL}(L)\right) = \frac{L}{\kappa_{RL}+L}, \qquad \pi_{\mathrm{rep}}^{L} \equiv \pi_{\mathrm{rep}}^{\alpha}\left(\nu_{RL}(L)\right) = \frac{1}{\kappa_{RL}+L}, \qquad \text{where } \kappa_{RL} = \alpha k_{RL} \quad (8)$$

These functions have the same form as the fraction of occupied receptors ν_{RL} and the fraction of unoccupied receptors $1-\nu_{RL}$ (see Eq. (6), except that the parameter κ_{RL} corresponds to a rescaling of the binding parameter k_{RL} by the shape parameter α. In this way, a cell is able to respond to a same receptor stimulus with different potencies in different behaviours. In the bone cell population model presented below, we will use the activator and repressor functions in the form (8). Table 3 in the appendix lists both the receptor-ligand binding parameters k_{RL} and the parameters κ_{RL} involved in the phenomenological relations between receptor occupancy and cell response.

3.2. Bone cell population model

In this section we give a summary of the bone cell population models developed by our group (Pivonka, Zimak et al. 2008; Pivonka, Zimak et al. 2010; Buenzli, Pivonka et al. 2012). The presented model takes into account the RANK-RANKL-OPG signalling pathway between bone cells, the action of TGF-β on different types of bone cells and pre-osteoblast proliferation. A revised formulation of competitive binding reactions in the RANK-RANKL-OPG system is presented, which is used to also include the effect of the anti-resorptive drug denosumab. A schematic picture of the model is presented in Figure 5.

Changes in porosity and bone matrix fraction due to cell activity

The activity of osteoclasts and osteoblasts leads to the removal and deposition of new bone. This activity modifies the volume fraction of bone matrix in the bone tissue ($f_{\mathrm{bm}}=V_{\mathrm{bm}}/V_{\mathrm{RVE}}$), or equivalently the bone porosity ($f_{\mathrm{vas}}=V_{\mathrm{pores}}/V_{\mathrm{RVE}}=1-f_{\mathrm{bm}}$). Osteoblasts deposit osteoid, a collagen-rich substance which later mineralizes into new bone. Primary mineralisation of osteoid is fast, i.e., 70% of the maximum mineral density is reached within a few days in humans (Parfitt 1983). Given the much larger time spans involved in evolution of bone diseases, we assume that osteoblasts "instantaneously" deposit "fully" mineralised new bone matrix as a first approximation. We further assume that the resorption rate of bone matrix k_{res} by an individual active osteoclast (in volume per unit time) and the formation rate of new bone matrix deposition k_{form} by an individual osteoblast (in volume per unit time) are constant. The evolution of the bone matrix volume fraction is thus given by

$$\partial_t f_{\mathrm{bm}} = k_{\mathrm{form}} \cdot \mathrm{OB_a} - k_{\mathrm{res}} \cdot \mathrm{OC_a} \qquad (9)$$

where OB_a and OC_a denote the density of active osteoblasts and active osteoclasts respectively (number of cells per unit volume). These densities are determined by biochemical and cellular processes such as differentiation from precursor cell types and apoptosis under the control of several signalling molecules. Such processes are governed in general by the material balance equation expressed for each cell type. The equations governing the evolution of the bone cell densities are presented in the following.

Governing equations for bone cell densities

From the osteoblastic lineage the following cell types are taken into account: uncommitted osteoblasts (OB_u) (representing a constant pool of mesenchymal stem cells or bone marrow stromal cells), osteoblast precursor cells (OB_p), and active osteoblasts (OB_a). From the osteoclastic lineage the following cell types are taken into account: osteoclast precursor cells (OC_p), and active osteoclasts (OC_a). The material balance of each cell type (in the representative volume element under consideration) is determined by specifying their production rate (source term) and elimination rate (sink term). The production and elimination of cells considered here are schematically represented in Figure 5 as flows between the different cell types. Cell differentiation accounts both for an elimination (outflux) from the pool of precursor cells and for a production (influx) into the pool of differentiated cells. In our model, OB_us differentiate into OB_ps in presence of TGF-β, but the differentiation of OB_ps in to OB_as is repressed by TGF-β. Osteoclasts become active upon RANKL binding to their receptor RANK. The availability of RANKL is in turn determined by the concentration of OPG and of the systemic hormone PTH (see below). The material balance equations for the densities of OB_p, OB_a and OC_a thus take the form of a system of rate equations with first-order production and elimination rates (i.e. rates proportional to the densities of cells) to account for the population size:

$$\partial_t \mathrm{OB_p} = D_{\mathrm{OB_u}} \cdot \mathrm{OB_u} \cdot \pi_{\mathrm{act,OB_u}}^{\mathrm{TGF\text{-}\beta}} + \mathcal{P}_{\mathrm{OB_p}} \cdot \mathrm{OB_p} - D_{\mathrm{OB_p}} \cdot \mathrm{OB_p} \cdot \pi_{\mathrm{rep,OB_p}}^{\mathrm{TGF\text{-}\beta}} \tag{10}$$

$$\partial_t \mathrm{OB_a} = D_{\mathrm{OB_p}} \cdot \mathrm{OB_p} \cdot \pi_{\mathrm{rep,OB_p}}^{\mathrm{TGF\text{-}\beta}} - A_{\mathrm{OB_a}} \cdot \mathrm{OB_a} \tag{11}$$

$$\partial_t \mathrm{OC_a} = D_{\mathrm{OC_p}} \cdot \mathrm{OC_p} \cdot \pi_{\mathrm{act,OC_p}}^{\mathrm{RANKL}} - A_{\mathrm{OC_a}} \cdot \mathrm{OC_a} \cdot \pi_{\mathrm{act,OC_a}}^{\mathrm{TGF\text{-}\beta}} \tag{12}$$

where D_{OB_u}, D_{OB_p} and D_{OC_p} denote the differentiation rates of uncommitted osteoblast progenitors, osteoblast precursor cells, uncommitted osteoclast progenitors and osteoclast precursor cells, respectively. $\mathcal{P}_{\mathrm{OB_p}}$ is the proliferation rate of osteoblast precursor cells, assumed to be self-limited by the current population of OB_ps: $\mathcal{P}_{\mathrm{OB_p}} = P_{\mathrm{OB_p}}(1 - \mathrm{OB_p} / \mathrm{OB_p^{sat}})$ (Buenzli, Pivonka et al. 2012). A_{OB_a} and A_{OC_a} denote the apoptosis rates of active osteoblasts and active osteoclasts. Note that OB_us and OC_ps are assumed to be constant and so are not part of the state variables. In the above equations, the signalling molecules TGF- β and RANKL influence cell differentiation and apoptosis through the activator and repressor regulatory functions introduced in Eq. (8). In particular, $\pi_{\mathrm{act,OB_u}}^{\mathrm{TGF\text{-}\beta}}$, $\pi_{\mathrm{rep,OB_p}}^{\mathrm{TGF\text{-}\beta}}$, and $\pi_{\mathrm{act,OC_a}}^{\mathrm{TGF\text{-}\beta}}$

represent an activation or repression by TGF-β of osteoblast differentiation and osteoclast apoptosis, whereas $\pi_{\mathrm{act,OC_p}}^{\mathrm{RANKL}}$ represents the activation by RANKL of osteoclast differentiation.

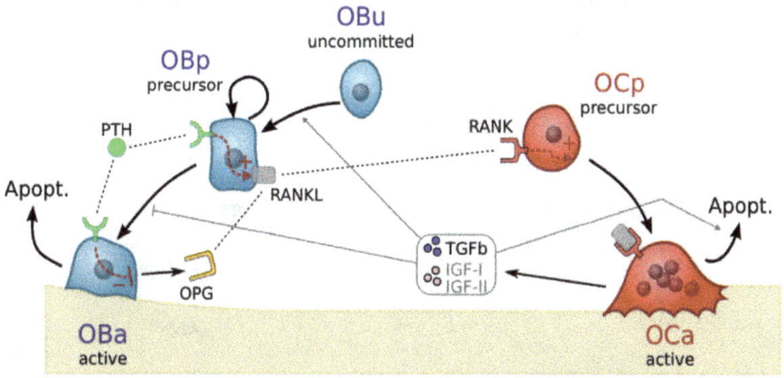

Figure 5. Mathematical model of bone cell interactions taking into account different bone cell types and biochemical regulatory mechanism including the RANK-RANKL-OPG signaling pathway, proliferation of osteoblast precursor cells and action of TGF-β on bone cells.

Binding reactions in the RANK-RANKL-OPG system

We now exemplify the general approach presented earlier for modelling competitive binding reactions between receptors and ligands in the RANK–RANKL–OPG system, including also the anti-resorptive drug denosumab. To simplify, we only consider one type of RANKL, expressed on the membrane of $\mathrm{OB_p}$s. However, we consider that RANKL may bind with three kinds of receptors, i.e. RANK (bound to the membrane of $\mathrm{OC_p}$s), OPG and denosumab (soluble molecules). The equations (5) governing the steady-state concentration of RANKL, RANK, OPG and denosumab are fully determined once the production rate P_{RANKL} of RANKL, as well as the production rate of its receptors, P_R, R=RANK, OPG, denosumab are specified. All degradation rates (D_{RANKL} and D_R) are assumed constant (see Table 3 in the appendix). The formulation of the equations governing the binding reactions in the RANK-RANKL-OPG system presented below differs from the formulation used in (Lemaire, Tobin et al. 2004) and (Pivonka, Zimak et al. 2008; Pivonka, Zimak et al. 2010). In these previous works, competitive binding was only partially implemented.[2] A fully consistent approach is considered here instead.

The expression of RANKL on $\mathrm{OB_p}$ is known to be increased by the systemic hormone PTH (see Table 1). Accordingly, we assume that PTH is activating the production rate $\beta_{\mathrm{OBp}}^{\mathrm{RANKL}}$ of RANKL by $\mathrm{OB_p}$s.[3] To account for a limited space per cell where RANKL can be expressed (i.e. a limited carrying capacity), the production rate of RANKL is also multiplied by a

[2] For example, while the concentration of RANKL was dependent on that of OPG, the concentration of OPG was not dependent on that of RANKL.

[3] In Lemaire et al. 2004 and Pivonka et al. 2008 regulation by PTH was assumed to limit the so-called 'carrying capacity' of RANKL on a cell instead, but there is no biological evidence for this fact. A more direct regulatory effect of PTH is assumed here.

saturation function such that production stops when the number $N_{\mathrm{OB_p}}^{\mathrm{RANKL}}$ of RANKL molecules on the $\mathrm{OB_p}$ reaches a maximum number, $N_{\mathrm{OB_p,max}}^{\mathrm{RANKL}}$. Finally, to account for the population size of $\mathrm{OB_p}$ expressing RANKL, the production rate of RANKL is multiplied by the density of $\mathrm{OB_p}$ cells, such that:

$$P_{\mathrm{RANKL}} = \beta_{\mathrm{OBp}}^{\mathrm{RANKL}} \pi_{\mathrm{act}}^{\mathrm{PTH}} \left(1 - \frac{N_{\mathrm{OBp}}^{\mathrm{RANKL}}}{N_{\mathrm{OBp,max}}^{\mathrm{RANKL}}}\right) \mathrm{OB_p} \tag{13}$$

where the total number of RANKL ligands (free and bound) present on the membrane of an $\mathrm{OB_p}$ cell, $N_{\mathrm{OBp}}^{\mathrm{RANKL}}$, is given by

$$N_{\mathrm{OBp}}^{\mathrm{RANKL}} = \frac{\mathrm{RANKL^{tot}}}{\mathrm{OB_p}}, \quad \text{where } \mathrm{RANKL^{tot}} = \mathrm{RANKL} + \sum_R \widehat{R\,\mathrm{RANKL}} = \mathrm{RANKL}\left(1 + \sum_R \frac{R}{k_{R,RANKL}}\right) \tag{14}$$

The expression of OPG by $\mathrm{OB_a}$s is known to be inhibited by PTH (see Table 1). Accordingly, we assume that PTH is repressing the production rate of OPG per OBa cell, $\beta_{\mathrm{OBa}}^{\mathrm{OPG}}$. Following (Pivonka, Zimak et al. 2008), the endogeneous production of OPG is also assumed to be self-limited, and so the production rate of OPG is multiplied by a saturation function, such that production stops when the concentration of OPG reaches a maximum concentration $\mathrm{OPG_{max}}$. Finally, to account for the population size of $\mathrm{OB_a}$ expressing OPG, the production rate of OPG is multiplied by the density of $\mathrm{OB_a}$ cells. Hence:

$$P_{\mathrm{OPG}} = \beta_{\mathrm{OBa}}^{\mathrm{OPG}} \pi_{\mathrm{rep}}^{\mathrm{PTH}} \left(1 - \frac{\mathrm{OPG}}{\mathrm{OPG_{max}}}\right) \mathrm{OB_a} \tag{15}$$

Following Refs (Lemaire, Tobin et al. 2004; Pivonka, Zimak et al. 2008), the receptor RANK is assumed be present on $\mathrm{OC_p}$s in constant numbers. Thus the production rate of RANK is proportional to that of $\mathrm{OC_p}$ cells:

$$P_{\mathrm{RANK}} = N_{\mathrm{OCp}}^{\mathrm{RANK}} P_{\mathrm{OCp}} \tag{16}$$

where $N_{\mathrm{OCp}}^{\mathrm{RANK}}$ is the number of RANK receptors per $\mathrm{OC_p}$ cell and P_{OCp} is the production rate of $\mathrm{OC_p}$s. The degradation rate of RANK receptors (whether free or bound) corresponds here to the differentiation rate of $\mathrm{OC_p}$s into $\mathrm{OC_a}$s, D_{OCp} (in the model, active osteoclasts are not assumed to express RANK receptors for simplicity). Therefore, one has $D_{\mathrm{RANK}} = D_{\mathrm{RANK,RANKL}} = D_{\mathrm{OC_p}}$ in the first equation in (5) such that only the ratio $P_{\mathrm{RANK}}/D_{\mathrm{RANK}}$ occurs in that equation. With Eq. (14), this ratio is equal to:

$$\frac{P_{\mathrm{RANK}}}{D_{\mathrm{RANK}}} = N_{\mathrm{OC_p}}^{\mathrm{RANK}} \mathrm{OC_p} \tag{17}$$

since the density of $\mathrm{OC_p}$s is equal to $P_{\mathrm{OCp}}/D_{\mathrm{OCp}}$.

The production rate of denosumab $P_{\text{denosumab}}^{\text{dosing}}$ will be set as given function of time according to the chosen dosing regime of the drug. All the binding and degradation properties of denosumab are assumed identical to those of OPG. In summary, the equations governing the concentrations of RANKL, RANK, OPG and denosumab are:

$$
\begin{aligned}
\text{RANKL} &= \frac{P_{\text{RANKL}}}{D_{\text{RANKL}} + \sum\limits_{R=\text{RANK,OPG,denosumab}} \frac{D_{R,RANKL}}{k_{R,RANKL}} R} \\
\text{RANK} &= \frac{N_{\text{OCp}}^{\text{RANK}}\, \text{OC}_{\text{p}}}{1 + \frac{1}{k_{\text{RANK, RANKL}}} \text{RANKL}} \\
\text{OPG} &= \frac{P_{\text{OPG}}}{D_{\text{OPG}} + \frac{D_{\text{OPG, RANKL}}}{k_{\text{OPG, RANKL}}} \text{RANKL}} \\
\text{denosumab} &= \frac{P_{\text{denosumab}}^{\text{dosing}}}{D_{\text{denosumab}}}
\end{aligned}
\tag{18}
$$

where P_{RANKL} and P_{OPG} are given by Eqs. (13)–(15) and all the parameters values are listed in Table 3 in the appendix. A closed form solution for these equations is not possible. Equations (10)–(12) governing the evolution of the cell populations and Eqs (18) form a system of differential algebraical equations (DAE) that need to be solved numerically.

4. Numerical simulations of bone diseases and therapeutic interventions

In this section we outline how bone diseases and drug treatments can be investigated in a computational modelling framework such as the bone cell population model of bone remodelling presented in Section 3. We refer to these computational experiments as "in-silico disease modelling" and "in-silico drug treatment modelling". We will then exemplify such in-silico experiments in our bone cell population model by considering a simple case of an osteoporotic condition, modelled by a decrease in the endogeneous OPG production rate. While osteoporosis is often associated with dysregulations in several pathways or components, such as increased osteocyte apoptosis due to reduced estrogen levels in post-menopausal osteoporosis, a decrease in endogeneous OPG production in our model leads to a high turnover rate with a catabolic bone imbalance, which is characterisitic of the first stage of age-related osteoporosis. Simple treatments of this disease will be first simulated by prescribing constant changes in model parameters in an attempt to restore bone volume (Section 4.1). The discussion of these simple in-silico treatments is largely based on the work by (Pivonka, Zimak et al. 2010). Finally, we will investigate a temporal in-silico treatment mimicking the anti-resorptive action of the drug denosumab with different dosing regimes (Section 4.2).

To model a disease in-silico, it is important that the computational model contains suitable parameters associated with the known pathophysiology of the disease. This emphasises the importance of including a comprehensive set of signalling molecules between bone cells when modelling osteoporosis and other disorders of bone remodelling. As has been pointed out in Section 2, many catabolic bone diseases are associated with disruptions in the RANK-RANKL-OPG pathway. However, several model parameters are involved in this pathway and so several model parameters may be modified to model such a disease. While this complicates the modelling of in-silico diseases, it also opens up the possibility for patient-specific disease modelling. The measurements of different bone properties from a patient, such as the temporal evolution of the bone matrix f_{bm} from micro-CT scans, bone turnover rates from histological estimates of bone cell numbers in biopsies or from serum levels of turnover markers, can be used to estimate the evolution of model parameters required to simulate the evolution of the disease in this particular patient (e.g. using optimisation algorithms). Often, however, the time evolution of such properties cannot be known with precision. In a situation where for example only two time points are known in a single bone property of a patient, one may assume a constant change in a single model parameter by lack of further information. This is the situation considered in the model of osteoporosis simulated below, where it is assumed that the bone matrix volume fraction is known only at the onset of the disease (t_0) and at one further time point after disease progression and before treatment (t_1) (see Figure 6). An in-silico therapy is then applied in a second step starting from the time point t_1 to treat the disease. The effect of this therapy is checked at a later time point (t_2).

To model realistic in-silico therapies, is it crucial to identify the specific action of a therapeutic drug on the various biochemical or cellular components of the system, and to know the pharmacokinetic properties of that drug, such as the rapidity of its clearance from the system. Similarly to a disease, a drug may affect a single component or multiple components in a biological system. Numerically simulating the effect of a drug can help understand the overall action of the drug, especially when responses of different characteristic times are involved in different affected components. However, simpler in-silico therapies, in which the effect of a constant change in a model parameter is studied in a sensitivity analysis, are also worthwhile to investigate. Such theoretical treatments can give insights into the most effective therapeutic strategies and potential drug targets.

The identification of successful treatment strategies is critical for the development of clinical applications of in-silico therapies. Bone therapeutic guidelines currently classify therapeutic agents into anabolic or anti-resorptive drugs based on their effect on bone volume (or bone mass) and bone turnover (see (Riggs and Parfitt 2005) and Section 2 for more details). Using the terminology of (Riggs and Parfitt 2005) anti-resorptive drugs (such as bisphosphonates and denosomab) increase bone volume, but lead to low bone turnover, whereas anabolic drugs (such as intermittent PTH) strongly increase bone volume while producing high bone turnover. Depending on the current bone volume and bone turnover state of a particular patient, a treatment regime combining both types of drugs may thus be advised. Optimisation algorithms could help a clinician find the "optimal" treatment regime for that patient. In Figure 6, three therapeutic treatments are schematically shown to successfully restore bone volume and bone turnover rate (bone cell numbers), however with different efficiencies. The "ideal" treatment (red curve) recovers all of the bone lost and normal cell

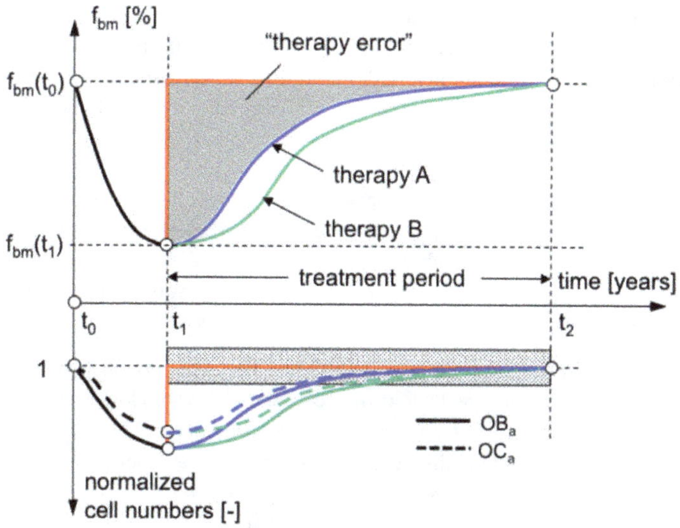

Figure 6. Schematic representation of catabolic bone diseases and applied therapy: (a) volume fraction of bone matrix [%] vs. time [years] and (b) normalized bone cell numbers [-] vs. time [years] (t_0 = onset of bone disease; t_1 = start treatment; t_2 ... end treatment; ideal therapy ... red curve; "example" therapy ... blue curve; therapy error ... gray shaded area between ideal therapy and typical therapy, Figure modified from Pivonka et al 2010).

numbers instantaneously. Clearly this is impossible, and realistic treatments (blue and green curves) may only restore the normal state after some period of time. However, different treatments will usually exhibit different time courses for bone volume and cell numbers. These differences may be used in the search for optimal treatment regimes. For example, one may want to minimise the area between the "ideal treatment" curve and the actual treatment curve, i.e. to minimise the "therapy error" (gray area in Figure 6). Weighted combinations of several criteria, for example involving turnover rate and cell numbers, can also be used to define an objective "therapy error" to minimise.

4.1. Simple in-silico treatment of osteoporosis in the bone cell population model: constant changes in parameter values

In this subsection we extend some of the results presented in (Pivonka, Zimak et al. 2010) using the improved model of bone cell interactions presented in Section 3, which includes proliferation of osteoblast precursor cells (Buenzli, Pivonka et al. 2012) and competitive receptor-ligand binding reactions in the RANK-RANKL-OPG system. In (Pivonka, Zimak et al. 2010), we showed that many bone disorders that act via the RANK-RANKL-OPG system induce more pronounced osteoclast responses than osteoblast responses. This indicates that targeting the RANK-RANKL-OPG pathway is not very effective at triggering bone formative responses. Furthermore, it was shown in this study that the severity of catabolic bone diseases strongly depended on how many components of the RANK-RANKL-OPG pathway were assumed to be affected. Changes of single parameters in the RANK–RANKL–OPG system led to less severe bone loss compared to multiple concurrent changes of parameters. In a subsequent study we have also shown that bone disorders related to excessive bone formation (such as van Buchem disease and sclerosteosis) may act through a different pathway, with the Wnt pathway being a major candidate (Buenzli, Pivonka et al. 2012). Regulation of osteoblast differentiation and proliferation by Wnt were shown to be very effective mechanisms to induce bone formative responses.

As mentioned above, modifying the value of a model parameter as an in-silico therapeutic treatment does not represent a realistic drug treatment scenario. However, evaluating the sensitivity of the bone response to such changes provides insights into the general mechanism of action of a potential therapeutic agent and can help identify the most effective pathways that should be targeted for disease intervention, even if such treatments may not currently exist. In Figure 7, we apply such constant changes in model parameters to restore the bone loss induced by a simulated osteoporotic condition. This osteoporotic condition is simulated by a reduction in endogeneous OPG production (i.e. a reduction in the value of the model parameter $\beta_{OB_a}^{OPG}$) assumed to develop instantaneously at time t_0=0 from a healthy state. The evolution of the bone volume fraction f_{bm} is followed in a representative volume element located in cortical bone with initial bone volume fraction $f_{bm}(t_0)$=95%. The reduction in OPG production is chosen such that a 5% of bone volume fraction is lost after one year, i.e. at t_1=365 days. At time t_1, different in-silico treatments are commenced by changing single model parameters with the objective to restore bone volume over a period of four years (i.e., at time t_2=(1+4)·365=1825 days). Changes in the model parameters D_{OBu}, P_{OBp}, D_{OBp}, A_{OBa} (targeting osteoblast development), in A_{OCa} and in the binding parameter between RANK and RANKL, $k_{RANK\text{-}RANKL}$ (targeting osteoclast development) have been investigated.

The numerical results shown in Figure 7 clearly show that model parameters associated with osteoblast development have a strong potential to restore bone volume within the prescribed treatment period of four years. In fact, the numbers of OB_as and OC_as are increased in these situations (not shown) while bone balance is positive. The combination of high bone turnover rate and positive bone balance enables a quick restoration of bone volume. According to the classification by (Riggs and Parfitt 2005), these model parameters can therefore be characterised as having a pro-anabolic potential.

On the other hand, parameters associated with osteoclast development (including parameters associated with the RANK-RANKL-OPG pathway) have a weak potential to restore bone volume. Increasing osteoclast apoptosis A_{OCa} (which can be associated with the action of a bisphosphonate) or reducing the binding affinity between RANK and RANKL

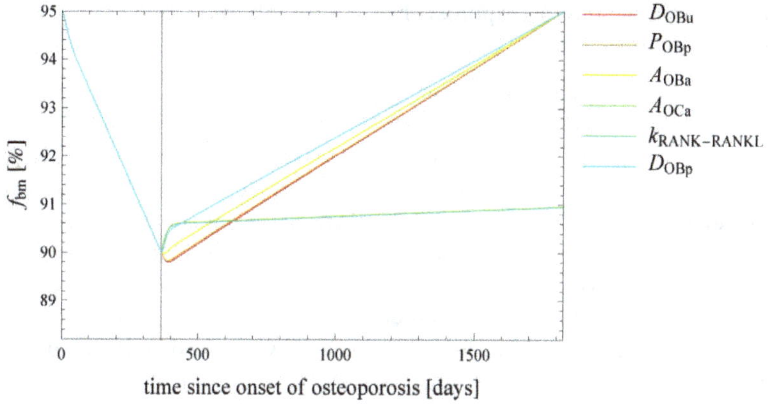

Figure 7. Single constant parameter therapies for treatment of catabolic bone disease due to constant reduction in OPG production rate leading to 5% bone loss over one year: (a) bone matrix volume fraction [%] vs. time [days] and (b) normalized cell numbers [-] vs. time [days] (D_{OBu}=differentiation of uncommitted osteoblasts, P_{OBp}=proliferation of osteoblast precursors, A_{OBa}=apoptosis of active osteoblasts, A_{OCa}=apoptosis of active osteoclasts, $k_{D,RANK-ARNKL}$=dissociation binding constant RANK-RANKL, D_{OBp}=differentiation of osteoblast precursors).

via the parameter $k_{RANK-RANKL}$ fail to restore bone volume within the prescribed treatment period of four years. In fact, the numbers of OB_as and OC_as are significantly reduced in these situations. Thus, although bone balance is positive, bone turnover rate is too low for this positive bone balance to build up bone volume fast enough. According to the classification by (Riggs and Parfitt 2005), these parameters can be characterised as having an anti-resorptive potential. The initial strong formative response is due to an immediate reduction of active osteoclasts without corresponding reduction in active osteoblasts. After a period of about 60 days, this initial reduction in active osteoclasts leads to a marked depletion of TGF-β and so to decreased osteoblastogenesis. This depletes RANKL which further enhances the initial reduction of active osteoclasts and leads to a low bone turnover state in the second part of the bone response.

According to the definition of "therapy error" presented in Figure 6, the most effective treatment is achieved via manipulation of the differentiation rate of osteoblast precursor cells (D_{OBp}). However, the ability of D_{OBp} to restore bone volume depends strongly on the proliferative potential of pre-osteoblasts, i.e. on the ratio P_{OBp}/D_{OBu} (Buenzli, Pivonka et al. 2012). At low ratios P_{OBp}/D_{OBu}, an increase in D_{OBp} depletes the pool of pre-osteoblasts enough to lead to a low bone turnover state unable to restore bone volume within the four years treatment period (data not shown). At higher ratios P_{OBp}/D_{OBu}, this depletion occurs at higher values of D_{OBp} and bone volume can still be restored within the treatment period as shown in Figure 7. The results of Figure 7 suggest that an alternative efficient treatment minimising therapy error in the evolution of bone volume may consist of first using an anti-resorptive treatment during the first few months, followed by a pro-anabolic treatment.

4.2. In-silico treatment of osteoporosis by denosumab in the bone cell population model

Finally, we investigate a therapeutic strategy that mimicks the effect of the anti-resorptive drug denosumab on the simulated osteoporotic condition of Section 4.1. Although a realistic model of denosumab treatment should take into account pharmacokinetic properties, here, we simply assume that denosumab is produced in the bone compartment according to its dosing regime and that it is eliminated at a constant rate (see Eq (18)). The explicit consideration of denosumab as a new biochemical component of the system associated with its own dynamics enables to investigate temporal dosing regimes consistently. Moreover, the explicit consideration of denosumab enables competitive binding effects between denosumab, RANK and OPG to bind RANKL to be fully accounted for.

Three different dosing regimes of denosumab are shown in Figure 8, all commencing at time t_1=365 days. In Figure 8a and 8b, a continuous infusion is assumed, i.e. the production rate $P_{\text{denosumab}}$ of denosumab is constant. The influence of the infusion rate ("dose", in concentration per unit time) on the evolution of the bone volume fraction is shown in Figure 8a. The influence of the infusion rate for the bone volume fraction reached after four years of treatment is shown in Figure 8b. As one may expect, the lowest infusion rate (3 pM/day) has the least effect to restore bone loss. Interestingly, a medium infusion rate (7 pM/day) is more efficient to restore bone loss than lower or higher infusion rates, an effect that would be difficult to predict without computational modelling. The similarity of the evolution of the bone volume fraction in Figure 8a and those seen in Figure 7 when modifying anti-resorptive parameters is due to the fact that denosumab acts via the RANK-RANKL-OPG pathway.

In Figure 8c, a single injection of denosumab is simulated. This situation is modelled by specifying the production rate $P_{\text{denosumab}}(t)$ as a function having a single peak at time t_1=365 days. The height of the peak is taken as a measure of the dose (in concentration per unit time) and the width of the peak at half height is taken to be two days. The dosing rate is seen to influence both the maximum bone volume fraction attained and the final gain in bone volume compared to the untreated case (dotted line). Higher denosumab doses lead to larger bone gain and reduce bone loss for a longer period, although a saturation of the dose dependence is quickly reached from about 100 pM/day upwards.

Finally, dosing regimes made of multiple injections spaced by regular time intervals are shown in Figure 8d. Each injection of these dosing regimes is simulated by the case 100 pM/day of Figure 8c. The dosing regime with a small administration interval (every 5 days) is comparable to constant denosumab infusion. Administration of denosumab every 10 days still has the potential to inhibit significant bone loss. However, an administration interval of 30 days at the same dose is not sufficient to efficiently inhibit bone loss.

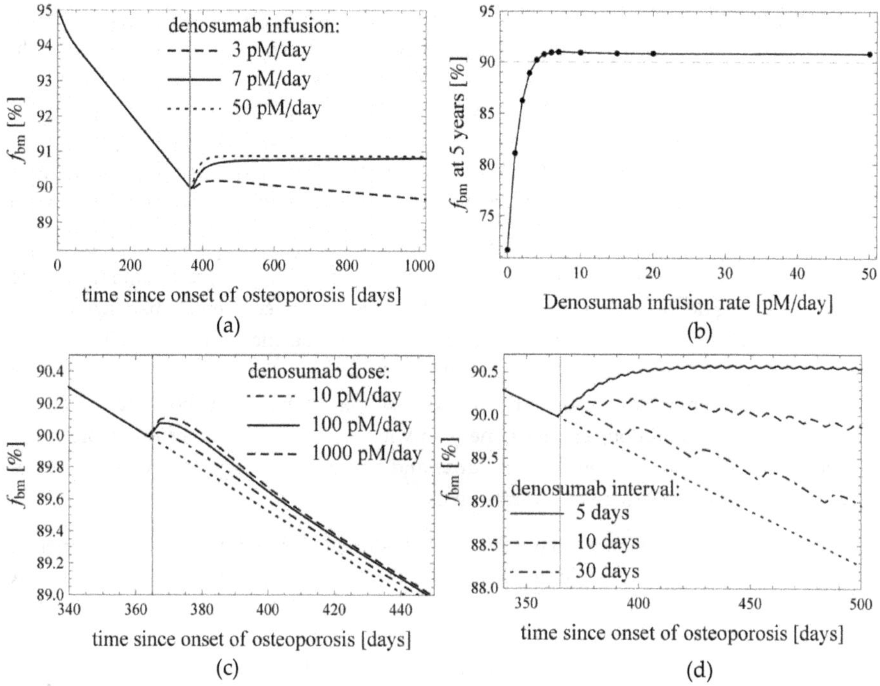

Figure 8. Single time dependent parameter therapy mimicking pharmacokinetic action of denosumab for treatment of catabolic bone disease due to constant reduction in OPG production rate leading to 5% bone loss over one year: (a) bone matrix f_{bm} [%] vs. time [days] for constant denosumab doses, (b) bone matrix f_{bm} [%] vs. denosumab infusion rate [pM/day], (c) bone matrix f_{bm} [%] vs. time [days] for singe injections of denosumab and (d) bone matrix f_{bm} [%] vs. time [days] for multiple injections of denosumab at a dose of 100 pM/day.

5. Summary and conclusions

In this contribution we presented a mathematical model describing bone remodelling taking into account complex bone cell interactions. A general approach to modelling cell-cell signalling and cell behaviour was proposed. This approach considers explicit extracellular signalling mechanisms and receptor-ligand binding reactions governed by the law of mass action and material balance, while intracellular signalling mechanisms are assumed to be captured by phenomenological response functions. We highlighted the fact that different patients may exhibit different bone disease patterns which can be captured in a so-called bone disease signature diagram. Within the presented framework patient specificity can be accounted for via adjustments of model parameters.

As an application of the model, we simulated an osteoporotic condition associated with a disruption in the RANK-RANKL-OPG pathway and considered various 'in-silico' therapeutic treatments. The notion of "therapy error" was introduced to identify the most efficient ways to restore normal bone mass and bone turnover from a diseased condition. In these therapeutic interventions, we could identify two groups of model parameters: (i) anti-resorptive parameters characterised by low bone turnover and (ii) pro-anabolic parameters characterised by high bone turnover. Anti-resorptive parameters corresponded to parameters of the RANK-RANKL-OPG pathway and parameters associated to osteoclast developement. Pro-anabolic parameters corresponded to parameters associated with osteoblast development.

Finally, temporal treatment regimes were investigated using a simple model mimicking the action of the drug denosumab. Competitive binding effects between denosumab, RANK and OPG to bind RANKL could be fully accounted for by the specification of explicit receptor-ligand binding reactions in the RANK-RANKL-OPG system. For constant administration regimes, there is an intermediate optimal dose leading to maximum bone gain. For the single and multiple injection regimes bone gain was stronger with higher doses or higher administration frequencies, however, the dose and administration frequency dependences of the bone response were seen to saturate quickly.

Future advancement of bone biology research will strongly rely on how well experimental and theoretical groups are able to communicate and collaborate with each other. A prerequisite for such collaborations is the mutual understanding of research methodologies employed. We hope that this chapter will provide some guidance on how theoretical tools such as mathematical modeling can be used in biomedical research in general, and particularly in bone biology research.

Appendix

Symbol	Value	Description
OC_p	$1x10^{-3}$ pM	pre-osteoclast density
OB_u	$1x10^{-3}$ pM	uncommitted osteoblast progenitors density
k_{res}	200 $pM^{-1}day^{-1}$	daily volume of bone matrix resorbed per osteoclast
k_{form}	40 $pM^{-1}day^{-1}$	daily volume of bone matrix formed per osteoblast
$n_{TGF-\beta}^{bone}$	$1x10^{-2}$ pM	concentration of TGF-β stored in the bone matrix
D_{OC_p}	2.1/day	$OC_p \rightarrow OC_a$ differentiation rate parameter
A_{OC_a}	5.65/day	OC_a apoptosis rate parameter
D_{OB_u}	0.7/day	$OB_u \rightarrow OB_p$ differentiation rate parameter
D_{OB_p}	0.166/day	$OB_p \rightarrow OB_a$ differentiation rate parameter
P_{OB_p}	0.054/day	OB_p proliferation rate parameter
A_{OB_a}	0.211/day	OB_a apoptosis rate
$\kappa_{OB_a,PTH}$	150 pM	Response parameter for RANKL activation by PTH on OB_a
$\kappa_{OB_p,PTH}$	0.2226 pM	Response parameter for OPG repression by PTH on OB_p

$K_{OC_p,RANKL}$	16.65 pM	Response parameter for activation of OC_p differentiation by RANKL
$K_{OB_u,TGF\text{-}\beta}$	5.63x10-4 pM	Response parameter for activation of OB_u differentiation by TGF-β
$K_{OB_p,TGF\text{-}\beta}$	1.75x10-4 pM	Response parameter for repression of OB_p differentiation by TGF-β
$N_{OC_p}^{RANK}$	10000	Number of RANK receptors on an OC_p cell
OPG_{max}	$2x10^{+8}$ pM	Concentration at which endogeneous production of OPG stops
$N_{OB_a}^{RANKL}$	$2.7\ 10^{+6}$	Carrying capacity of RANKL on an OB_a cell
β_{PTH}	250 pM/day	Systemic production rate of PTH
$\beta_{OB_a}^{RANKL}$	$1.64245x10^{+5}$/day	Production rate of RANKL by an OB_a cell
$\beta_{OB_p}^{OPG}$	$1.625x10^{+8}$/day	Production rate of OPG by an OB_p cell
D_{PTH}	86/day	Degradation rate of PTH
D_{TGFb}	1/day	Degradation rate of TGF-β
D_{RANKL}	10.1325/day	Degradation rate of RANKL
D_{OPG}	0.35/day	Degradation rate of OPG
$D_{denosumab}$	0.35/day	Degradation rate of denosumab
$D_{RANK,RANKL}$	10.1325/day	Degradation rate of the bound complex $\widehat{RANK\,RANKL}$
$D_{OPG,RANKL}$	10.1325/day	Degradation rate of the bound complex $\widehat{OPG\,RANKL}$
$D_{denosumab,RANKL}$	10.1325/day	Degradation rate of the bound complex $\widehat{denosumab\,RANKL}$
$k_{RANK,RANKL}$	29.3 pM	Binding parameter between RANK and RANKL
$k_{OPG,RANKL}$	1000 pM	Binding parameter between OPG and RANKL
$k_{denosumab,RANKL}$	1000 pM	Binding parameter between denosumab and RANKL

Table 3. Parameters of the model

Author details

P. Pivonka and P.R. Buenzli

Faculty of Engineering, Computing and Mathematics, University of Western Australia, Australia

C.R. Dunstan

Department of Biomedical Engineering, University of Sydney, Australia

Acknowledgement

The authors are grateful to Prof T.J. Martin (St. Vincent's Institute of Medical Research, University of Melbourne) for his feedback in preparing this manuscript and to Dr S. Scheiner (Vienna University of Technology) for fruitful discussions. The work of A/Prof. Pivonka is supported by the Australian Research Council (ARC) for discovery project funding (DP0988427).

6. References

Abe, E., M. Yamamoto, et al. (2000). "Essential Requirement of BMPs-2/4 for Both Osteoblast and Osteoclast Formation in Murine Bone Marrow Cultures from Adult Mice: Antagonism by Noggin." *Journal of Bone and Mineral Research* 15(4): 663-673.

Babuska, I. and J. T. Oden (2004). "Verification and validation in computational engineering and science: basic concepts." *Computer Methods in Applied Mechanics and Engineering* 193(36-38): 4057-4066.

Brändström, H., K. B. Jonsson, et al. (1998). "Regulation of Osteoprotegerin mRNA Levels by Prostaglandin E2in Human Bone Marrow Stroma Cells." *Biochemical and Biophysical Research Communications* 247(2): 338-341.

Buenzli, P. R., J. Jeon, et al. (2012). "Investigation of bone resorption within a cortical basic multicellular unit using a lattice-based computational model." *Bone* 50(1): 378-389.

Buenzli, P. R., P. Pivonka, et al. (2012). "Modelling the anabolic response of bone using a cell population model." *Journal of Theoretical Biology* 37: 42–52.

Buenzli, P. R., P. Pivonka, et al. (2011). "Spatio-temporal structure of cell distribution in cortical Bone Multicellular Units: A mathematical model." *Bone* 48(4): 918-926.

Cacciagrano, D., F. Corradini, et al. (2010). Bone Remodelling: A Complex Automata-Based Model Running in BioShape, Springer Berlin / Heidelberg. 6350: 116-127.

Calvi, L. M., G. B. Adams, et al. (2003). "Osteoblastic cells regulate the haematopoietic stem cell niche." *Nature* 425(6960): 841-846.

Caplan, A. I. and D. Correa (2011). "PDGF in bone formation and regeneration: New insights into a novel mechanism involving MSCs." *Journal of Orthopaedic Research* 29(12): 1795-1803.

Cassman, M., A. Arkin, et al. (2005). WTEC Panel Report on International Research and Development in Systems Biology, World Technology Evaluation Centre (WTEC): 1-184.

Chambers, T. J. (1982). "Osteoblasts relase osteoclasts from calcitonin-induced quiescence " *J. Cell Sci.* 57(Oct.): 247-260.

Chavassieux, P., E. Seeman, et al. (2007). "Insights into Material and Structural Basis of Bone Fragility from Diseases Associated with Fractures: How Determinants of the Biomechanical Properties of Bone Are Compromised by Disease." *Endocr Rev* 28(2): 151-164.

Clemens, T. L. and G. Karsenty (2011). "The osteoblast: An insulin target cell controlling glucose homeostasis." *Journal of Bone and Mineral Research* 26(4): 677-680.

Dallas, S. L., J. L. Rosser, et al. (2002). "Proteolysis of Latent Transforming Growth Factor-Î² (TGF-Î²)-binding Protein-1 by Osteoclasts." *Journal of Biological Chemistry* 277(24): 21352-21360.

Dunstan, C. R., R. Boyce, et al. (1999). "Systemic Administration of Acidic Fibroblast Growth Factor (FGF-1) Prevents Bone Loss and Increases New Bone Formation in Ovariectomized Rats." *Journal of Bone and Mineral Research* 14(6): 953-959.

Eastell, R., S. Nagase, et al. (2011). "Safety and efficacy of the cathepsin K inhibitor ONO-5334 in postmenopausal osteoporosis: The OCEAN study." *Journal of Bone and Mineral Research* 26(6): 1303-1312.

Gazzerro, E. and E. Canalis (2006). "Bone morphogenetic proteins and their antagonists " *Reviews in Endocrine & Metabolic Disorders* 7(1-2): 51-65.

Gutteridge, D. H., R. I. Price, et al. (1990). "Spontaneous hip fractures in fluoride-treated patients: Potential causative factors." *Journal of Bone and Mineral Research* 5(S1): S205-S215.

Gutteridge, D. H., G. O. Stewart, et al. (2002). "A randomized trial of sodium fluoride (60 mg) +/- estrogen in postmenopausal osteoporotic vertebral fractures: increased vertebral fractures and peripheral bone loss with sodium fluoride; concurrent estrogen prevents peripheral loss, but not vertebral fractures." *Osteoporos Int.* 13(2): 158-170.

Hofbauer, L. C., C. R. Dunstan, et al. (1998). "Osteoprotegerin Production by Human Osteoblast Lineage Cells Is Stimulated by Vitamin D, Bone Morphogenetic Protein-2, and Cytokines." *Biochemical and Biophysical Research Communications* 250(3): 776-781.

Hofbauer, L. C., F. Gori, et al. (1999). "Stimulation of Osteoprotegerin Ligand and Inhibition of Osteoprotegerin Production by Glucocorticoids in Human Osteoblastic Lineage Cells: Potential Paracrine Mechanisms of Glucocorticoid-Induced Osteoporosis." *Endocrinology* 140(10): 4382-4389.

Hofbauer, L. C., S. Khosla, et al. (1999). "Estrogen Stimulates Gene Expression and Protein Production of Osteoprotegerin in Human Osteoblastic Cells*." *Endocrinology* 140(9): 4367-4370.

Hofbauer, L. C., D. L. Lacey, et al. (1999). "Interleukin-1beta and tumor necrosis factor-alpha, but not interleukin-6, stimulate osteoprotegerin ligand gene expression in human osteoblastic cells." *Bone* 25(3): 255-259.

Horwood, N. J., J. Elliott, et al. (1998). "Osteotropic Agents Regulate the Expression of Osteoclast Differentiation Factor and Osteoprotegerin in Osteoblastic Stromal Cells." *Endocrinology* 139(11): 4743.

Hurwitz, S. (1996). "Homeostatic Control of Plasma Calcium Concentration." *Critical Reviews in Biochemistry and Molecular Biology* 31(1): 41-100.

Jaworski, Z. F. G. (1984). "Coupling of Bone Formation to Bone Resorption." *Calcif Tissue Int* 36: 531-535.

Kavanagh, K. L., K. Guo, et al. (2006). "The molecular mechanism of nitrogen-containing bisphosphonates as antiosteoporosis drugs." *Proceedings of the National Academy of Sciences* 103(20): 7829-7834.

Kawaguchi, H., C. C. Pilbeam, et al. (1995). "The role of prostaglandins in the regulation of bone metabolism." *Clin Orthop Relat Res.* 313(Apr.): 36-46.

Kawano, Y. and R. Kypta (2003). "Secreted antagonists of the Wnt signalling pathway." *Journal of Cell Science* 116(13): 2627-2634.

Kearns, J. D., S. Basak, et al. (2006). "IκBε provides negative feedback to control NF-κB oscillations, signaling dynamics, and inflammatory gene expression." *The Journal of Cell Biology* 173(5): 659-664.

Keller, H. and M. Kneissel (2005). "SOST is a target gene for PTH in bone." *Bone* 37(2): 148-158.

Kidd, L. J., A. S. Stephens, et al. (2010). "Temporal pattern of gene expression and histology of stress fracture healing." *Bone* 46(2): 369-378.

Komarova, S. V., R. J. Smith, et al. (2003). "Mathematical model predicts a critical role for osteoclast autocrine regulation in the control of bone remodeling." *Bone* 33(2): 206-215.

Kulkarni, R., A. Bakker, et al. (2010). "Inhibition of Osteoclastogenesis by Mechanically Loaded Osteocytes: Involvement of MEPE." *Calcified Tissue International* 87(5): 461-468.

Lam, J., S. Takeshita, et al. (2000). "TNF-α induces osteoclastogenesis by direct stimulation of macrophages exposed to permissive levels of RANK ligand." *J Clin Invest.* 106(12): 1481–1488.

Lanyon, L. E. (1993). "Osteocytes, strain detection, bone modeling and remodeling." *Calcified Tissue International* 53(0): S102-S107.

Lauffenburger, D. A. and J. J. Linderman (1996). *Receptors: Models for Binding, Trafficking, and Signaling* Oxford, Oxford University Press.

Lauffenburger, D. A. and J. J. Linderman (1996). *Receptors: models for binding, trafficking, and signaling*. Oxford, Oxford University Press.

Lee, S.-K. and J. A. Lorenzo (1999). "Parathyroid Hormone Stimulates TRANCE and Inhibits Osteoprotegerin Messenger Ribonucleic Acid Expression in Murine Bone Marrow Cultures: Correlation with Osteoclast-Like Cell Formation." *Endocrinology* 140(8): 3552-3561.

Lemaire, V., F. L. Tobin, et al. (2004). "Modeling the interactions between osteoblast and osteoclast activities in bone remodeling." *Journal of Theoretical Biology* 229(3): 293-309.

Li, Z., M. Q. Hassan, et al. (2008). "A microRNA signature for a BMP2-induced osteoblast lineage commitment program." *Proceedings of the National Academy of Sciences* 105(37): 13906-13911.

Liu, X. S., L. Ardeshirpour, et al. (2012). "Site-specific changes in bone microarchitecture, mineralization, and stiffness during lactation and after weaning in mice." *Journal of Bone and Mineral Research* 27(4): 865-875.

Lu, Z., G. Wang, et al. (2012). "Short-Term Exposure to Tumor Necrosis Factor-Alpha Enables Human Osteoblasts to Direct Adipose Tissue-Derived Mesenchymal Stem Cells into Osteogenic Differentiation " *Stem Cells and Development*: (accepted for publication).

Martin, T. J. (2005). "Osteoblast-derived PTHrP is a physiological regulator of bone formation." *The Journal of Clinical Investigation* 115(9): 2322-2324.

Mazzuoli, G. F., M. Passeri, et al. (1986). "Effects of salmon calcitonin in postmenopausal osteoporosis: a controlled double-blind clinical study." *Calcif Tissue Int.* 38(1): 3-8.

McClung, M. R., E. M. Lewiecki, et al. (2006). "Denosumab in Postmenopausal Women with Low Bone Mineral Density." *New England Journal of Medicine* 354(8): 821-831.

Miao, D., B. He, et al. (2005). "Osteoblast-derived PTHrP is a potent endogenous bone anabolic agent that modifies the therapeutic efficacy of administered PTH 1â€"34." *The Journal of Clinical Investigation* 115(9): 2402-2411.

Michael, H., P. L. Härkönen, et al. (2005). "Estrogen and Testosterone Use Different Cellular Pathways to Inhibit Osteoclastogenesis and Bone Resorption." *Journal of Bone and Mineral Research* 20(12): 2224-2232.

Nakagawa, N., M. Kinosaki, et al. (1998). "RANK Is the Essential Signaling Receptor for Osteoclast Differentiation Factor in Osteoclastogenesis." *Biochemical and Biophysical Research Communications* 253(2): 395-400.

Nakashima, T., M. Hayashi, et al. (2011). "Evidence for osteocyte regulation of bone homeostasis through RANKL expression." *Nat Med* 17(10): 1231-1234.

Onyia, J. E., L. M. Helvering, et al. (2005). "Molecular profile of catabolic versus anabolic treatment regimens of parathyroid hormone (PTH) in rat bone: An analysis by DNA microarray." *Journal of Cellular Biochemistry* 95(2): 403-418.

Parfitt, A. M. (1976). "The actions of parathyroid hormone on bone: relation to bone remodeling and turnover, calcium homeostasis, and metabolic bone diseases. II. PTH and bone cells: bone turnover and plasma calcium regulation." *Metabolism: clinical and experimental* 25(8): 909-55.

Parfitt, A. M. (1983). The physiological and clinical significance of bone histomorphometric data. *Bone histomorphometry: Techniques and interpretation*. R. R. Recker. Boca Raton, CRC Press: 143–223.

Paschalis, E. P., E. V. Glass, et al. (2005). "Bone Mineral and Collagen Quality in Iliac Crest Biopsies of Patients Given Teriparatide: New Results from the Fracture Prevention Trial." *Journal of Clinical Endocrinology & Metabolism* 90(8): 4644-4649.

Peterson, M. C. and M. M. Riggs (2010). "A physiologically based mathematical model of integrated calcium homeostasis and bone remodeling." *Bone* 46(1): 49-63.

Pivonka, P. and S. V. Komarova (2010). "Mathematical modeling in bone biology: From intracellular signaling to tissue mechanics." *Bone* 47(2): 181-189.

Pivonka, P., J. Zimak, et al. (2008). "Model structure and control of bone remodeling: A theoretical study." *Bone* 43(2): 249-263.

Pivonka, P., J. Zimak, et al. (2010). "Theoretical investigation of the role of the RANK-RANKL-OPG system in bone remodeling." *Journal of Theoretical Biology* 262(2): 306-316.

Raposo, J. F., L. G. Sobrinho, et al. (2002). "A Minimal Mathematical Model of Calcium Homeostasis." *Journal of Clinical Endocrinology & Metabolism* 87(9): 4330-4340.

Reginster, J.-Y. (2002). "Strontium Ranelate in Osteoporosis " *Curr. Pharm. Des.* 8(21): 1907-1916.

Riggs, B. L. and A. M. Parfitt (2005). "Drugs Used to Treat Osteoporosis: The Critical Need for a Uniform Nomenclature Based on Their Action on Bone Remodeling." *Journal of Bone and Mineral Research* 20(2): 177-184.

Rodan, G. and T. Martin (1981). "Role of osteoblasts in hormonal control of bone resorption—A hypothesis." *Calcified Tissue International* 33(1): 349-351.

Rodda, S. J. and A. P. McMahon (2006). "Distinct roles for Hedgehog and canonical Wnt signaling in specification, differentiation and maintenance of osteoblast progenitors." *Development* 133(16): 3231-3244.

Ryser, M. D., S. V. Komarova, et al. (2010). "The Cellular Dynamics of Bone Remodeling: A Mathematical Model " *SIAM J. on Applied Mathematics* 70(6): 1899-1921.

Shelburne, K. B. and M. G. Pandy (1997). "A musculoskeletal model of the knee for evaluating ligament forces during isometric contractions." *Journal of Biomechanics* 30(2): 163-176.

Shim, V. B., P. J. Hunter, et al. (2011). "A Multiscale Framework Based on the Physiome Markup Languages for Exploring the Initiation of Osteoarthritis at the Bone-Cartilage Interface." *Biomedical Engineering, IEEE Transactions on* 58(12): 3532-3536.

Simonet, W. S., D. L. Lacey, et al. (1997). "Osteoprotegerin: A Novel Secreted Protein Involved in the Regulation of Bone Density." *Cell* 89(2): 309-319.

Smallwood, R. (2011). Understanding the dynmaics of biological systems: lessons learned from integrative systems biology. *Cell-centred modeling of tissue behaviour*. W. Dubitzky, J. Southgate and H. Fuss, Springer.

Strogatz, S. H. (2001). *Non-linear dynmaics and chaos: applications to physics, biology, chemistry and engineering*, Westview Press.

Suva, L. J., G. A. Winslow, et al. (1987). "A parathyroid hormone-related protein implicated in malignant hypercalcemia: cloning and expression." *Science* 237(4817): 893-896.

Talmage, R. V. and H. T. Mobley (2008). "Calcium homeostasis: Reassessment of the actions of parathyroid hormone." *General and Comparative Endocrinology* 156(1): 1-8.

Tanaka, H., H. Ogasa, et al. (1999). "Actions of bFGF on mitogenic activity and lineage expression in rat osteoprogenitor cells: effect of age." *Molecular and Cellular Endocrinology* 150(1â€"2): 1-10.

Teng, A. Y.-T., H. Nguyen, et al. (2000). "Functional human T-cell immunity and osteoprotegerin ligand control alveolar bone destruction in periodontal infection." *J. Clin. Invest.* 106(6): R59–R67

Tsuda, E., M. Goto, et al. (1997). "Isolation of a Novel Cytokine from Human Fibroblasts That Specifically Inhibits Osteoclastogenesis." *Biochemical and Biophysical Research Communications* 234(1): 137-142.

Tsukii, K., N. Shima, et al. (1998). "Osteoclast Differentiation Factor Mediates an Essential Signal for Bone Resorption Induced by 1alpha,25-Dihydroxyvitamin D3, Prostaglandin E2, or Parathyroid Hormone in the Microenvironment of Bone." *Biochemical and Biophysical Research Communications* 246(2): 337-341.

Vega, D., N. M. Maalouf, et al. (2007). "The Role of Receptor Activator of Nuclear Factor-{kappa}B (RANK)/RANK Ligand/Osteoprotegerin: Clinical Implications." *J Clin Endocrinol Metab* 92(12): 4514-4521.

Vidal, O. N. A., K. Sjögren, et al. (1998). "Osteoprotegerin mRNA Is Increased by Interleukin-1 alpha in the Human Osteosarcoma Cell Line MG-63 and in Human Osteoblast-Like Cells." *Biochemical and Biophysical Research Communications* 248(3): 696-700.

Webster, D. and R. Müller (2011). "In silico models of bone remodeling from macro to nano—from organ to cell." *Wiley Interdisciplinary Reviews: Systems Biology and Medicine* 3(2): 241-251.

Wiktor-Jedrzejczak, W., A. Bartocci, et al. (1990). "Total absence of colony-stimulating factor 1 in the macrophage-deficient osteopetrotic (op/op) mouse." *PNAS* 87(12): 4828-32.

Woodrow, J. P., C. J. Sharpe, et al. (2006). "Calcitonin Plays a Critical Role in Regulating Skeletal Mineral Metabolism during Lactation." *Endocrinology* 147(9): 4010-4021.

Zhou, H., W. Mak, et al. (2008). "Osteoblasts Directly Control Lineage Commitment of Mesenchymal Progenitor Cells through Wnt Signaling." *Journal of Biological Chemistry* 283(4): 1936-1945.

Section 5

Methods in Cell Interactions

Chapter 9

Development of Dioctadecyldimethylammonium Bromide/Monoolein Liposomes for Gene Delivery

J. P. Neves Silva, A. C. N. Oliveira, A. C. Gomes and M. E. C. D. Real Oliveira

Additional information is available at the end of the chapter

1. Introduction

The electrostatic repulsion between the phosphodiester anionic charges of nucleic acids (NA) and the negatively-charged headgroups of cell membrane phospholipids hinders naked NA to permeate the plasma membrane [1, 2]. Additionally, nucleases present in the cells and in biological fluids enzymatically degrade NA, limiting their biofunctionality. Although many alternative methods have been developed to deliver NA to cells, factors such as versatility [3, 4], applicability [5, 6] and efficiency [7, 8] have discouraged their disseminated use in gene therapy.

Given these limitations, the intracellular delivery of genetic material can only be achieved through the use of physical, biological or chemical methods that promote gene insertion into cells. Physical methods have generally low *in vivo* applicability and include direct injection of NA into organs of live animals, micro projectile biolistics with a gene-propelling gun, cell sonication using an ultrasonic transducer and cell electroporation by exposure to an electric field. Biological methods rely on attenuated or inactivated versions of adenoviruses, lentiviruses and retroviruses, whose deactivated components can be used as gene vectors to obtain relatively high *in vivo* transfection efficiencies (transfection mediated by viral vectors) [9, 10]. Viral vectors pose nevertheless important safety, toxicity and immunogenicity issues, which greatly limit their use in humans. Chemical methods are based on the use of chemical adjuvants with relatively high levels of biocompatibility, such as synthetic polymers (polyfection) [11, 12] and cationic liposomes (lipofection) [13-15]. These molecules self-assemble in highly organized structures capable of complexing the genetic material and later releasing it inside the cells. Whereas viruses generally impose issues of mutagenicity and immunogenicity [16, 17], polymers such as polyethylenimine (PEI) are known to be highly cytotoxic [18, 19]. By exclusion of alternatives, cationic liposomes have emerged as the carriers of excellence for intracellular delivery of nucleic acids due to their high versatility [20], reduced cytotoxicity [21], and high transfection efficiency [22].

Cationic liposomes are spherical vesicles composed of one or more cationic lipid or phospholipid bilayers [23, 24]. They include both cationic and neutral surfactants in their composition and may differ in size [25], lamellarity [26] or charge [27]. The cationic amphiphiles (which are mainly of synthetic nature) share two common features: the net cationic charge on the hydrophilic headgroup, and the hydrophobic tail that anchors the molecule to the liposome lipid bilayer [28]. The chemical structure of the cationic lipids varies markedly and each molecule can have a single (monovalent surfactant) or multiple cationic charges (multivalent surfactant) [29]. The neutral *helper* lipid also plays an important role in lipoplex fate by promoting the formation of inverted non-lamellar structures. These structures facilitate lipoplex fusion with the cell membrane and the subsequent release of the genetic material in the cytoplasm [30]. In addition, the presence of *helper* lipid reduces the amount of cationic lipid required for NA condensation, which reflects itself on a reduction of the toxic effects towards the cells, by decreasing the number of positively-charged headgroups in the lipoplex formulation [31, 32].

The driving force for lipoplex formation is the electrostatic interaction between the net positive charge of the cationic liposomes and the negatively charged DNA at an optimal ratio (+/-). This fact also enables the resulting complex to adsorb to the negatively charged cell surface [33-36]. After adsorption, cellular uptake of the complexed DNA facilitates intracellular DNA delivery and subsequent transgene expression [37]. In the case of DODAB/MO formulations, in which DODAB acts as a monovalent cationic surfactant and MO as *helper* lipid, the inclusion of MO leads to a dual-lipoplex phase diagram with lamellar structures prevalent at DODAB molar fractions above 0.5 and inverted bicontinuous cubic mesophases below 0.5 [38].

The high structural dependence of the system on MO content and temperature [39, 40] could reveal itself useful for optimizing lipoplex resistance against deleterious interactions with biological fluids and cell components, while remaining biocompatible and efficient as delivery agent. The presence of MO in these formulations also reduces the net positive charge necessary for successful NA complexation, thus reducing transfection-associated cytotoxicity [41]. In summary, a multidisciplinary approach to lipofection vectors will lead to the development of formulations with the most appropriate characteristics. Careful design of liposomal composition is essential for overcoming biological barriers, in order to achieve optimal transfection efficiency *in vitro* and *in vivo.*

2. Cationic lipid-mediated gene transfection

2.1. DODAB:MO liposomes

When assessing the potential of a new lipofection reagent, it is fundamental to study the physicochemical properties of the base liposomal formulation, to better adjust lipoplex morphology (lipoplex size, charge ratio (+/-), fluidity and structure) for optimal transfection conditions. In this way, the behaviour of the lipofection reagent (resistance to extracellular components, cytotoxicity) *in vitro and in vivo* is more predictable [42-44].

MO was first proposed as *helper* lipid for non-viral transfection [45, 46] in a new liposomal formulation also including synthetic surfactant Dioctadecyldimethylammonium Bromide (DODAB) [41]. DODAB is a bilayer-forming cationic lipid that tends to form large unilamellar vesicles (LUV's) in excess water [47, 48]. It features a hydrophobic moiety consisting of a double acyl chain (C18:0) attached to a quaternary ammonium headgroup (one single positive charge per molecule) [49, 50]. DODAB's phase behaviour has been extensively studied [51, 52], thus its physicochemical characteristics can be easily controlled, making it straightforward to design DODAB-based formulations with specific molecular structures. DODAB's main limitation is the relatively high gel-to-liquid crystalline phase transition temperature (T_M = 45°C) [53-56], superior to the human physiological temperature (T_M = 37°C), meaning that DODAB's bilayers display a strong rigidity at normal body temperature which greatly limits its use as a delivery agent. This limitation can be counteracted by including a co-lipid with a lower T_M value, such as DOPE [57, 58], cholesterol [59, 60] or MO [40], which will lower the T_M of the lipid mixture.

The use of MO in liposomal formulations brings other advantages apart from the fluidization of DODAB's membranes. MO is a natural-occurring neutral surfactant that has the particularity of forming two inverted bicontinuous cubic phases (Q_{II}^D and Q_{II}^G) in excess water [61, 62]. It possesses a single unsaturated acyl chain (C18:1) attached to a glycerol headgroup [63]. Its tendency to form inverted bicontinuous cubic phases has been explored in the past for different applications such as protein crystallization [64, 65] or matrix for gel electrophoresis [66], and justifies the structural richness of the liposomal system formed with DODAB [40].

The aggregation behaviour of concentrated DODAB/MO mixtures has been studied through different techniques including phase scan imaging (Fig. 1) that reveals a two-region phase diagram consisting of either DODAB or MO enriched zones [40]. If $X_{DODAB} \geq$ 0.5, bilayer-based structures dominate (Fig. 1A', 1B') and their size and fluidity depend on the molar composition of the mixture, with DODAB gel phase appearing as hydrated crystals [40]. When $X_{DODAB} < 0.5$, aggregates are dominated by densely packed cubic-oriented particles, visible as a cubic isotropic phase (Q) associated with high MO contents (Fig. 1D', 1E') [40].

This dual phase behaviour of DODAB/MO lipid mixtures confers a structural complexity to the system that extends itself to lipoplex organization, which can be fine-tuned to suit the biological application. Additionally, results show that MO has a similar effect on aggregate morphology than an increase in temperature, which can be modulated to produce formulations more suitable for gene transfection [39, 40].

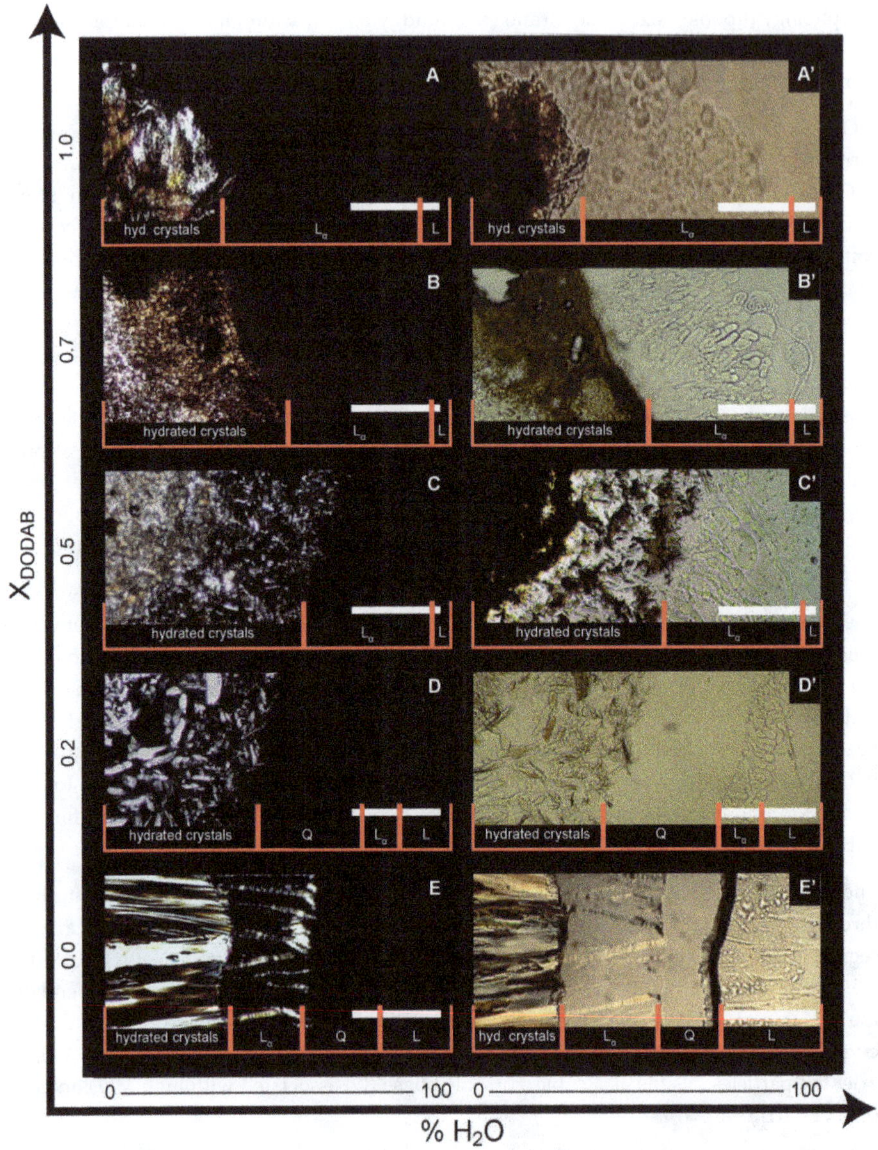

Figure 1. Phase scan imaging of neat DODAB (A, A′); X_{DODAB} = 0.7 (B, B′); X_{DODAB} = 0.5 (C, C′); X_{DODAB} = 0.2 (D, D′) and neat MO (E, E′) at 25°C. The images on the left side were obtained under polarized light and the images on the right side were obtained with normal light, using DIC lenses. Scale Bar: 200 μm. Abbreviations: L_α, lamellar liquid crystalline phase; Q, cubic isotropic phase; L, isotropic phase. Adapted from [40].

2.2. Role of MO as helper lipid in pDNA/DODAB/MO lipoplexes

The incubation of nucleic acids with DODAB/MO mixtures or other cationic vesicle formulation leads to the formation of lipoplexes [67, 68]. The electrostatic interaction between opposite charges is the key factor that determines the adsorption of the cationic vesicles to the DNA molecules, a transient state that ends when a critical cationic vesicle concentration is reached. This leads to the disruption of the lipid vesicles which allows the formation of highly organized structures where the DNA molecules are tightly condensed between adjacent bilayers – the so-called lipoplexes [69-71]. The excess of cationic lipid is required for lipoplex binding to the cell surface but any subsequent addition of cationic lipid to the complex does not enhance DNA delivery and only increases toxicity in the exposed cells [72].

Lipoplexes such as the pDNA/DODAB/MO system can be directly visualized by techniques such as cryo-TEM imaging (Fig. 2), which also gives information on the structural properties of the system (size, compactation, organization) [38]. Cryo-TEM imaging reveals that pDNA/DODAB/MO lipoplexes present the same dual phase diagram as obtained for DODAB/MO lipid mixtures [38]. pDNA/DODAB/MO lipoplexes at $X_{DODAB} > 0.5$ (Fig. 2A) exhibit a multilamellar structure consisting of stacked alternating lipid bilayers and pDNA monolayers. The analysis using Fast Fourier Transforms (FFT) corroborates this observation, by denoting a mono-orientated organization pattern at repeating distances of about 5 nm (Fig. 2A″, 2A‴ and 2A⁗) [38].

In contrast, pDNA/DODAB/MO lipoplexes at $X_{DODAB} \leq 0.5$ (Fig. 2B) show high-curvature zones where lipid bilayers intercross each other with pDNA monolayers stacked between them. These high-curvature zones have been interpreted as MO-rich domains that alternate with DODAB-rich domains presenting multilamellar organization. The FFT diagrams show that these MO-rich domains possess a distinct structural organization with bi-orientated patterns in angles of 90% between them, consistent with the existence of cubic inverted bicontinuous mesophases (Fig. 2B″, 2B‴ and 2B⁗) [38].

The DODAB/MO aggregate organization influences the final structural properties of the resulting pDNA/DODAB/MO lipoplexes, with MO content having a dramatic effect on how DNA is condensed and protected within the membrane.

By definition, a *helper* lipid (also termed as co-lipid or simply adjuvant) is any neutral surfactant not directly contributing to NA condensation or to targeting of the cell membrane by the lipoplex. *Helper* lipids enhance transfection efficiency by forming non-lamellar structures that intervene in several steps of the transfection process [73, 74]. These non-lamellar structures influence transfection efficiency in at least two ways: i) lipoplex-cell membrane fusion promoted by the fusogenic character of the *helper*; ii) improved endosomal escape of NA due to the disruption of the endosomal membrane by these structures prior to endosome/lysosome fusion, which would lead to NA degradation [75].

Dioleoylphosphatidylethanolamine (DOPE) is the most established *helper* used in non-viral vectors and is known to enhance transfection mediated by different cationic liposomal formulations [76-79]. DOPE stimulates the formation of inverted hexagonal structures (H_{II} –

Fig 3A) which represent a major structural variation from the classic multilamellar sandwich model of lipoplex organization (L_{α}^{C} – Fig. 3B) [80]. However, its application in gene therapy has been strongly limited because of the strong cytotoxicity associated with it [81, 82].

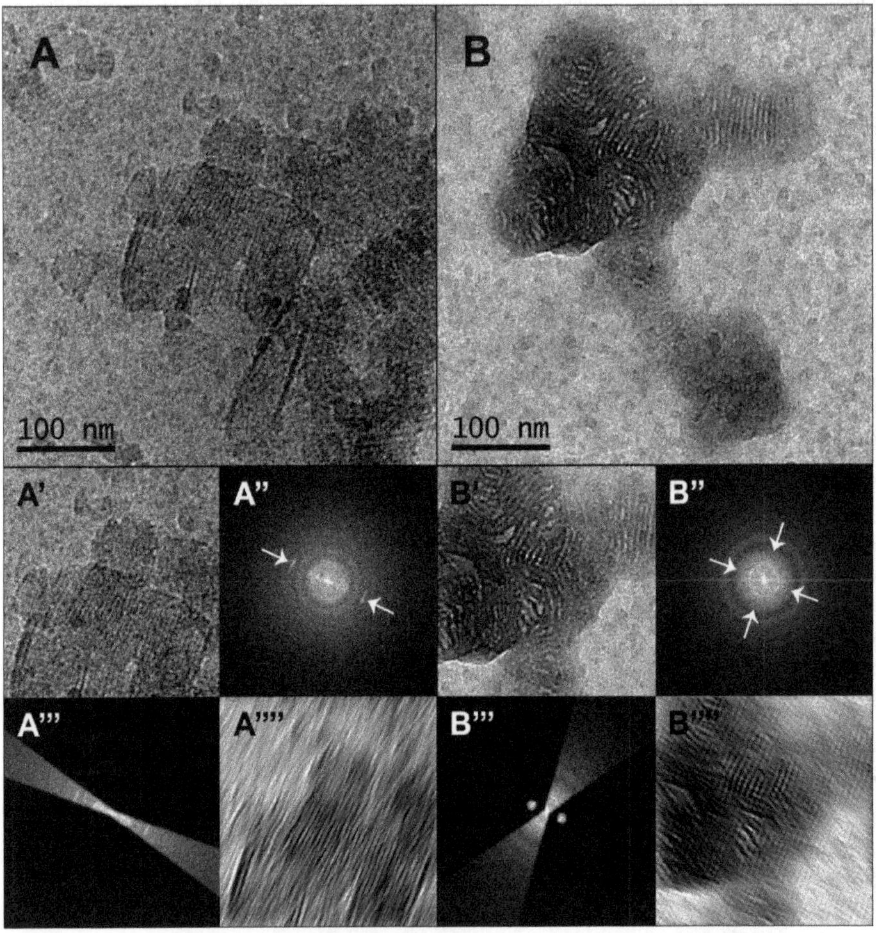

Figure 2. Cryo-TEM imaging of pDNA/DODAB:MO lipoplexes at C.R. (+/-) 4.0 (1mM total lipid). Panels A and B represent two different DODAB:MO molar fractions (2:1 and 1:1, respectively) from which have been selected two distinct zones A′ and B′. The corresponding FFT diagrams A″ and B″ are shown after the appliance of "Mask" tool in the Digital Micrograph™ (GATAN) software (A‴ and B‴). The inverse FFT diagrams of the previous images allow the emergence of distinct structural patterns: mono-orientated organization consistent with the existence of lamellar structures for high DODAB contents (A⁗, 2:1) and 90° bi-orientated organization associated with inverted bicontinuous cubic mesophases for high MO contents (B⁗, 1:1). Magnification: 50 000x. Adapted from [38].

This evidence has motivated the search for new *helpers* with higher levels of biocompatibility while maintaining the same efficiency as DOPE. Cholesterol is one of such molecules due to its ability to modify bilayer fluidity [83]. Inclusion of cholesterol results in the formation of complexes that are more stable but less efficient *in vitro* compared to DOPE-containing lipoplexes. In contrast, addition of cholesterol results in more efficient complexes for *in vivo* application [84, 85].

MO is another promising alternative to common *helper* lipids, as it seems to combine positive aspects of both DOPE and cholesterol: tendency to promote inverted non-lamellar structures similarly to DOPE (although different from the common inverted hexagonal structures – Fig. 3C) and the fluidizing effect of cholesterol, which increases the fusogenicity of the lipoplexes.

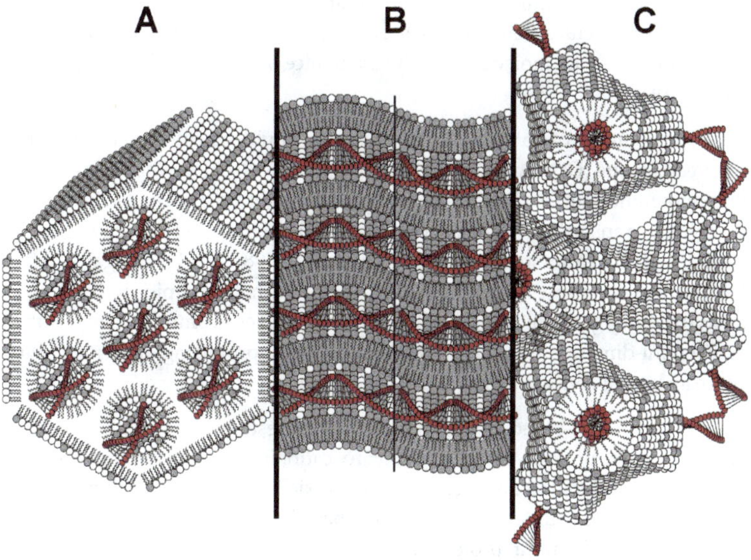

Figure 3. Different types of pDNA/cationic lipid structural organizations: A - inverted non-lamellar hexagonal structure characteristic of cationic vesicles containing DOPE at $X_{DOPE} \geq 0.5$); B - lamellar structural characteristic of cationic vesicles containing $X_{Helper} \leq 0.5$; and C - inverted bicontinuous cubic structure characteristic of cationic vesicles containing MO at $X_{MO} \geq 0.5$. Double-tailed surfactant with grey-headgroup represents cationic lipid, double-tailed surfactant with white-headgroup represents DOPE and single-tailed surfactant with white-headgroup represents MO. Grey-coloured regions represent cationic lipid rich-domains and white-coloured regions represent MO or DOPE rich-domains.

The fluidizing effect of MO contributes favourably to the complexation efficiency of DNA, quickening lipoplex formation [41]. At the same time, the formed inverted bicontinuous cubic mesophases improve the resistance of aggregates to extracellular component destabilization, thereby potentially enhancing transfection efficiency [38]. MO-based aggregates induce a relatively low cytotoxicity level, which further reinforces its use as a new *helper* lipid in this type of non-viral systems. *In vivo* evaluation of MO-based lipoplexes shall confirm the potential for this neutral surfactant to replace classic *helpers* in lipofection formulations, although some promising results have already been obtained for other cationic lipid formulations that also form inverted bicontinuous cubic structures [86-88]. Ethylphosphatidylcholines are a family of positively charged membrane lipid derivatives that promote the formation of Q_{II}^G and Q_{II}^P structures, having been linked to high levels of transfection efficiency with low cytotoxicity in several animal cell lines [89, 90], consubstantiating MO's potential role in gene delivery.

2.3. Recent progress in gene delivery with cationic lipids

The quest for the perfect cationic liposome formulation has been based on empirical testing of novel surfactant molecules that had never been previously used for NA delivery [91-93]. The only goals for candidate molecules are the attainance of high transfection efficiency with low cytotoxicity [94, 95].

After the first generation of cationic lipids based on double-chain surfactants with plain ammonium headgroups (DODAB, DOTAP, DOTMA or DMRIE) [96, 97], soon came cationic lipids with poly-ammonium and multivalent functional radicals (DOGS, DOSPA). The latter exhibited higher transfection efficiencies but also higher cytotoxicity due to the immunogenicity of the cationic ammonium headgroups [98, 99]. This negative effect was balanced with the appearance of *helpers* (DOPE) [74] and natural lipid-derivatives such as cholesterol [100] or glycerol [101], although sometimes compromising transfection efficiency. Gemini-dimeric surfactants also presented promising potential but with significant toxicological consequences [102-106].

Polyethylene glycol (PEG)-based lipids emerged as interesting hydrophilic polymer-based surfactants that could provide steric stability to cationic liposomes, increasing lipoplex lifetime in the bloodstream and also decreasing the toxic effects observed *in vitro* and *in vivo* [107-109]. The polymeric counterpart of the PEG-based surfactants (variable both in chain length and branching) forms a protective surface coating that inhibits the adhesion of plasma components which could promote NA release and particle aggregation [109]. This protective effect is enhanced by including up to 5-10% of PEG in the liposomal formulation, with no visible effects on lipoplex structure [110]. PEG addition reduces net electric charge and increases hydration of the liposome surface, decreasing immunogenicity and cytotoxicity elicited by the particles. Nevertheless, at high concentrations, these polymers are known to be toxic and of difficult clearance from the organism. Therefore, when developing PEGylated particles, one must weigh advantages and disadvantages of including PEG, especially when aiming for long-term therapeutic administration [111].

Inclusion of pH-sensitive molecules in the formulations has been shown to improve transgene expression by favouring DNA release from the endosomal compartment. Examples of pH-sensitive molecules used in non-viral gene delivery include polyhistidine, dioleoyldimethylammonium propane (DODAP) or cholesteryl hemisuccinate (CHEMS) [112, 113].

Another major breakthrough with impact in gene therapy was the possibility of specific cell targeting by liposomes. Amphiphiles with hydrophilic headgroups could be chemically linked to molecules such as folate, transferrin or the epidermal growth factor that potentiate specific delivery to cancer cells, markedly increasing the therapeutic benefits achieved with lipoplexes, with little secondary effects [114-118].

More recently, cationic lipids with amino acid headgroup (serine, alanine) [119, 120] and sugar-based cationic lipids (D-galactose) have appeared as promising families of cationic surfactants [121, 122]. Small molecular weight peptides (glutamate, cysteine) augment the hydrophilicity of the lipoplex surface, as with small surface sugars (galactose, mannose) that additionally allow targetability of the lipoplexes.

3. Lipoplex interaction with extracellular milieu

3.1. Resistance to components of biological fluids

An effective delivery system must confer stability to complexed NA in physiological conditions [123, 124]. Systemic delivery of NA requires a stealth carrier that protects NA from indiscriminate interaction with complement and coagulation pathways that lead to rapid removal from blood circulation of the lipoplexes by opsonization [125-127]. pDNA/DODAB/MO lipoplexes were therefore tested regarding their sensitivity when simulating their interaction with the body (temperature, salt, exposure to serum, nucleases and membrane lipases), to be validated for systemic applications [128].

Fig. 4 shows the variation of free pDNA fraction after incubation of pDNA/DODAB:MO lipoplexes (2:1, 1:1 and 1:2) with different constituents of the plasma. Increasing the temperature from 25°C to physiological temperature (37°C) leads to a reduced but visible release of pDNA from the lipoplex, more evident for lower MO contents. The gel phase of DODAB ($X_{DODAB}> 0.5$) is clearly more disturbed by incubation at higher temperature than the liquid-crystalline phase of DODAB/MO lipid mixtures ($X_{DODAB} \leq 0.5$). This tendency is maintained upon NaCl addition at physiological concentration (150mM), showing the protective role of MO upon the electrostatic imbalance provoked by salt addition.

DODAB/MO formulations with varying MO content behave very differently when exposed to serum (Fig. 4). Serum may strongly interfere with lipoplexes, both *in vitro* and *in vivo*, causing lipoplex-protein aggregation that lead to degradation of the genetic material and possibly clogging the blood vessels in intravenous application [129]. pDNA/DODAB/MO lipoplexes release up to 30% of the initially complexed pDNA when incubated for 30min with bovine serum albumin, the major constituent of bovine serum, particularly in the case of formulations with low MO content. MO contributes, in fact, to a better resistance of pDNA/DODAB/MO to extracellular components, eventually related to the inverted bicontinuous cubic structures present that reduce the exposure of DNA molecules to the plasma constituents. In fact, the results suggest a direct correlation between lipoplex stability and MO content.

Some authors have managed to transiently overcome this inhibitory effect of serum on lipofection by increasing the charge ratio (+/-) of cationic liposome to DNA [130, 131]. Significantly enhanced gene transfer has also been achieved by pre-incubating the delivery system with serum proteins prior to NA complexation [132, 133].

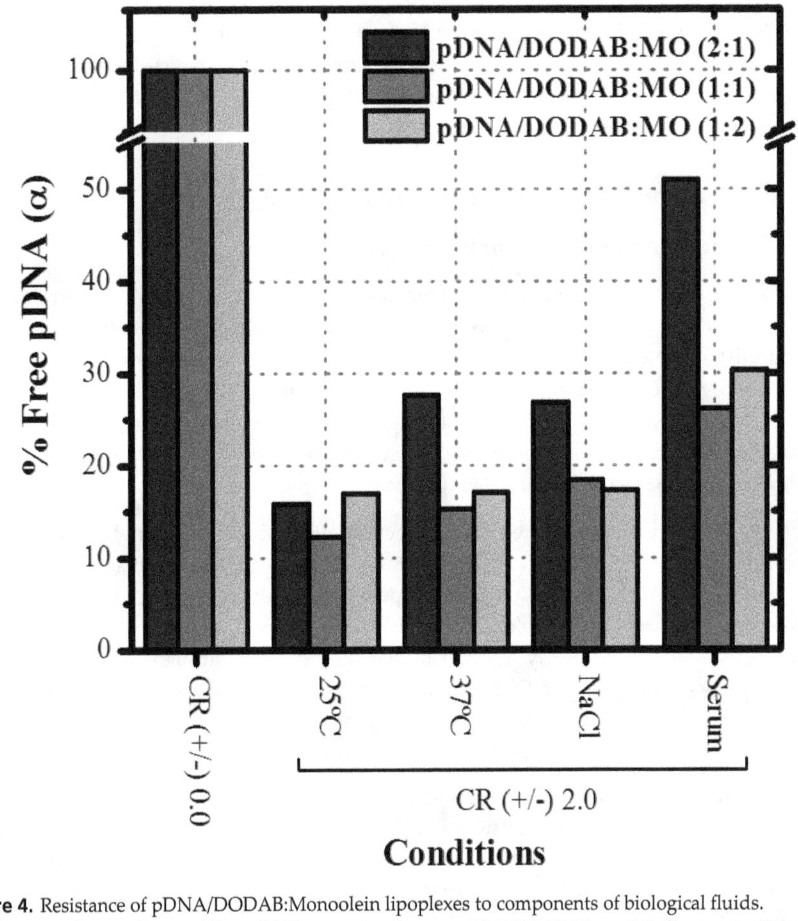

Figure 4. Resistance of pDNA/DODAB:Monoolein lipoplexes to components of biological fluids. Variation on the percentage of free pDNA upon incubation with DODAB:Monoolein liposomes (2:1, 1:1, and 1:2) at CR (+/-) 2.0, and subsequently exposed to a temperature increase from 25°C to 37°C in the presence of NaCl salt (150mM) and BSA (0.5g/L) at incubation times of 30 min. The values were calculated through spectral decomposition of ethidium bromide steady-state fluorescence, as described elsewhere [38].

3.2. Lipoplex adhesion to the cell surface

The adsorption and uptake of lipoplexes may be affected by the presence of proteoglycans at the plasma cell membrane surface. It is therefore important to study how lipoplexes interact with these extracellular matrix components during cell transfection. Association of lipoplexes with negative polyelectrolytes free in solution might also be useful to evaluate eventual loss of pDNA at the cell surface [134].

Proteoglycans (membrane receptors consisting of a protein core and one or more anionic glycosaminoglycan chains including heparin, dermatan and chondroitin sulphates) were identified as the mediating agents for cationic liposome/DNA cellular uptake both *in vitro* and *in vivo* [135]. Lipoplex/proteoglycan interaction is suggested to depend upon three major aspects: the ionic strength, the effect of *helper* lipids and of the glycosaminoglycan structure [134, 135].

When the lipoplexes interact with heparin and heparin sulphate, the negative charge of the polyelectrolytes determines NA release from the lipoplex through the same type of cooperative process that is responsible for lipoplex formation [136-138].

On subjecting pDNA/DODAB:MO (2:1 and 1:1) lipoplexes to increasing amounts of heparin (HEP), the improved resistance and stability of the lipoplexes obtained with increasing amounts of MO could be confirmed (Fig. 5). The fact that the system with higher MO content (X_{DODAB}= 0.5) shows enhanced resistance to heparin relatively to pDNA/DODAB:MO (2:1) lipoplexes suggests that pDNA dissociation is mainly dependent on structural properties (Fig. 2) rather than physicochemical properties of the lipoplexes.

4. Modulation of cell behaviour by lipoplexes

4.1. Cytotoxicity

In addition to the efficiency of MO based lipoplexes, patient tolerability is determinant for therapeutic application of these systems. *In vitro* toxicity tests are a useful, time and cost-effective first approach in the validation process of a therapeutic agent. The adverse effects of liposomes on cells can be identified through different assays that look at particular aspects of cell behavior, such as metabolism, proliferation or cell membrane integrity. To determine if liposomal formulations will be well tolerated by all cells it will contact with, it is important to test cytotoxicity using different cell types. The cell lines should be selected: i) to evaluate how target cells will react; ii) to predict eventual toxicity for the heart and liver, by using cardiomyocytes and hepatocytes, respectively; iii) to screen if the liposomes can be applied to all types of cells. In the case of DODAB:MO liposomes, four different mammalian cell lines (HEK 293, BJ5ta, L929 and C2C12) were exposed for two days to increasing concentrations of these systems, after which different analytical methods were applied (Figs. 6 to 8).

The cell lines presented here are routinely used for toxicity studies and are commercially available. The human Embryonic Kidney (HEK) 293 cell line was originally derived from human embryonic kidney cells grown in tissue culture, from which 293T cell line is derived. BJ5ta cells are normal human foreskin fibroblasts immortalized with telomerase. Murine cell lines L929 and C2C12 are fibroblasts and myoblasts, respectively.

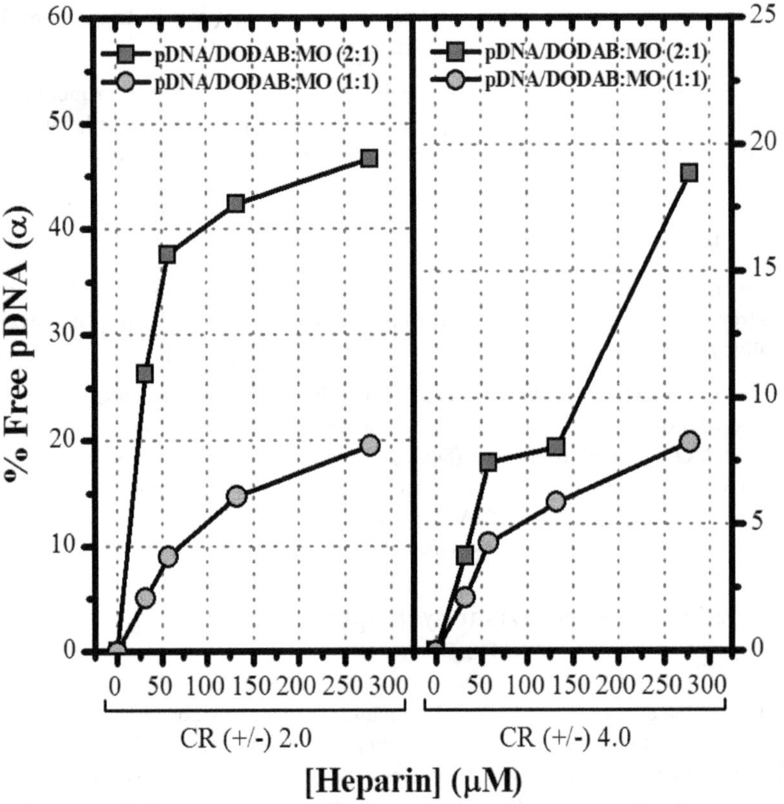

Figure 5. Resistance of pDNA/DODAB:Monoolein lipoplexes to model proteoglycans. Variation on the percentage of free pDNA upon incubation of pDNA/DODAB:Monoolein lipoplexes (2:1, 1:1, and 1:2) at CRs (+/-) 2.0/4.0 with increasing amounts of heparin (HEP) at incubation times of 30 min. The values were calculated through spectral decomposition of ethidium bromide steady-state fluorescence, as described elsewhere [38]. Adapted from [38].

Another aspect to be taken into account is the possibility that the liposomes and lipoplexes may differently affect parameters such as metabolism, cell membrane structure and chemistry, cell proliferation and mobility. For a comprehensive study, a minimum of three different methodologies, monitoring at least two of these parameters, should be used. From our own results, it was observable that the metabolism of L929 and C2C12 cells was more pronouncedly affected by the contact with DODAB:MO liposomes compared to the other cell lines, especially with a lipid concentration ≥ 20 μg/ml (Fig. 6). At these higher concentrations, DODAB:MO (1:1) induced lower levels of cytotoxicity in all the cell types, which probably reflects the higher content of MO and concomitant lower content of cationic lipid. Interestingly, the cell membrane integrity assay did not reveal such obvious, concentration-dependent variations in cytotoxicity (Fig. 7). The results obtained with the proliferation test (Fig. 8) were quite concordant with those from the metabolic assay (Fig. 6), indicating again the L929 and C2C12 cells as more sensitive, while BJ5ta proliferation was clearly increased when incubated with up to 20 μg/ml lipid (Fig. 8).

Lipoplexes prepared from these liposomal formulations, at concentrations typically used in transfection experiments, constantly leading to slightly lower viability rates compared to the base DODAB:MO liposomes (data not shown).

The fact that MO-based aggregates cause reduced levels of cytotoxicity for concentrations typically used on transfection assays, reinforcing the use of MO as a new *helper* lipid in this type of non-viral systems. Even if there is general agreement in the reduced toxicity of liposomes as non-viral vectors, these results emphasize the need for accurate liposome/lipoplex evaluation to better assess human risk prior to using them as lipofection vectors.

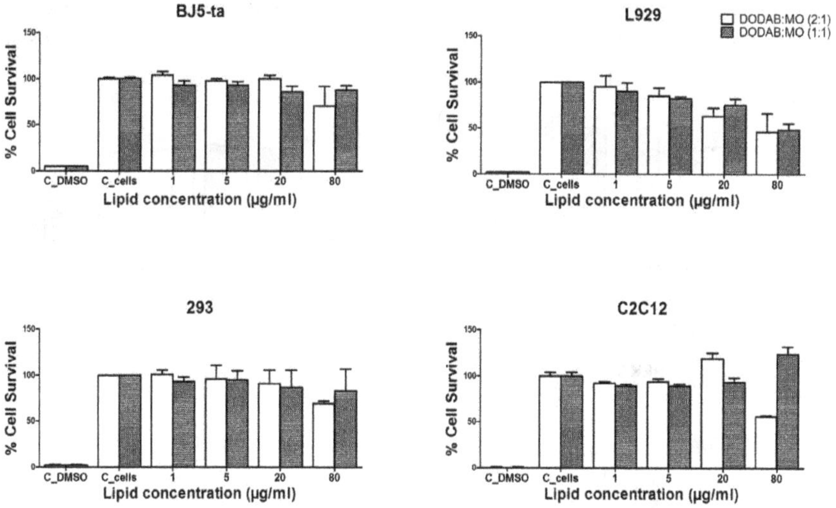

Figure 6. Evaluation of the cytotoxicity (metabolic assay) in four different mammalian cell lines (BJ5-ta, L929, 293 and C2C12) induced by varying concentrations of DODAB:MO-based liposomes after 48 h of incubation. C_DMSO: cells incubated with 30 % DMSO; C_Cells: cells alone. The mean (+/−) SD was obtained from two independent experiments. MTT assay can be used to estimate cell viability, specifically as marker of the cell metabolic capacity. The soluble tetrazolium MTT is reduced by metabolically active cells, thus the developed purple color proportional to the number of viable cells.

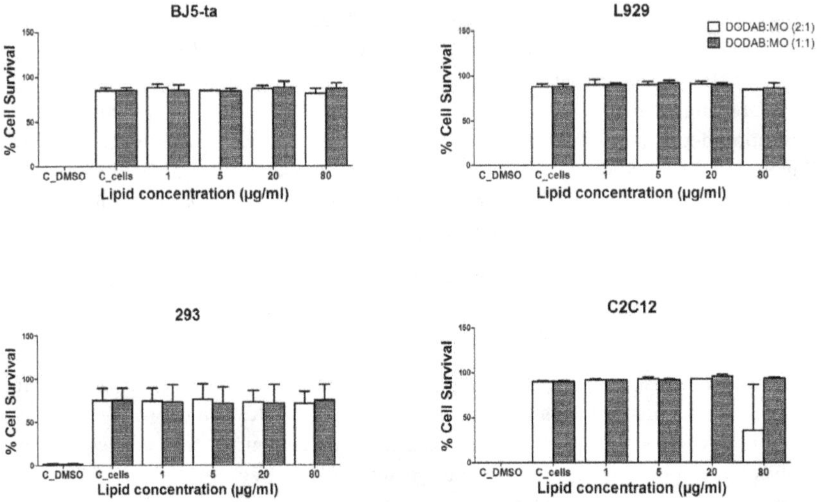

Figure 7. Evaluation of the cytotoxicity (cell membrane integrity) in four different mammalian cell lines (BJ5-ta, L929, 293 and C2C12) induced by varying concentrations of DODAB:MO-based liposomes after 48 h of incubation. C_DMSO: cells incubated with 30 % DMSO; C_Cells: cells alone. The mean (+/−) SD was obtained from two independent experiments. The LDH assay is used to estimate cell viability, as the intracellular enzyme LDH is released into the extracellular medium when cell membranes are damaged.

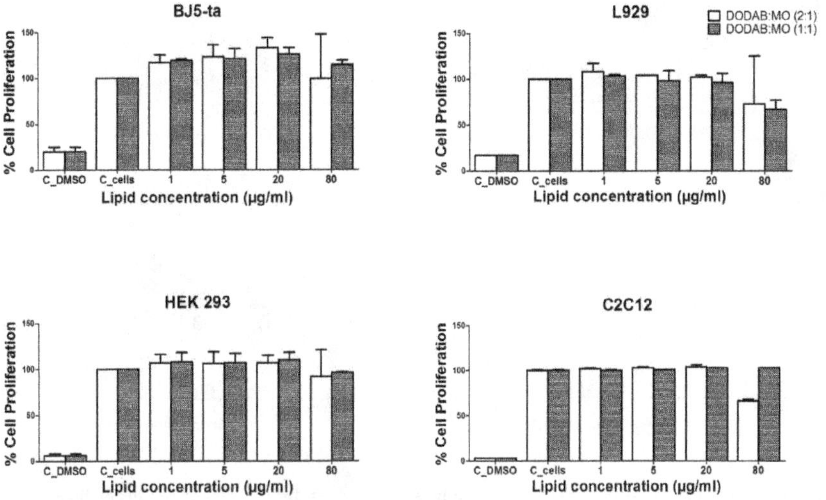

Figure 8. Evaluation of the cytotoxicity (proliferation) in four different mammalian cell lines (BJ5-ta, L929, 293 and C2C12) of varying concentrations of DODAB:MO-based liposomes after 48 h of incubation. C_DMSO: cells incubated with 30 % DMSO; C_Cells: cells alone. The mean (+/−) SD was obtained from two independent experiments. Sulforhodamine B (SRB) is considered a proliferation assay, used for cell density determination, based on the determination of the cellular protein content.

4.2. Cellular uptake and intracellular trafficking

In spite of extensive efforts to unravel the *in vitro/in vivo* mechanisms of internalization of lipoplexes, doubts remain as to whether the topology of lipoplexes facilitates the entry of DNA by fusion with the plasma membrane or with endosomal vesicles. Other studies have indicated that endocytosis is possibly the preferred mechanism of lipoplex internalization by cells [139, 140]. After the formation of the endocytic vesicle containing the lipoplexes, the internal pH of the endosomes decreases to about 5.5 [141].The endosomes then fuse with the lysosomes, in which the condensed NA component may be hydrolysed by lysosomal enzymes [142, 143]. Endosomal release of the NA should occur, avoiding the lysosomal lythic pathway, leading to successful transfection.

Different mechanisms for complex internalization have been proposed, in particular for lipoplexes and polyplexes. Endocytosis at the plasma membrane may be clathrin-dependent or -independent. Clathrin-independent mechanisms include fusion of lipoplexes with the plasma membrane, phagocytosis, macropinocytosis and caveolae-mediated uptake [144]. *In vitro* cell culture systems provide the opportunity to experimentally address how lipoplexes interact with the plasma membrane. Although it is widely accepted that endocytosis is the most important route for lipoplex entry, different endocytic pathways may be used in parallel. The most likely explanation is that different cell types prefer a particular mechanism but use more than one. Therefore, optimization remains largely dependent of trial and error.

Intracellular trafficking of lipoplexes can be followed by co-localization studies of labeled particle components and dyes, or antibodies that recognize cell organelles or molecules playing a role in the process (e.g. clathrin coating endocytic pits in the plasma membrane) [139, 145] (Fig. 9). Cell lines harboring mutations in some of these molecules may also be used to evaluate their importance for the internalization process of specific formulations. The use of inhibitors of endocytosis has also been widely used but has two major limitations: the significant toxicity induced by the inhibitors themselves and the evidence corroborating that internalization can be simultaneously mediated by different pathways.

The endosomal escape is thought to be the major limitation for efficient gene transfection [146]. A number of strategies have been explored to enhance NA endosomal release. For example, the incorporation of a non-lamellar forming lipid such as DOPE that disrupts the endosome membrane or inclusion of a pH-dependent molecule that senses the acidification in the endosome compartment leading to disruption of its membrane [147].

Modulation of the endosomal escape during lipoplex intracellular trafficking was replicated by exposing pDNA/DODAB:MO (2:1, 1:1 and 1:2) lipoplexes to acidic conditions in the presence of increasing amounts of hydrochloric acid (pH ranging from 7.5 to 2.5) (Fig. 10). The percentage of released DNA steadily increased upon milieu acidification from pH 7.4 to 4.5, which is the pH range typical in the endosome. This trend correlates negatively with the MO content in the formulation, suggesting that MO's inverted bicontinuous cubic structures may protect more efficiently the lipoplex structure in this environment. More stringent acidification of the environment (pH 4.5 to pH 2.5) inverts the release tendency, which can be related to degradation of naked pDNA in solution.

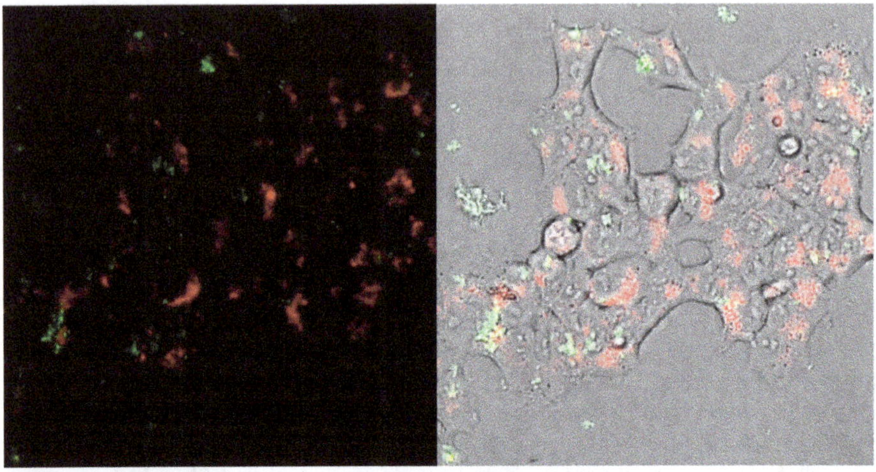

Figure 9. Visualization of cellular uptake of DODAB:MO lipoplexes by HEK 293T cells. Liposomes are labelled with Bodipy-PE (green) and endo-lisosomes with dextran (red). Co-localization (yellow) indicates sites of active endocytosis of lipoplexes. DODAB:MO (2:1) (+/-) 4.0, 1 μg pDNA/well, amplification 200x.

Lipoplex charge ratio (+/-) also affects the intensity of pDNA release. Using the same DODAB:MO base formulation, increasing charge ratio (+/-) seems to prevent pDNA release from the lipoplex. This effect was already visible in the destabilization of pDNA/DODAB/MO lipoplexes by plasma constituents such as serum and salt, and probably reflects more efficient pDNA condensation in presence of excess cationic lipid.

Increasing ammonium/phosphate ratio carries the risk of increased cytotoxicity. One possible solution may be using increasing amounts of MO in lipoplex formulation for better protection of pDNA integrity without imposing major toxic effects to the target cell.

Non-viral vectors, although less toxic than viral vectors, may still elicit a strong, nonspecific immune response. Toxicity frequently results from characteristics of the encapsulating polymer or lipid such as the length, saturation, or branching of the polymer. Efforts to reduce the toxicity of nonviral vectors have largely resulted in attempts to make the vectors more biodegradable and biocompatible. Many of the aforementioned systems (i.e. triggered release with disulfides, PEG copolymers) incorporated more biologically active components, thereby reducing the elicited immune response. For example, the incorporation in liposomes of molecules known to suppress the production of the cytokine tumor necrosis factor (TNF-α), as compared to lipoplex alone, succeeded in maintain its levels low while achieving comparable levels of transgene expression [148]. Another method explored by Tan [149] significantly reduced toxicity through the sequential injection of liposome and later of DNA, as opposed to using formed lipoplexes. With this approach, cytokine levels (IL-12, TNF-α) were reduced by greater than 80% compared to lipoplex delivery [149]. Thus, significant advances have been made towards decreasing the toxicity of these non-viral vectors.

Interestingly, DODAB:MO based liposomes and lipoplexes were found to induce production of low levels of TNF-α by macrophages, comparable or lower than DOTMA/DOPE and DOTMA/cholesterol lipoplexes (data not shown) [150].

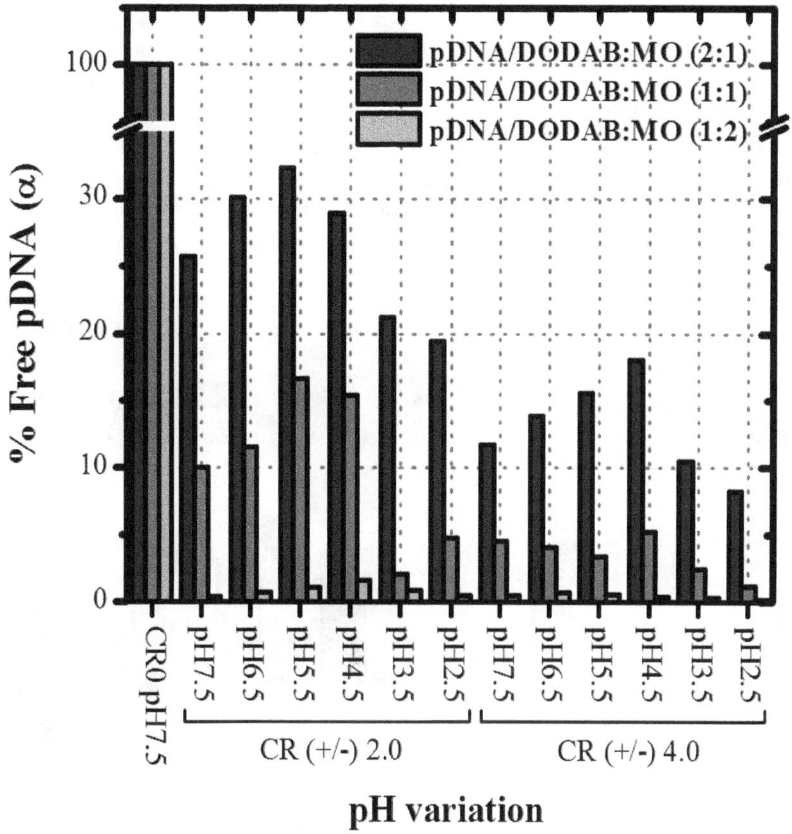

Figure 10. Resistance of pDNA/DODAB:Monoolein lipoplexes to pH decrease (modulation of endosomal escape). Variation on the percentage of free pDNA upon incubation of pDNA/DODAB:Monoolein lipoplexes (2:1, 1:1, and 1:2) at CRs (+/-) 2.0/4.0 with increasing amounts of hydrochloric acid at incubation times of 30 min. The values were calculated through spectral decomposition of ethidium bromide steady-state fluorescence, as described elsewhere [38].

4.3. Transfection efficiency

Transfection efficiency of plasmid DNA can be directly evaluated by detecting the protein encoded by the reporter gene. Examples of reporter genes are: green fluorescent protein (GFP) and similar, detectable by techniques as microscopy or flow cytometry; β-galactosidase, whose activity can be evaluated by a colorimetric assay; luciferase, whose activity can be measured with a luminometer, after a substrate is converted into a luminescent form by luciferase. In Figure 11 is depicted an experiment that allows to identify the effect on transfection efficiency of varying the content of MO in the liposomal formulations, lipid:DNA charge ratio in the lipoplexes and also the quantity of pDNA added to the cells, as pDNA dosage is known to affect transfection efficiency. It can be observed that the incorporation of MO in the liposomes resulted in a transfection efficiency improvement when compared to the cationic lipid DODAB alone. When using 1 μg DNA/well, the transfection levels of pDNA/DODAB:MO systems are of the same order of magnitude as Lipofectamine™ LTX. For a lower MO content (pDNA/DODAB:MO (2:1) formulation), a dose effect response (0.5 μg and 1 μg of pDNA) was observed. For higher MO content (pDNA/DODAB:MO (1:1) formulation), the transfection efficiencies remained constant at both CRs. This result strengthens the role of MO as *helper* lipid in the transfection agent.

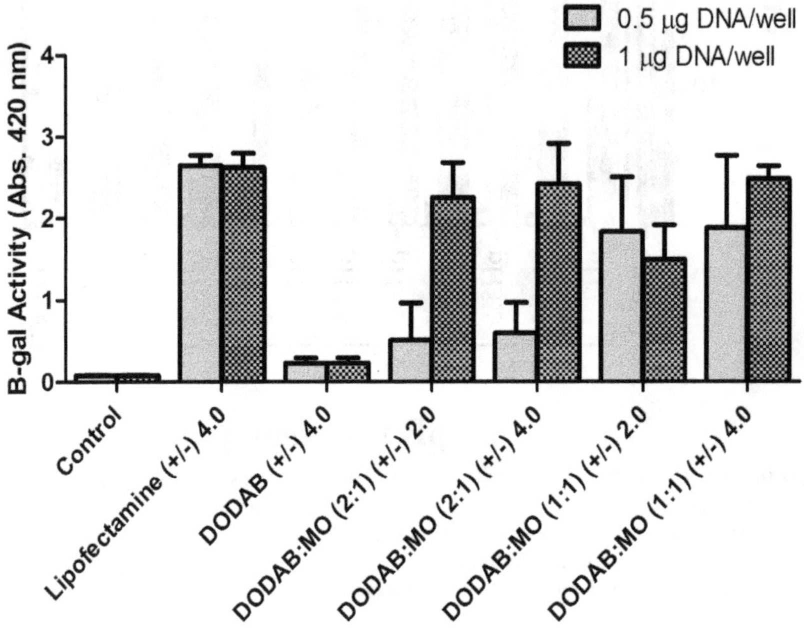

Figure 11. Transfection efficiency of HEK 293T cells by MO-based lipoplexes. Transfected pDNA encoded the β-galactosidase gene whose activity was evaluated by a colorimetric assay after 48 h of incubation. Lipoplexes prepared at charge ratio (+/-) 4.0 or 2.0, 0.5 or 1.0 μg pDNA/well. Controls: cells incubated with free pDNA; cells transfected using Lipofectamine® as lipofection agent. The mean (+/−) SD was obtained from three independent experiments. Adapted from [38].

5. Conclusions

The identification of the most important formulation parameters and how they influence macromolecule delivery and bioactivity will give direction towards the development of novel therapeutic solutions. The morphology and structure of the lipoplex is influenced by the surrounding environment and the chemical nature of its constituents. Physicochemical properties of the systems define the course of most events when lipoplex interact with the body, tissues and cells. The effectiveness of vector internalization, its intracellular trafficking and successful transgene expression in target cells, is directly dependent on the *helper* lipid features, net charge of the lipoplex and the degree of NA compactation within the complex. Different target cells may impose specific challenges to transfection and many inherent factors are unknown. The advent of controlled cell targeting for improved specificity holds great promise for application of these formulations in nanomedicine.

A good lipofection system must protect NA from deleterious interaction with biological fluids and cell components, while remaining biocompatible and efficient as delivery agent. In summary, with this work we intend to demonstrate that MO can be used safely and efficiently as *helper* lipid in the preparation of non-viral vectors for transfection. The presence of this natural lipid in the formulations reduces the net positive charge necessary for successful NA complexation, thus decreasing transfection associated cytoxicity.

Author details

J. P. Neves Silva and M. E. C. D. Real Oliveira*
Centre of Physics, University of Minho, Campus of Gualtar, Braga, Portugal

A. C. N. Oliveira and A. C. Gomes
Centre of Molecular & Environmental Biology - CBMA, University of Minho, Campus of Gualtar, Braga, Portugal

Acknowledgement

This work was supported by FCT research project PTDC/QUI/69795/2006, which is co-funded by the program COMPETE from QREN with co-participation from the European Community fund FEDER; CFUM [PEst-C/FIS/UI0607/2011]; CBMA [Pest C/BIA/UI4050/2011]; J.P.N. Silva holds a PhD Grant (SFRH/BD/46968/2008); A. C.N. Oliveira holds a PhD grant (SFRH/BD/68588/2010).

6. References

[1] Patil, S.D., D.G. Rhodes, and D.J. Burgess, *DNA-Based Therapeutics and DNA Delivery Systems - A Comprehensive Review.* American Association of Pharmaceutical Scientists Journal, 2005. 7: p. E61-E77.

* Corresponding Author

[2] Levin, Y. and J.J. Arenzon, *Kinetics of Charge Inversion.* Journal of Physics A: Mathematical and General, 2003. 36: p. 5857-5863.

[3] Pitt, W.G., G.A. Husseini, and B.J. Staples, *Ultrasonic Drug Delivery - A General Review.* Expert Opinion on Drug Delivery, 2004. 1(37-56).

[4] Miller, D.L. and C. Dou, *Induction of Apoptosis in Sonoporation and Ultrasonic Gene Transfer.* Ultrasound in Medicine and Biology 2009. 35: p. 144-154.

[5] Rebersek, M. and D. Miklavcic, *Advantages and Disadvantages of Different Concepts of Electroporation Pulse Generation.* Automatika, 2011. 52: p. 12-19.

[6] Gehl, J., *Electroporation Theory and Methods, Perspectives for Drug Delivery, Gene Therapy and Research.* Acta Physiologica Scandinavica, 2003. 177: p. 437-447.

[7] André, F.M., et al., *Variability of Naked DNA Expression After Direct Local Injection - The Influence of the Injection Speed.* Gene Therapy, 2006. 13: p. 1619-1627.

[8] Herweijer, H. and J.A. Wolff, *Progress and Prospects - Naked DNA Gene Transfer and Therapy.* Gene Therapy, 2003. 10: p. 453-458.

[9] Machida, C.A., *Viral Vectors for Gene Therapy - Methods and Protocols.* 1st ed, ed. H. Press. 2003, New Jersey (U.S.A). 606.

[10] Heiser, W.C., *Gene Delivery to Mammalian Cells - Viral Gene Transfer Techniques.* 1st ed, ed. H. Press. 2004, New Jersey (U.S.A.). 584.

[11] Arima, H., *Polyfection ad Nonviral Gene Transfer Method - Design of Novel Nonviral Vector using Ciclodextrin.* Journal of the Pharmaceutical Society of Japan, 2004. 124: p. 451-464.

[12] Eliyahu, H., Y. Barenholz, and A.J. Domb, *Polymers for DNA Delivery.* Molecules, 2005. 10: p. 34-64.

[13] Felgner, P.L., et al., *Lipofection Procedure - A Highly, Efficient, Lipid-mediated DNA-transfection Procedure.* Proceedings of the National Academy of Sciences U.S.A., 1987. 84: p. 7413-7417.

[14] Pampinella, F., et al., *Analysis of Differential Lipofection Efficiency in Primary and Established Myoblasts.* Molecular Therapy, 2002. 5: p. 161-169.

[15] Felgner, P.L., et al., *Nomenclature for Synthetic Gene Delivery Systems.* Human Gene Therapy, 1997. 8: p. 511-512.

[16] Douglas, K.L., *Toward Development of Artificial Viruses for Gene Therapy A Comparative Evaluation of Viral and Non-Viral Transfection.* Biotechnology Progress, 2008. 24: p. 871-883.

[17] Stone, D., et al., *Viral Vectors for Gene Delivery and Gene Therapy Within the Endocrine System.* Journal of Endocrinology, 2000. 164: p. 103-118.

[18] Florea, B.I., et al., *Transfection Efficiency and Toxicity of Polyethylenimine in Differentiated Calu-3 and Nondifferentiated COS-1 Cell Cultures.* American Association of Pharmaceutical Scientists Journal, 2002. 4: p. 1-11.

[19] Lee, M., *Apoptosis Induced by PolyethylenimineDNA Complex in Polymer Mediated Gene Delivery.* Bulletin of the Korean Chemical Society, 2007. 28: p. 95-98.

[20] Egilmez, N.K., Y. Iwanuma, and R.B. Bankert, *Evaluation And Optimization Of Different Cationic Liposome Formulations.* Biochemical and Biophysical Research Communications, 1996. 221: p. 169-173.

[21] Choi, W., et al., *Low Toxicity Of Cationic Lipid-based Emulsion For Gene Transfer.* Biomaterials, 2004. 25: p. 5893-5903.

[22] Yingyongnarongkul, B., et al., *High Transfection Efficiency and Low Toxicity Cationic Lipids with Aminoglycerol-Diamine Conjugate.* Bioorganic & Medicinal Chemistry, 2009. 17: p. 176-188.

[23] Lasic, D.D. and Y. Barenholz, *Nonmedical Applications of Liposomes - From Gene Delivery and Diagnostics to Ecology.* 1st ed, ed. C. Press. 1996, Boca Raton (USA).

[24] Lasic, D.D., *Liposomes.* Science & Medicine, 1996. 3: p. 34-43.

[25] Gonçalves, E., R.J. Debs, and T.D. Heath, *The Effect of Liposome Size on the Final Lipid-DNA Ratio of Cationic Lipoplexes.* Biophysical Journal, 2004. 86: p. 1554-1563.

[26] Zuidam, N.J., et al., *Lamellarity of Cationic Liposomes and Mode of Preparation of Lipoplexes Affect Transfection Efficiency.* Biochimica et Biophysica Acta, 1999. 1419: p. 207-220.

[27] Claessens, M.M.A.E., et al., *Charged Lipid Vesicles Effects of Salts on Bending Rigidity, Stability,and Size.* Biophysical Journal, 2004. 87: p. 3882-3893.

[28] Labas, R., et al., *Nature as a Source of Inspiration for Cationic Lipid Synthesis.* Genetica, 2010. 138: p. 153-168.

[29] Ahmad, A., et al., *New Multivalent Cationic Lipids Reveal Bell Curve for Transfection Efficiency Versus Membrane Charge Density Lipid-DNA Complexes for Gene Delivery.* Journal of Gene Medicine, 2005. 7: p. 739-748.

[30] Zuhorn, I.S., et al., *Phase Behavior Of Cationic Amphiphiles And Their Mixtures With Helper Lipids.* Biophysical Journal, 2002. 83: p. 2096-2108.

[31] Lasic, D.D. and D. Papahadjopoulos, *Medical Applications of Liposomes.* 1st ed, ed. Elsevier. 1998, Amsterdam (Netherlands). 795.

[32] Madeira, C., et al., *Liposome Complexation Efficiency Monitored by FRET - Effect of Charge Ratio, Helper Lipid and Plasmid Size.* European Biophysical Journal, 2007. 36: p. 609-620.

[33] Zelphati, O., et al., *Stable and Monodisperse Lipoplex Formulations for Gene Delivery.* Gene Therapy, 1998. 5: p. 1272-1282.

[34] Gershon, H., et al., *Mode of Formation and Structural Features of DNA-Cationic Liposome Complexes Used for Transfection.* Biochemistry, 1993. 32: p. 7143-7151.

[35] Akao, T., et al., *Conformational Change in DNA Induced by Cationic Bilayer Membranes.* Federation of European Biochemical Societies Letters, 1996. 391: p. 215-218.

[36] Dan, N., *Formation of Ordered Domains in Membrane-Bound DNA.* Biophysical Journal, 1996. 71: p. 1267-1272.

[37] Cruz, M.T.G., et al., *Kinetic Analysis of the Initial Steps Involved in Lipoplex-Cell Interactions.* Biochimica et Biophysica Acta, 2001. 1510: p. 136-151.

[38] Silva, J.P.N., et al., *DODAB:Monoolein-based Lipoplexes as Non-viral Vectors for Transfection of Mammalian Cells.* Biochimica et Biophysica Acta, 2011. 1808: p. 2440-2449.

[39] Silva, J.P.N., P.J.G. Coutinho, and M.E.C.D.R. Oliveira, *Characterization of Mixed DODAB-Monoolein Aggregates Using Nile Red as a Solvatochromic and Anisotropy Fluorescent Probe.* Journal of Photochemistry and Photobiology A: Chemistry, 2009. 203: p. 32-39.

[40] Oliveira, I.M.S.C., et al., *Aggregation Behavior of Aqueous Dioctadecyldimethylammonium Bromide/Monoolein Mixtures: A Multitechnique Investigation on the Influence of Composition and Temperature.* Journal of Colloid and Interface Science, 2011. 374: p. 206-217.

[41] Silva, J.P.N., P.J.G. Coutinho, and M.E.C.D.R. Oliveira, *Characterization of Monoolein-Based Lipoplexes Using Fluorescence Spectroscopy.* Journal of Fluorescence, 2008. 18: p. 555-562.

[42] Kearns, M.D., A.M. Donkor, and M. Savva, *Structure-Transfection Activity Studies of Novel Cationic Cholesterol-Based Amphiphiles.* Molecular Pharmaceutics, 2008. 5: p. 128-139.

[43] Tarahovsky, Y.S., R. Koynova, and R.C. MacDonald, *DNA Release from Lipoplexes by Anionic Lipids Correlation with Lipid Mesomorphism, Interfacial Curvature, and Membrane Fusion.* Biophysical Journal, 2004. 87: p. 1054-1064.

[44] Koynova, R., et al., *Lipoplex Formulation of Superior Efficacy Exhibits High Surface Activity and Fusogenicity, and Readily Releases DNA.* Biochimica et Biophysica Acta, 2007. 1768: p. 375-386.

[45] Real-Oliveira, M.E.C.D., et al., *Aplicação da Monooleína como Novo Lípido Adjuvante em Lipofecção,* in *Portuguese Patent n. PT104158,* I.N.d.P. Industrial, Editor. 2011. p. 1-27.

[46] Real-Oliveira, M.E.C.D., et al., *Use of Monoolein as a New Auxiliary Lipid in Lipofection,* in *International Patent n. WO2010/020935 A2,* W.I.P. Organization, Editor. 2010. p. 1-27.

[47] Tsuruta, L.R., M.M. Lessa, and A.M. Carmona-Ribeiro, *Interactions Between Dioctadecyldimethylammonium Chloride or Bromide Bilayers in Water.* Langmuir, 1995. 11: p. 2938-2943.

[48] Nascimento, D.B., et al., *Counterion Effects on Properties of Cationic Vesicles.* Langmuir, 1998. 14: p. 7387-7391.

[49] Okuyama, K., et al., *Molecular and Crystal Structure of the Lipid-Model Amphiphile Dioctadecyldimethylammonium Bromide Monohydrate.* Bulletin of the Chemical Society of Japan, 1988. 61: p. 1485-1490.

[50] Feitosa, E. and W. Brown, *Fragment and Vesicle Structures in Sonicated Dispersions of Dioctadecyldimethylammonium Bromide.* Langmuir, 1997. 13: p. 4810-4816.

[51] Schulz, P.C., et al., *Phase Behaviour of the Dioctadecylammonium Bromide-Water System.* Journal of Thermal Analysis, 1998. 51: p. 49-62.

[52] Feitosa, E., P.C.A. Barreleiro, and G. Olofsson, *Phase Transition in Dioctadecyldimethylammonium Bromide and Chloride Vesicles Prepared by Different Methods.* Chemistry and Physics of Lipids, 2000. 105: p. 201-213.

[53] Benatti, C.R., et al., *Structural and Thermal Characterization of Dioctadecyldimethylammonium Bromide Dispersions by Spin Labels.* Chemistry and Physics of Lipids, 2001. 111: p. 93-104.

[54] Proverbio, Z.E., P.C. Schulz, and J.E. Puig, *Aggregation of the Aqueous Dodecyltrimethylammonium Bromide - Didodecyldimethylammonium Bromide System at Low Concentration.* Colloid and Polymer Science, 2002. 280: p. 1045-1052.

[55] Feitosa, E., et al., *Cationic Liposomes in Mixed Didodecyldimethylammonium Bromide and Dioctadecyldimethylammonium Bromide Aqueous Dispersions Studied by Differential*

Scanning Calorimetry, Nile Red Fluorescence and Turbidity. Langmuir, 2006. 22: p. 3579-3585.

[56] Feitosa, E., G. Karlsson, and K. Edwards, *Unilamellar Vesicles Obtained by Simply Mixing Dioctadecyldimethylammonium Chloride and Bromide with Water.* Chemistry and Physics of Lipids, 2006. 140: p. 66-74.

[57] Rodriguez-Pulido, A., et al., *A Theoretical and Experimental Approach to the Compaction Process of DNA by Dioctadecyldimethylammonium Bromide-Zwitterionic Mixed Liposomes.* Journal of Physical Chemistry B, 2009. 113: p. 15648-15661.

[58] Mel'nikov, Y.S., S.M. Mel'nikova, and J.E. Lofroth, *Physico-chemical Aspects of the Interaction between DNA and Oppositely Charged Mixed Liposomes.* Biophysical Chemistry, 1999. 81: p. 125-141.

[59] Hungerford, G., et al., *Domain Formation in DODAB-Cholesterol Mixed Systems Monitored Via Nile Red Anisotropy.* Journal of Fluorescence, 2005. 15: p. 835-840.

[60] Hungerford, G., et al., *Interaction of DODAB with Neutral Phospholipids and Cholesterol Studied Using Fluorescence Anisotropy.* Journal of Photochemistry and Photobiology A: Chemistry, 2006. 181: p. 99-105.

[61] Briggs, J., H. Chung, and M. Caffrey, *The Temperature-Composition Phase Diagram and Mesophase Structure Characterization of the Monoolein-Water System.* Journal de Physique II, 1996. 6: p. 723-751.

[62] Czeslik, C., et al., *Temperature- and Pressure-Dependent Phase Behavior of Monoacylglycerides Monoolein and Monoelaidin.* Biophysical Journal, 1995. 68: p. 1423-1429.

[63] Vacklin, H., et al., *The Bending Elasticity of 1-Monoolein upon Relief of Packing Stress.* Langmuir, 2000. 16: p. 4741-4748.

[64] Caffrey, M., *A Lipid's Eye View of Membrane Protein Crystallization in Mesophases.* Current Opinion in Structural Biology, 2000. 10: p. 486-497.

[65] Caffrey, M., *Membrane Protein Crystallization.* Journal of Structural Biology, 2003. 142: p. 108-132.

[66] Carlsson, N., et al., *Bicontinuous Cubic Phase of Monoolein and Water as Medium for Electrophoresis of Both Membrane-bound Probes and DNA.* Langmuir, 2006. 22: p. 4408-4414.

[67] Dias, R. and B. Lindman, *DNA Interactions with Polymers and Surfactants.* 1st ed, ed. J.W.S. Incorporated. 2008, Hoboken (USA). 425.

[68] Bielke, W. and C. Erbacher, *Nucleic Acid Transfection.* 1st ed, ed. S. Verlag. 2010, Berlin (Germany). 316.

[69] Hofland, H.E.J., L. Shepard, and S.M. Sullivan, *Formation of Stable Cationic Lipid/DNA Complexes.* Proceedings of the National Academy of Sciences U.S.A., 1996. 93: p. 7305-7309.

[70] Oberle, V., et al., *Lipoplex Formation under Equilibrium Conditions Reveals a Three-Step Mechanism.* Biophysical Journal, 2000. 79: p. 1447-1454.

[71] Barreleiro, P.C.A., R.P. May, and B. Lindman, *Mechanism of Formation of DNA-Cationic Vesicle Complexes.* Faraday Discussions, 2002. 122: p. 191-201.

[72] Masotti, A., et al., *Comparison of Different Commercially Available Cationic Liposome-DNA Lipoplexes Parameters Influencing Toxicity and Transfection Efficiency.* Colloids and Surfaces B: Biointerfaces, 2009. 68: p. 136-144.

[73] Hui, S.W., et al., *The Role of Helper Lipids In Cationic Liposome-Mediated Gene Transfer.* Biophysical Journal, 1996. 71: p. 590-599.

[74] Farhood, H., N. Serbina, and L. Huang, *The Role of Dioleoyl Phosphatidylethanolamine in Cationic Liposome Mediated Transfer.* Biochimica et Biophysica Acta, 1995. 1235: p. 289-295.

[75] Hirsch-Lerner, D., et al., *Effect of Helper ''Lipid'' on Lipoplex Electrostatics.* Biochimica et Biophysica Acta, 2005. 1714: p. 71-84.

[76] Ciani, L., et al., *DOTAP-DOPE and DC-Chol-DOPE Lipoplexes for Gene Delivery Zeta Potential Measurements and Electron Spin Resonance Spectra.* Biochimica et Biophysica Acta, 2004. 1664: p. 70-79.

[77] Esposito, C., et al., *The Analysis of Serum Effects on Structure, Size and Toxicity of DDAB-DOPE and DC-Chol-DOPE Lipoplexes Contributes to Explain their Different Transfection Efficiency.* Colloids and Surfaces B: Biointerfaces, 2006. 53: p. 187-192.

[78] Maitani, Y., et al., *Cationic Liposome (DC-Chol-DOPE 1-2) and a Modified Ethanol Injection Method to Prepare Liposomes, Increased Gene Expression.* International Journal of Pharmaceutics, 2007. 342: p. 33-39.

[79] Rodriguez-Pulido, A., et al., *A Physicochemical Characterization of the Interaction between DC-Chol/DOPE Cationic Liposomes and DNA.* Journal of Physical Chemistry B, 2008. 112: p. 12555-12565.

[80] Smisterová, J., et al., *Molecular Shape of the Cationic Lipid Controls the Structure of Cationic Lipid-DOPE-DNA Complexes and the Efficiency of Gene Delivery.* Journal of Biological Chemistry, 2001. 276: p. 47615-47622.

[81] Yuan, X.B., *Non-Viral Gene Therapy.* 1st ed, ed. Intech. 2011, Rijeka (Croatia). 707.

[82] Kiefer, K., J. Clement, and P. Garidel, *Transfection Efficiency and Cytotoxicity of Non-viral Gene Transfer Reagents in Human Smooth Muscle and Endothelial Cells.* Pharmaceutical Research, 2004. 21: p. 1009-1017.

[83] Tenchov, B.G., R.C. MacDonald, and D.P. Siegel, *Cubic Phases in Phosphatidylcholine-Cholesterol Mixtures - Cholesterol as Membrane Fusogen.* Biophysical Journal, 2006. 91: p. 2508-2516.

[84] Huang, L., M. Hung, and E. Wagner, *Nonviral Vectors for Gene Therapy - Part I (1st Edition).* 1st ed, ed. E.A. Press. 1999, California (U.S.A.). 442.

[85] Huang, L., M. Hung, and E. Wagner, *Nonviral Vectors for Gene Therapy - Part I (2nd Edition).* 2nd ed, ed. E.A. Press. 2005, California (U.S.A.). 377.

[86] Koynova, R., L. Wang, and R.C. MacDonald, *Cationic Phospholipids Forming Cubic Phases - Lipoplex Structure and Transfection Efficiency.* Molecular Pharmaceutics, 2008. 5: p. 739-744.

[87] Koynova, R., L. Wang, and R.C. MacDonald, *An Intracellular Lamellar-Nonlamellar Phase Transition Rationalizes the Superior Performance of Some Cationic Lipid Transfection Agents.* Proceedings of the National Academy of Sciences U.S.A., 2006. 103: p. 14373-14378.

[88] Koynova, R. and R.C. MacDonald, *Mixtures of Cationic Lipid O-Ethylphosphatidylcholine with Membrane Lipids and DNA.* Biophysical Journal, 2003. 85: p. 2449-2465.
[89] Koynova, R., *Lipid Phases Eye View to Lipofection - Cationic Phosphatidylcholine Derivatives as Efficient DNA Carriers for Gene Delivery.* Lipid Insights, 2008. 2: p. 41-59.
[90] Tenchov, B.G., et al., *Modulation of a Membrane Lipid Lamellar-Nonlamellar Phase Transition by Cationic Lipids - A Measure for Transfection Efficiency.* Biochimica et Biophysica Acta, 2008. 1778: p. 2405–2412.
[91] Taira, K., K. Kataoka, and T. Niidome, *Non-viral Gene Therapy - Gene Design and Delivery.* 1st ed, ed. S. Verlag. 2005, Tokyo (Japan). 490.
[92] Wickstrom, E., *Clinical Trials of Genetic Therapy with Antisense DNA and DNA Vectors.* 1st ed, ed. M.D. Incorporated. 1998, New York (U.S.A). 421.
[93] Templeton, N.S., *Gene and Cell Therapy: Therapeutic Mechanisms and Strategies.* 2nd ed, ed. M.D. Incorporated. 2005, New York (U.S.A.). 896.
[94] Villiers, M.M., P. Aramwit, and G.S. Kwon, *Nanotechnology in Drug Delivery.* 1st ed, ed. S. Verlag. 2009, Berlin (Germany). 663.
[95] Lamprecht, A., *Nanotherapeutics - Drug Delivery Concepts in Nanoscience.* 1st ed, ed. P. Stanford. 2009, Danvers (USA). 293.
[96] Rose, J.K., *Liposomal Transfection of Nucleic Acids into Animal Cells*, in *U.S. Patent n. 005279833*, U.S.I.P. Organization, Editor. 1994. p. 1-13.
[97] Ashley, G.W., R.C. MacDonald, and M. Shida, *Cationic Phospholipids for Transfection*, in *U.S. Patent n. 005651981*, U.S.I.P. Organization, Editor. 1997. p. 1-14.
[98] Niedzinski, E.J. and M. Bennet, *Multi-Functional Polyamines for Delivery of Biologically-Active Polynucleotides*, in *International Patent n. 2003/102150*, W.I.P. Organization, Editor. 2003. p. 1-37.
[99] Barenholz, Y. and D. Simberg, *Sphingolipids Polyakylamine Conjugates for Use in Transfection*, in *International Patent n. 2004/110499*, W.I.P. Organization, Editor. 2004. p. 1-49.
[100] Reszka, R., *Cholesterol Derivative for Liposomal Gene Transfer*, in *U.S. Patent n. 005888821*, U.S.I.P. Organization, Editor. 1999. p. 1-4.
[101] Mori, H. and N. Nishikawa, *Glycerol Derivatives*, in *U.S. Patent n. 005221796*, U.S.I.P. Organization, Editor. 1993. p. 1-14.
[102] Camilleri, P., et al., *A Novel Class of Cationic Gemini Surfactants Showing Efficient In Vitro Gene Transfection Properties.* Chemical Communications, 2000. 31: p. 1253-1254.
[103] Kirby, A.J., et al., *Gemini Surfactants: New Synthetic Vectors for Gene Transfection.* Angewandte Chemie International Edition, 2003. 42: p. 1448-1457.
[104] Sekhon, B.S., *Gemini (Dimeric) Surfactants.* Resonance, 2004. 9: p. 42-49.
[105] Wettig, S.D., R.E. Verrall, and M. Foldvari, *Gemini Surfactants - A New Family of Building Blocks for Non-Viral Gene Delivery Systems.* Current Gene Therapy, 2008. 8: p. 9-23.
[106] Bombelli, C., et al., *Gemini Surfactant Based Carriers in Gene and Drug Delivery.* Current Medicinal Chemistry, 2009. 16: p. 171-183.
[107] Palmer, L.R., et al., *Transfection Properties of Stabilized Plasmid-Lipid Particles Containg Cationic PEG Lipids.* Biochimica et Biophysica Acta, 2003. 1611: p. 204-216.

[108] Buyens, K., et al., *Elucidating the Encapsulation of Short Interfering RNA in PEGylated Cationic Liposomes.* Langmuir, 2009. 25: p. 4886-4891.

[109] Kim, J.Y., et al., *The Use of PEGylated Liposomes to Prolong Circulation Lifetimes of Tissue Plasminogen Activator.* Biomaterials, 2009. 30: p. 5751-5756.

[110] Harvie, P., F.M.P. Wong, and M.B. Bally, *Use of Poly(ethylene glycol)-Lipid Conjugates to Regulate the Surface Attributes and Transfection Activity of Lipid-DNA Particles.* Journal of Pharmaceutical Sciences, 2000. 89: p. 652-663.

[111] Veronese, F.M., *PEGylated Protein Drugs: Basic Science and Clinical Applications (Milestones in Drug Therapy).* 1st ed, ed. B. Basel. 2009, Basel (Switzerland). 297.

[112] Sakaguchi, N., et al., *The Correlation Between Fusion Capability and Transfection Activity in Hybrid Complexes of Lipoplexes and pH-Sensitive Liposomes.* Biomaterials, 2008. 29: p. 4029-4036.

[113] Wasungu, L., et al., *Transfection Mediated by pH-Sensitive Sugar-Based Gemini Surfactants; Potential for In Vivo Gene Therapy Applications.* Journal of Molecular Medicine, 2006. 84: p. 774-784.

[114] Chang, E.H., L. Xu, and K. Pirollo, *Targeted Liposome Gene Delivery,* in *U.S. Patent n. 006749863,* U.S.I.P. Organization, Editor. 2004. p. 1-25.

[115] Gorman, C.M. and M. McClarrinon, *Cationic Lipid-DNA Complexes for Gene Targeting,* in *U.S. Patent n. 005830878,* U.S.I.P. Organization, Editor. 1998. p. 1-24.

[116] Hattori, Y. and Y. Maitani, *Folate-Linked Lipid-Based Nanoparticle for Targeted Gene Delivery.* Current Drug Delivery, 2005. 2: p. 243-252.

[117] Petrak, K., *Essential Properties of Drug-Targeting Delivery Systems.* Drug Discovery Today, 2005. 10: p. 1667-1673.

[118] Russ, V. and E. Wagner, *Cell and Tissue Targeting of Nucleic Acids for Cancer Gene Therapy.* Pharmaceutical Research, 2007. 24: p. 1047-1057.

[119] Yang, P., et al., *Enhanced Gene Expression in Epithelial Cells Transfected With Aminoacid-Substituted Gemini Nanoparticles.* European Journal of Pharmaceutics and Biopharmaceutics, 2010. 75: p. 311-320.

[120] Rosa, M., et al., *DNA Pre-Condensation with an Amino Acid-based Cationic Amphiphile - A Viable Approach for Liposome-Based Gene Delivery.* Molecular Membrane Biology, 2008. 25: p. 23-34.

[121] Letrou-Bonneval, E., et al., *Galactosylated Multimodular Lipoplexes for Specific Gene Transfer into Primary Hepatocytes.* Journal of Gene Medicine, 2008. 10: p. 1198-1210.

[122] Higuchi, Y., et al., *Effect of the Particle Size of Galactosylated Lipoplex on Hepatocyte-Selective Gene Transfection after Intraportal Administration.* Biological and Pharmaceutical Bulletin, 2006. 29: p. 1521-1523.

[123] Audouy, S. and D. Hoekstra, *Cationic Lipid-Mediated Transfection In Vitro and In Vivo (Review).* Molecular Membrane Biology, 2001. 18: p. 129-143.

[124] Bihan, O.L., et al., *Probing the In Vitro Mechanism of Action of Cationic Lipid-DNA Lipoplexes at a Nanometric Scale.* Nucleic Acids Research, 2010. 1: p. 1-15.

[125] Gregoriadis, G., *Liposome Technology - I - Liposome Preparation and Related Techniques.* 3rd ed, ed. I. Healthcare. 2007, New York (USA). 660.

[126] Gregoriadis, G., *Liposome Technology - II - Entrapment of Drugs and Other Materials Into Liposomes.* 3rd ed, ed. I. Healthcare. 2007, New York (USA). 424.

[127] Gregoriadis, G., *Liposome Technology - III - Interactions of Liposomes with the Biological Milieu.* 3rd ed, ed. I. Healthcare. 2007, New York (USA). 464.

[128] Real-Oliveira, M.E.C.D., et al., *Use of Monoolein as a New Auxiliary Lipid in Lipofection,* in *European Patent n. EP2335687 A2,* E.P. Office, Editor. 2011. p. 1-27.

[129] Zuhorn, I.S., et al., *Interference of Serum with Lipoplex-cell Interaction.* Biochimica et Biophysica Acta, 2002. 1560: p. 25-36.

[130] Yang, J.P. and L. Huang, *Overcoming the Inhibitory Effect of Serum on Lipofection by Increasing the Charge Ratio of Cationic Liposome to DNA.* Gene Therapy, 1997. 4: p. 950-960.

[131] Zhang, Y. and T.J. Anchordoquy, *The Role of Lipid Charge Density in the Serum Stability of Cationic Lipid-DNA Complexes.* Biochimica et Biophysica Acta, 2004. 1663: p. 143-157.

[132] Conwell, C.C., F. Liu, and L. Huang, *Gene Transfer Activity is Enhanced Significantly by Allowing Cationic Polymer to Interact With Serum Proteins Prior to DNA Addition.* Molecular Therapy, 2005. 11: p. pp. S80.

[133] Arpke, R.W. and P.W. Cheng, *Characterization of Human Serum Albumin-Facilitated Lipofection Gene Delivery Strategy.* Journal of Cell Science & Therapy, 2011. 2: p. 108-114.

[134] Mounkes, L.C., et al., *Proteoglycans Mediate Cationic Liposome-DNA Complex-Based Gene Delivery In Vitro and In Vivo.* Journal of Biological Chemistry, 1998. 27: p. 26164-26170.

[135] Wiethoff, C.M., et al., *The Potential Role of Proteoglycans in Cationic Lipid-mediated Gene Delivery.* Journal of Biological Chemistry, 2001. 276: p. 32806-32813.

[136] Silva, M.E. and C.P. Dietrich, *Structure of Heparin.* Journal of Biological Chemistry, 1975. 25: p. 6841-6846.

[137] Rosenberg, R.D., G. Armand, and L. Lam, *Structure-Function Relationships of Heparin Species.* Proceedings of the National Academy of Sciences U.S.A, 1978. 75: p. 3065-3069.

[138] Mascotti, D.P. and T.M. Lohman, *Thermodynamics of Charged Oligopeptide-Heparin Interactions.* Biochemistry, 1995. 34: p. 2908-2915.

[139] Zuhorn, I.S., R. Kalicharan, and D. Hoekstra, *Lipoplex-Mediated Transfection of Mammalian Cells Occurs Through the Cholesterol-Dependent Clathrin-Mediated Pathway of Endocytosis.* Journal of Biological Chemistry, 2002. 27: p. 18021-18028.

[140] Habib, N.A., *Cancer Gene Therapy - Past Achievements and Future Challenges.* 1st ed, ed. K.A. Publishers. 2002, New York (U.S.A.). 458.

[141] Ilarduya, C.T., Y. Sun, and N. Düzgünes, *Gene Delivery by Lipoplexes and Polyplexes.* European Journal of Pharmaceutical Sciences, 2010. 40: p. 159-170.

[142] Resina, S., P. Prevot, and A.R. Thierry, *Physico-Chemical Characteristics of Lipoplexes Influence Cell Uptake Mechanisms and Transfection Efficacy.* Public Library of Science ONE, 2009. 4: p. e6058 (1-11).

[143] Ferrari, M.E., et al., *Trends in Lipoplex Physical Properties Dependent on Cationic Lipid Structure, Vehicle and Complexation Procedure do ot Correlate with Biological Activity.* Nucleic Acids Research, 2001. 29: p. 1539-1548.

[144] Hoekstra, D., et al., *Gene Delivery by Cationic Lipids In and Out of an Endosome.* Biochemical Society Transactions, 2007. 35: p. 68-71.

[145] Elouahabi, A. and J. Ruysschaert, *Formation and Intracellular Trafficking of Lipoplexes and Polyplexes*. Molecular Therapy, 2005. 11: p. 336-347.
[146] Bhushan, B., *Handbook of Nanotechnology*. 2nd ed, ed. S. Verlag. 2010, Berlin (Germany). 1968.
[147] Caracciolo, G., et al., *Efficient Escape from Endosomes Determines the Superior Efficiency of Multicomponent Lipoplexes*. Journal of Physical Chemistry B, 2009. 113: p. 4995-4997.
[148] Liu, F., et al., *Effect of Non-Ionic Surfactants on the Formation of DNA-Emulsion Complexes and Emulsion-Mediated Gene Transfer*. Pharmaceutical Research, 1996. 13: p. 1642-1646.
[149] Tan, Y., et al., *Sequential Injection of Cationic Liposome and Pasmid DNA Effectively Transfects the Lung with Minimal Inflammatory Toxicity*. Molecular Therapy, 2001. 3: p. 673-682.
[150] Yasuda, S., et al., *Comparison of the Type of Liposome Involving Cytokine Production Induced by Non-CpG Lipoplex in Macrophages*. Molecular Pharmaceutics, 2010. 7: p. 533-542.

Chapter 10

Enhancement of Homologous Recombination Efficiency by Homologous Oligonucleotides

Hidekazu Kuwayama

Additional information is available at the end of the chapter

1. Introduction

Gene targeting is a powerful technology to achieve gene ablation and modification in the study of gene function by phenotypic analysis. This method is widely recognized as being useful; however, its application is not yet versatile. Its main limitation is the small number of cells in which gene replacement occurs efficiently after endogenous homologous recombination. Artificial enhancement of homologous recombination has been hardly successful. Recently, however, several techniques have been reported that can increase the efficiency of homologous recombination. In this chapter, I first summarize the principles and applications of those techniques. Next, I focus on a simple technique, in which the addition of oligonucleotides, homologous to the targeted locus, significantly increases the efficiency of homologous recombination and, subsequently, the number of genetically targeted clones. The greatest benefit of oligonucleotide-aided homologous recombination is its versatility, i.e., its applicability to virtually any cell type. Finally, the presumed molecular mechanisms underlying oligonucleotide-aided homologous recombination are presented.

2. Gene targeting technology

The ultimate goal of genetic molecular biology is to modulate the activity of genes at will. In gene targeting technology, in vivo homologous recombination enables the replacement of a target genomic region with an exogenous DNA fragment that contains a region homologous to the targeted locus (Fig. 1). This technology is indispensable for the analysis of gene function. To acknowledge its importance, the discovery of the principles of gene targeting in mice was awarded the Noble Prize in 2007. Recently, applications of this technology have been expanded to gene therapy and transgenic plants. However, the success of the technique greatly depends on the efficiency of homologous recombination; therefore, it cannot be successfully applied to cells with low homologous recombination efficiency, such as mammalian somatic cells and higher plant cells. Furthermore, in multiploid somatic cells, simultaneous gene targeting is required, making complete gene replacement extremely difficult. Even in embryonic stem (ES) cells, where homologous recombination occurs with high efficiency, some genes are difficult to target and subject to homologous recombination. To overcome these problems, the development of efficient and versatile methods that can artificially increase the efficiency of homologous recombination is needed.

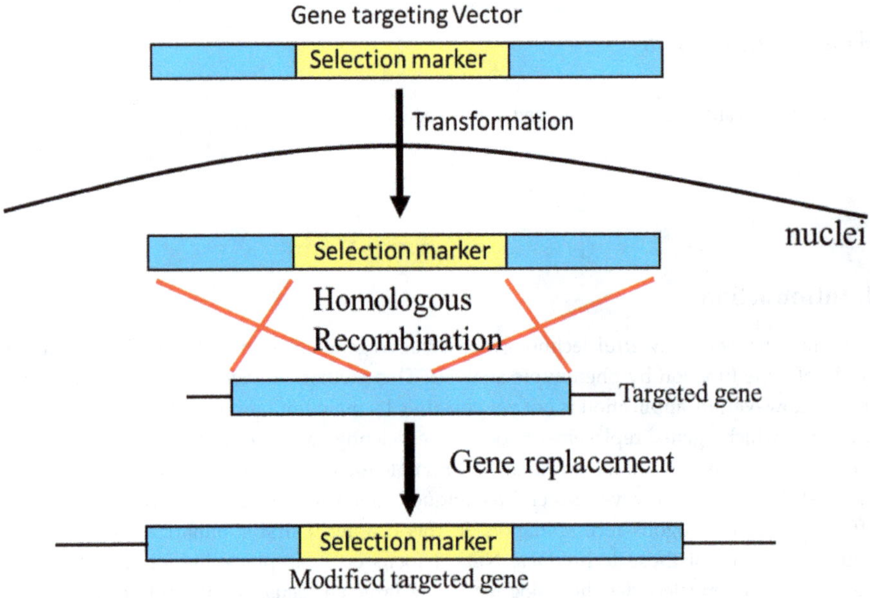

Figure 1. Schematic diagram of principle of Gene Targeting

When extracellular DNA fragment is introduced into cell, gene replacement rarely occurs by homologous recombination. Utilizing this phenomenon, one can 'target' a gene of interest and change its DNA alignment at will.

3. Introduction of vector DNA into cells

Introduction of a targeting vector (DNA fragment) into a cell is the primary step to targeting genes. In general, there are 3 types of strategies to introduce a DNA fragment into a cell: biological, chemical, and physical. The biological method, using a virus vector, yields high transformation efficiencies. However, there remains the risk of insertion of viral vector genes into the host genome with selectivity for the virus. The chemical method, usually utilizing a polycationic polymer, is easy to perform. However, the polymer can be recognized as a foreign substance and become enclosed within endosomes, where it is digested along with the transformation vector. It is possible to make the polymer escape this digesting pathway, but the strategy is generally not very efficient (Colosimo et al., 2003).

Physical methods such as microinjection and electroporation are relatively versatile and most frequently used in gene-targeted cell transformation (Niidome et al., 2002).

4. Construction of gene targeting vectors

In gene targeting, a linear DNA fragment is used as targeting vector. The targeting vector consists of 2 homologous regions to the genome and a drug resistance gene for the selection of transformants (Fig. 1). Drug resistance gene products can degrade drugs such as G418, hygromycin B, puromycin, and blasticidin S, which are toxic to untransformed cells, thus facilitating the survival of the transformed cells in drug-containing medium. The targeting vector is constructed by fusing the 3 DNA fragments in tandem by using basic molecular biological techniques, such as PCR, restriction enzyme digestion, and cloning techniques. Interestingly, Kuwayama et al. have described the construction of a gene targeting vector by PCR only, but unfortunately not in sufficient detail for reproduction (Kuwayama et al., 2002).

Briefly, in the first step, three separate PCR syntheses of a selectable marker cassette and the 5'- and 3'-homologous regions of a target gene. Of the four primers used in amplification of the 5'- and 3'-regions of the target gene, two primers placed proximal to the site of the marker cassette are designed to have sequence tags complementary to the 5'- or 3'-side of the marker cassette. The two primers used in PCR synthesis of the marker cassette are complementary to the tagged primers. By fusion PCR, the 5' and 3' PCR products are connected to the marker cassette via the regions of tagged primers that overlap. And then, a sufficient amount of the disruption construct can be directly amplified with the outermost primers (Fig. 2).

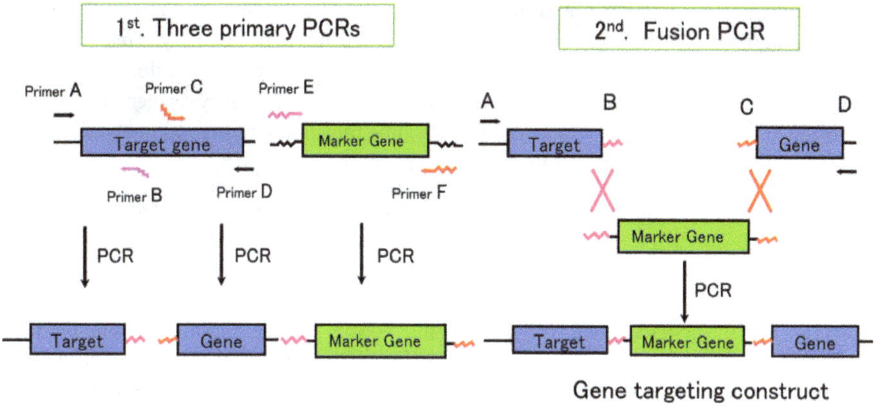

Figure 2. Schematic diagram of PCR-dependent construction of gene targeting vector.

Step 1, the three primary PCR reactions. The 5'- and 3'-flanking regions are amplified with primers (primers A - F) specific for the sequence of the target gene. The primers distal to the selectable marker insertion site are simple primers complementary to the target gene (primers A and D). The primers directly adjacent to the marker cassette are chimeric (primers B and C). Primers complementary to primers B and C are used to amplify the selection marker gene (primers E and F). Step 2, fusion PCR. The 5´- and 3´-flanking regions are joined to the marker gene and the final PCR product is amplified with the outermost primers A and D. The order in which the flanking sequences are joined to the ends of the selectable marker cassette is discretionary. The final PCR reaction mix containing the targeting vector is subjected to ethanol precipitation and can be directly used to transform cells.

For the transformation, the targeting vector must be linearized by enzymatic digestion at a site outside the insert region; alternatively, the insert region may be amplified by PCR before transformation. One of the critical points in constructing targeting vectors is the length of the homologous regions. Optimum length of the homologous regions for gene targeting varies and depends on the organism, cell type, and targeted locus. In practice, the length of the homologous DNA region, which is cloned or amplified by PCR, is limited. In addition, an excessively long DNA fragment is difficult to introduce into cells. Therefore, the homologous DNA fragment is generally designed to be 5–8 kb in length (Hasty et al., 1993).

5. Principles and applications of gene targeting methods

Introduced DNA fragments containing homologous sequences to genomic DNA rarely induce target gene replacement (Hasty & Bradley, 1993; Rouet et al., 1994; Smih et al., 1995; Jasin, 1996; Mamsour et al, 1988). Notwithstanding its low occurrence, this reaction (so-called homologous recombination) is considered the driving force of evolution and diversity in species. In a test tube, it is possible to manipulate the genomic DNA at will by means of gene targeting in organisms and/or cells with high transformation and endogenous homologous recombination efficiencies. However, even in such model systems, targeting vector transformation results in a high proportion of non-targeted (randomly inserted) transformants. This is because homologous recombination efficiency is generally much lower than genomic insertion efficiency. To overcome this problem, several methods, which eliminate non-targeted transformants, have been proposed.

One such method is positive/negative selection (Mamsour et al., 1988). This method, based on the addition of a negative selection marker gene in one or both ends of the gene targeting vector, aims at the enrichment of the small fraction of cells in which homologous recombination took place. In case of random insertion, the negative selection marker is integrated into the genome along with the targeting vector. The negative marker gene product eliminates non-targeted host cells, and only homologously recombined clones can survive by removal of the negative selection marker gene during homologous recombination. However, the added negative selection marker gene may reduce homologous recombination efficiency. In another method, the "promoterless" method, the promoter region of the selection marker gene is removed and the marker gene can be translated only when homologous recombination occurs. However, when the endogenous promoter activity is low, the marker gene is not expressed at levels high enough to degrade the selection drug.

Any of the 2 abovementioned methods can reduce the number of non-targeted clones, and minimize the time and effort required for selection of the targeted clones. However, none of them achieve the ideal removal of non-targeted clones. In fact, the ratio of targeted clones to non-targeted clones can be as low as 0.1%; in cultured mammalian cells, this percentage can reach up to 20% (Sedivy & Dutriaux, 1999). To increase the proportion of targeted clones, a method has been developed that suppresses non-homologous recombination by mutagenizing host cells. This method targets ku70/ku80 genes, which encode proteins that bind the ends of DNA linear fragments (Kooistra et al., 2004 ; Ninomiya et al., 2004). Without these genes, host cells suppress non-homologous recombination and, as a result, homologous recombination efficiency increases. However, application of this method is restricted to host cells possessing ku70/ku80 orthologous genes. Moreover, the effects of elimination of these genes should be carefully examined in each particular case.

6. Increased homologous recombination efficiency by artificial methods

Some methods have been proposed that artificially increase endogenous homologous recombination efficiency. The common basis of these methods relies on the observation that the occurrence of a DNA double-strand break (DSB) in the targeted region dramatically increases homologous recombination efficiency. Hence, artificial induction of DSB would effectively increase gene targeting efficiency, to (Hasty & Bradley, 1993; Rouet et al., 1994; , Jasin, 1996). At least 3 different DSB-inducing methods have been reported. *I-SceI* is a highly specific restriction enzyme that recognizes an 18-bp-long DNA sequence (Fig. 3). When the recognition site exists within the targeted domain, co-transformation of *I-SceI* with targeting vector results in specific digestion of the genomic DNA at the recognition site, increasing homologous recombination efficiency. The limitations of the method are the necessary pre-existence of a *I-SceI* site in the target region and the absence of that same site from the targeting vector. Moreover, there should be ideally no other *I-SceI* restriction site in the host genome. Therefore, application of the method is restricted to certain organisms and cells.

Another method utilizes a nuclease fused with engineered C2H2 zinc finger protein-based DNA-binding domains, which bind sequences specifically at the targeting region and cause site-directed DSBs (Fig. 4). A limitation of this method is the effort required to accurately and carefully design the zinc finger DNA-binding domain, because binding specificity and affinity are the critical determinants of recombination efficiency. Furthermore, simultaneous expression of zinc finger nucleases (ZFNs) with the gene-targeting construct is also indispensable (Urnov et al., 2005).

Triplex-forming oligonucleotides (TFOs) are known to induce a DNA DSB and repair system and, thus, are expected to increase homologous recombination efficiency *in vivo* (Fig. 5). However, TFOs that bind to double-stranded DNA are restricted to the polypurine or polypyrimidine tract; therefore, this technology is limited to segments with unique target sites (Demidov, 2003).

When the recognition site of I-SceI inserted into the targeted domain, co-transformation of I-SceI enzyme with targeting vector results in specific digestion of the genomic DNA at the recognition site, increasing homologous recombination efficiency.

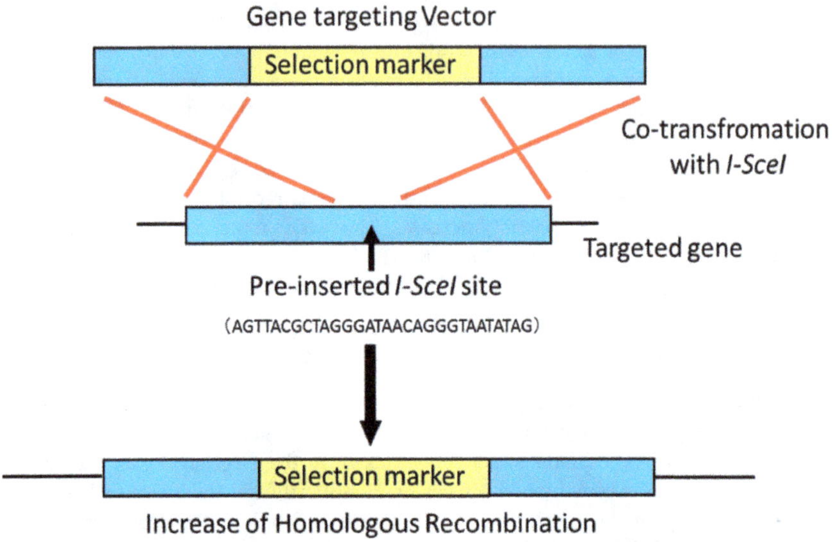

Figure 3. Schematic diagram of Restriction Enzyme (I-SceI) dependent increase of homologous recombination efficiency.

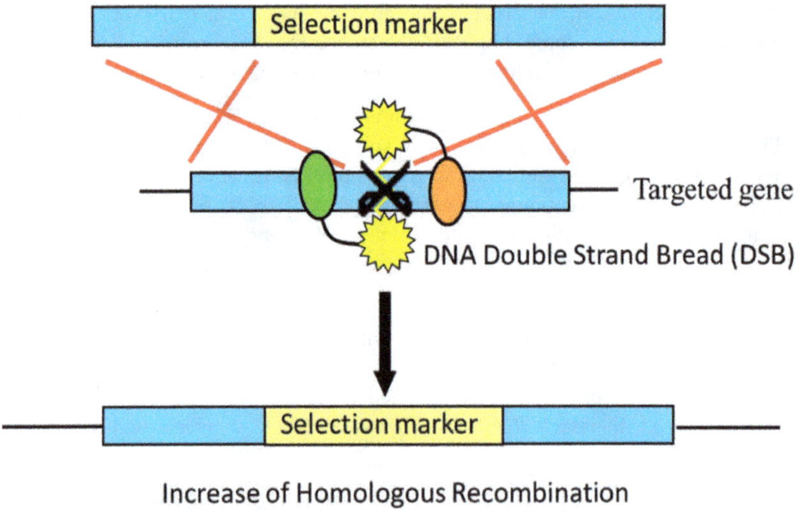

Figure 4. Schematic diagram of zinc finger nucleases (ZFNs) dependent increase of homologous recombination efficiency.

By expressing engineered C2H2 zinc finger protein-based DNA-binding domains which bind sequences specifically at the targeting region, a site-directed DSBs occurs and, thus, increases homologous recombination efficiency.

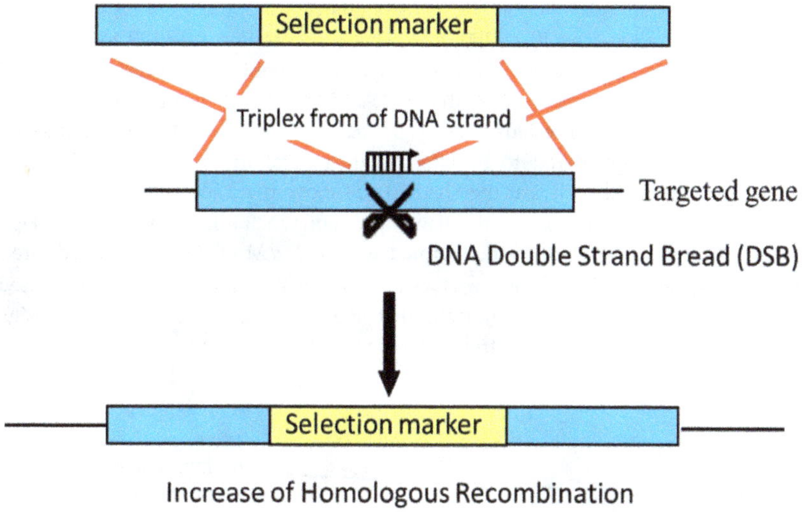

Figure 5. Schematic diagram of Triplex-forming oligonucleotides (TFOs) dependent increase of homologous recombination efficiency.

When a TFO is formed in genome, a DNA DSB is induced at the site and, thus, homologous recombination efficiency increases.

7. Enhancement of homologous recombination efficiency by homologous oligonucleotides

Recently, I described a general gene targeting method in which co-transformation of DNA oligonucleotides (oligomers) could significantly increase homologous recombination frequency and transformation efficiency (Kuwayama et al., 2008). This method is based on the experience that a high concentration of gene-targeting construct generally provides considerably better transformation and homologous recombination efficiencies. However, the amount of gene targeting construct that can be used at each transformation is limited and, furthermore, preparing a large amount of vector DNA is demanding. In order to overcome this limitation, I tested whether addition of only a part of the homologous region of a gene-targeting construct was as effective as the entire construct (Fig. 6).

A single-stranded DNA oligomer is much smaller than the gene targeting vector, and thus, it can be introduced into cells in much larger amounts than the targeting vector. Using the cellular slime mold model organism, *Dictyostelium discoideum*, and mammalian Hela cells, the effect of co-transformation of short homologous DNA oligomers was tested. In D. discoideum, the gbfA gene locus was chosen to be targeted because this gene was reportedly difficult to replace with an endogenous targeting vector. By electroporation, the gene targeting vector and homologous strand of short DNA oligomers were simultaneously transformed into *D. discoideum* cells. The DNA oligomer was about 20 bases in length, and the added concentration was 10 to 100 μM. This concentration was 100 to 10,000 times higher than that of the gene targeting vector. As a result, homologous recombination as well as transformation efficiencies significantly increased. Since this positive effect was also observed with all the genes tested—*pkaC, gbfB, ctxA,* and *ctxB*—addition of homologous DNA oligomer was considered to be effective in general in *D. discoideum*. The tested oligomers were designed such that they had 20 - 24 monomers, and the sequences at both the ends were identical to those of the flanking regions in the inward direction (Fig. 6). When the wild-type cells were co-transformed with 100 μM of the two inward-directed oligomers, the gene targeting efficiency as well as the transformation efficiency increased in all cases (Fig. 7). These results indicate that the co-transformation of the designed homologous oligomers increases the transformation efficiency.

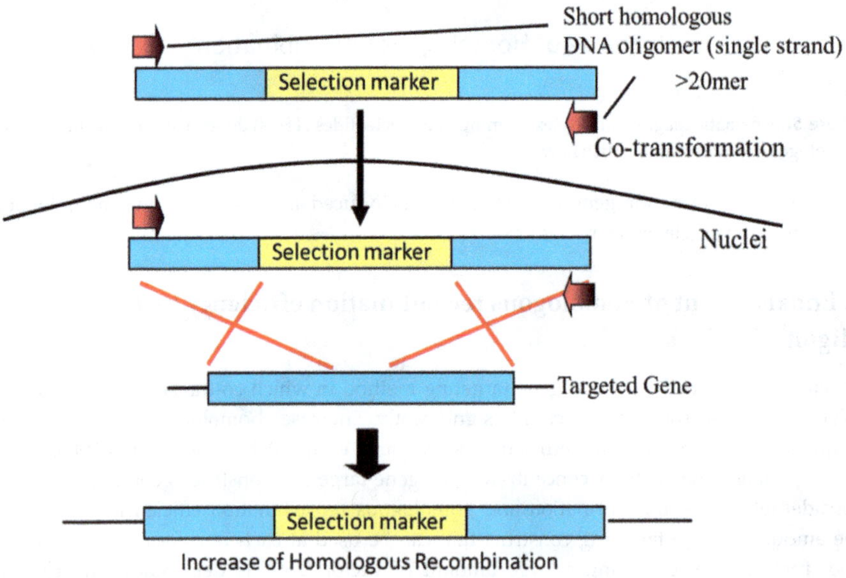

Figure 6. Schematic diagram of increase of homologous recombibation efficiency by short hologous DNA oligomers.

Furthermore, we also observed that, in diploid cells, sister alleles were simultaneously recombined with the targeting vectors. As this effect was also observed in the human cell line, Hela, it was suggested that this effect is not specific to *D. discoideum* cells but is general to all eukaryotic cells.

Why does the simple addition of homologous short DNA oligomers increase specific homologous recombination efficiency? It is not likely that TFOs are responsible because oligomers lacking TFO signature sequences are also effective. Hence, site-specific DSB does not seem to occur. One possibility is that short DNA oligomers can easily be introduced into the nuclei, affecting chromatin structure at the targeting locus and enhancing the interaction between the genomic targeted region and the targeting vector. Another possibility is that the added DNA oligomers contribute to the increase in the number of targeting vector molecules in the cell.

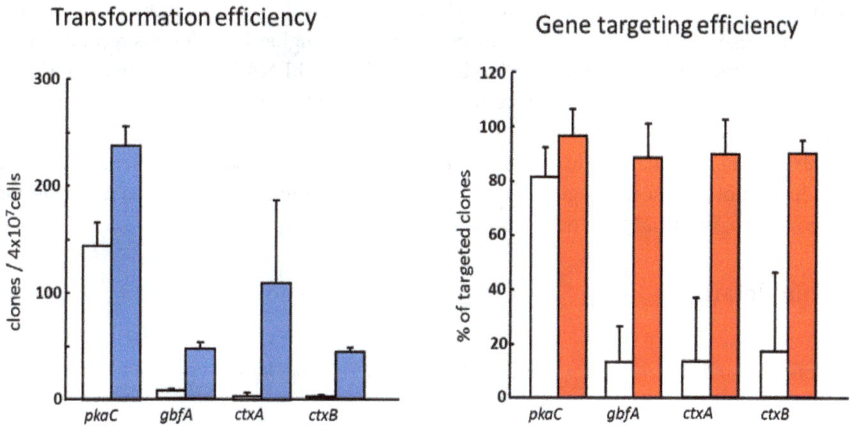

Figure 7. Increase of the homologous oligomers on the transformation and homologous recombination efficiencies.

Co-transformation of high concentration of short homologous DNA oligomers results in significant increase of homologous recombination. This technique also increases transformation efficiency, although the detailed mechanism is not unraveled, yet.

Transformation efficiencies were examined for the pkaC, gbfA, ctxA, and ctxB gene - targeting constructs. The data are represented as the number of primary transformants per transformation (2×10^7 cells). White bars represent transformation without oligomers. Blue bars represent transformation with 100 μM inward oligomers. Homologous recombination efficiencies of the gene- targeting construct without the oligomers (white bar) and with 100 μM oligomers (Red bar) are represented as an average percentage of the gene targeted transformants to the total number of transformants. Bars represent standard deviation (SD) of 3 independent transformations.

8. Conclusion

The DNA oligomer-aided homologous recombination herein presented can, in principle, be applied to any general transformation method, including microinjection and lipofection, and cell line, including mammalian transformed and primary somatic cells. Furthermore, no cytotoxicity and no random insertion of DNA oligomers into the genome were observed (unpublished data). Although the reason underlying the increase in the homologous recombination efficiency after oligomer addition remains unknown at present, it may become possible in the future to design oligomers to target the most effective position at the locus of interest.

Further improvements in this method are expected to result in practical and clinically safe therapeutic modifications of human cells, in particular, by using artificial nucleic acid analogues such as peptide nucleic acid (PNA) and 2´-O or 4´-C locked nucleic acid (LNA). The use of these analogues is expected to provide higher homologous recombination frequency at low oligomer concentration because PNA and LNA have an increased affinity to native DNA and a high resistance to nucleases, thereby imparting higher biostability (Demidv, 2003). Furthermore, these analogues are low in toxicity (Wahlestedt et al., 2000; Kaihatsu et al., 2004). In the future, it is hoped that this method will contribute to development of genetically engineered high-efficiency yielding transformation methods and to innovation of epochal gene delivery systems.

Author details

Hidekazu Kuwayama
Faculty of Life and Environmental Sciences, University of Tsukuba, Japan

Acknowledgement

I thank Dr. Hideko Urushihara for the helpful discussions and encouragements. This work was supported by Grants-in-Aid for Scientific Research (S) (no. 15109003) from the Japan Society for the Promotion of Science and by the special fund for tenure-track faculty members of the Institute of Biological Sciences at the University of Tsukuba.

9. References

Cohen-Tannoudji, M. Robine, S. Choulika, A. Pinto, D. El Marjou, F. Babinet, C. Louvard, D. & Jaisser. F, (1998) I-SceI-induced gene replacement at a natural locus in embryonic stem cells. Mol Cell Biol Vol. 18, No. 3, pp. 1444–1448

Colosimo A, Goncz KK, Holmes AR, Kunzelmann K, Novelli G, Malone RW, Bennett MJ, & Gruenert DC. (2000) Transfer and expression of foreign genes in mammalian cells. Biotechniques Vol. 29, No. 2, pp. 314–331

Demidv, V.V. (2003) PNA and LNA throw light on DNA. Trends Biotechnol Vol.21, No. 1, pp. 4–7

Hasty, P. & Bradley, A. (1993) Gene targeting vectors for mammalian cells. In: Joyner AL, editor. Gene targeting: a practical approach, Oxford:IRL Press. pp. 1–31

Jasin, M. Genetic manipulation of genomes with rare-cutting endonucleases. (1996) Trends Genet Vol, 12, No 6, pp. 224–228

Kaihatsu, K. Huffman, K.E. & Corey, D.R. (2004) Intracellular uptake and inhibition of gene expression by PNAs and PNA-peptide conjugates. Biochemistry Vol. 43, No. 45, pp. 14340-14347.

Kooistra, R. Hooykaas, P. J. & Steensma, H.Y. Efficient gene targeting in Kluyveromyces lactis. (2004) Yeast Vol. 21, No. 9, pp. 781–792

Kuwayama, H. Obara, S. Morio, T. Katoh, M. Urushihara, H. & Tanaka, Y. (2002) Nucleic Acids Res Vol. 30, e2

Kuwayama, H. Yanagida, T. & Ueda, M. (2008) DNA oligonucleotide-assisted genetic manipulation increases transformation and homologous recombination efficiencies: Evidence from gene targeting of *Dictyostelium discoideum*. J Biotechnol Vol. 133, No. 4, pp. 418–423

Mamsour, S. L. Thomas, K. R. & Capecchi, M.R. (1988) Disruption of the proto-oncogene int-2 in mouse embryo-derived stem cells: a general strategy for targeting mutations to non-selectable genes. Nature Vol. 336, No. 6197, pp. 348–352

Niidome, T. & Huang, L. (2002) Gene therapy progress and prospects: nonviral vectors. Gene Ther Vol. 9, No. 24, pp. 1647–1652

Ninomiya, Y. Suzuki, K. Ishii, C. & Inoue, H. Highly efficient gene replacements in Neurospora strains deficient for nonhomologous end-joining. (2004) Proc Natl Acad Sci USA Vol. 101, No. 33, pp. 12248–12253

Rouet, P. Smih, F. & Jasin, M. (1994) Introduction of double-strand breaks into the genome of mouse cells by expression of a rare-cutting endonuclease. Mol Cell Biol Vol. 14, No, 12, pp. 8096–8106

Sedivy, J. M. & Dutriaux, A. (1999) Gene targeting and somatic cell genetics--a rebirth or a coming of age? Trends Genet Vol. 15, No. 3, pp. 88–90

Smih, F. Rouet, P. Romanienko, P. J. & Jasin, M. (1995) Double-strand breaks at the target locus stimulate gene targeting in embryonic stem cells. Nucleic Acids Res Vol. 23, No. 24, pp. 5012–5019

Urnov, F. D. Miller, J. C. Lee, Y.L. Beausejour, C. M. Rock, J. M. Augustus, S. Jamieson, A.C. Porteus, M. H. Gregory, P. D. & Holmes, M.C. (2005) Highly efficient endogenous human gene correction using designed zinc-finger nucleases. Nature Vol. 435, No. 7042, pp. 646–651

Wahlestedt, C. Salmi, P. Good, L. Kela, J. Johnsson, T. Hökfelt, T. Broberger, C. Porreca, F. Lai, J. Ren, K. Ossipov, M. Koshkin, A. Jakobsen, N. Skouv, J. Oerum, H. Jacobsen, M.H. & Wengel, J. (2000) Potent and nontoxic antisense oligonucleotides containing locked nucleic acids. Proc Natl Acad Sci USA Vol. 97, No, 10. pp. 5633-5638

Chapter 11

Micropatterned Coatings for Guided Tissue Regeneration in Dental Implantology

A. Pelaez-Vargas, D. Gallego-Perez, N. Higuita-Castro, A. Carvalho, L. Grenho, J.A. Arismendi, M.H. Fernandes, M.P. Ferraz, D.J. Hansford and F.J. Monteiro

Additional information is available at the end of the chapter

1. Introduction

Dento-alveolar trauma and congenital absences are the most important causes of edentulism that are not associated with bacteria. However, the World Health Organization reports show that dental caries and periodontitis, two conditions of bacterial origin, are the most frequent oral diseases in humans [1]. These conditions might be avoided if an adequate oral preventive health policy is implemented, including preventive and educational measures that, regardless of the population´s socioeconomic factors, have shown their effectiveness. Despite these facts, tooth extraction[1], defined as the surgical removal of a tooth, is currently the most frequent surgical procedure in the world [1].

Tooth loss or edentulism affects the aesthetics and function of the stomatognathic system. The missing interproximal contact produces an intra-arch imbalance that is visible as dental misalignment and the formation of anterior/posterior diastema. Additionally, the tooth distal to the extraction site will drift mesially into the space, thus creating an oclusal collapse. Inter-arch disharmonies are observed as occlusal collapse, supereruption of antagonist teeth, and alteration of the vertical dimension of occlusion. The synergy of inter-/ intra- arch disequilibrium is associated with ATM dysfunction, muscle hyperactivity, nutritional imbalances, tooth wear, mobility, and potential harmful contact areas during mandibular eccentric movements or otologic symptoms [2].

Recovering aesthetics and function is only possible using some oral rehabilitation procedures such as fixed or removable prostheses. In the fixed restorations, titanium implant-based therapy appears as the "gold standard', considering that successful rates of ~95% after 5 years have been reported.

[1] Tooth extraction was indexed (1965) in the MESH of the National Library of Medicine (NLM). Some synonyms are dental extraction, exodontia or pulling teeth.

The history of implantology has been divided in two different phases, namely the pre-osseointegration and post-osseointegration eras. Many audacious designs were developed to be used in sub- or endosteal areas in the former. The most commonly used implants were blades and plaques based on metallic alloys such as cobalt-chromium-molybdenum and stainless steel (Figure 1). These types of devices have been associated to doubtful long-term clinical success. Initial clinical research reported by Branemark was criticized, but the basis for a new philosophy in dental therapy had been initiated. After completing training in Sweden, Zarb and colleagues from Toronto started a longitudinal study to verify the

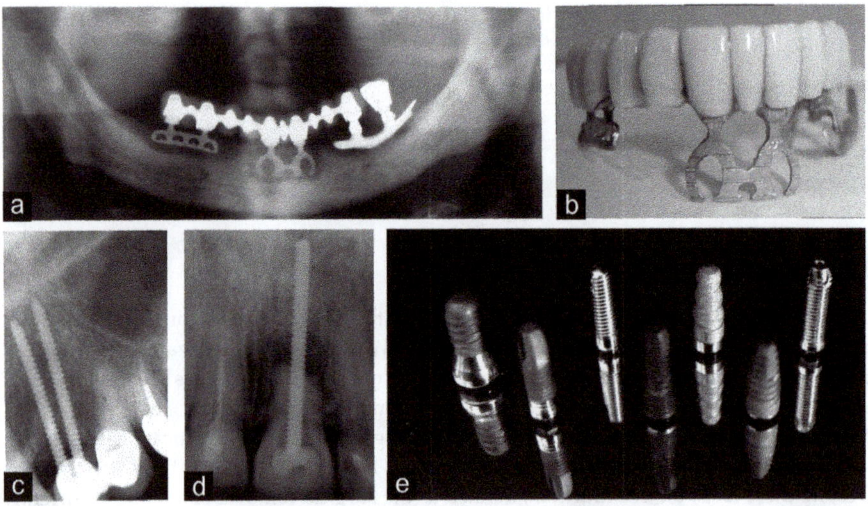

Figure 1. History of dental implants. (a,b) Panoramic radiography and general view of a failed total rehabilitation supported on blade/plate-shaped implants. (c) Parallel pins implant, (d) Endosteal endodontic stabilizers and (e) different designs and surfaces of screw dental implants.

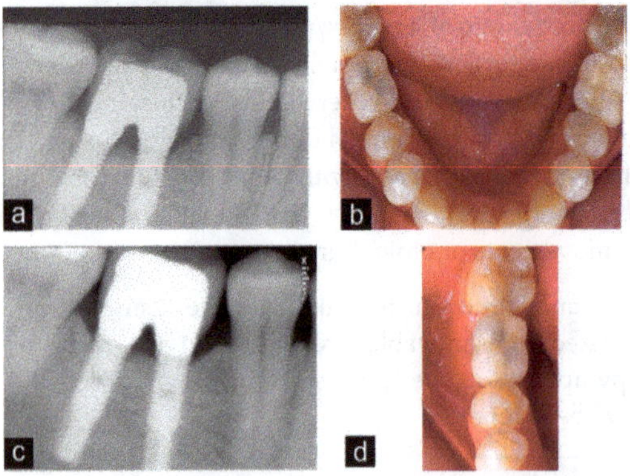

Figure 2. Double implants to replace a mandibular molar. (a,b) Periapical radiography and occlusal view of the final rehabilitation. (c,d) Periapical radiography and detailed occlusal view after 10 years of function.

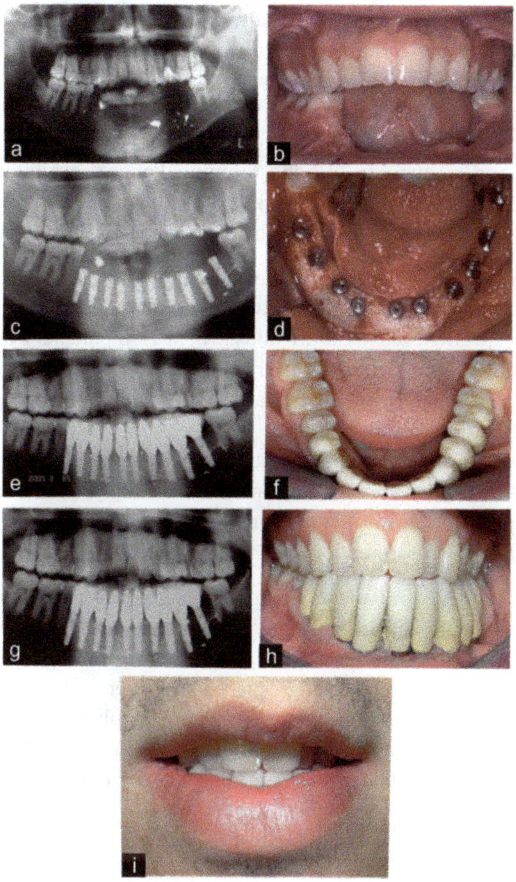

Figure 3. Ten-year follow-up of a successful implant-based rehabilitation after trauma. (a) Initial panoramic radiography, where radiopaque zones compatible with metallic residues are observed. (b) Initial clinical view. Note the extensive loss of mandibular teeth and their alveolar process. (c, d) Surgical phase: ten implants were placed in two surgical treatment phases after a procedure of distraction-based osteogenesis for vertical bone augmentation. (e,f) Final restorations exhibited an increased length to obtain a compensation of the vertically lost bone dimension. (g,h) Panoramic radiography and intraoral view after ten years of function. Note that the restoration is maintained in a good aesthetic condition. (i) Clinical view in labial resting position.

possibilities of osseointegrated implants [3]. In 1982, the concept of osseointegration was discussed in a meeting in Toronto attended by people from the most important dental schools in North America. For more detailed information about the first steps of osseointegrated implants, the reader is referred to more specific literature [3, 4]. Branemark's work described osseointegration as a biological phenomenon involving direct contact between bone and Ti surfaces, allowing for a new philosophy of therapy.

In general, it is accepted that implant therapy is a predictable treatment without any contraindications for partially and fully edentulous patients. However, several factors, such as implant design, surgical procedure, anatomic and osseous conditions, systemic diseases, prosthetic design (Figure 2), and two-stage or immediate loading may affect the prognosis and long-term success (Figure 3). A poor prognosis was observed in patients with insufficient quality and/or quantity of bone receptor. Patients exhibiting poor quality of bone (type IV) in the posterior area of the maxilla had a 35% implant failure. This retrospective study indicated that patients with type I, II and III showed only 3% failure [5].

Systemic diseases are potential factors that affect the prognosis. The available information is derived from empirical observational studies where multifactorial risks are considered. Moy et al. showed a 10-year retrospective cohort study where 1140 patients were evaluated. Their results presented evidence that smoking, history of diabetes, head and neck radiation and postmenopausal estrogen therapy were correlated with a significantly increased failure rate, while gender, hypertension, coronary arterial disease, pulmonary disease, steroid therapy, chemotherapy and absence of hormone replacement therapy for post-menopausal women were not associated with increased risk in implant failure [6]. This study concluded that there are no absolute medical contraindications to dental implant rehabilitation, although they must be individually considered for each new patient.

The decision to put implants in patients treated for aggressive periodontitis is a challenge since controlled studies with large sample sizes are not available. Some case reports showed approximately 8% failure in patients previously treated for aggressive periodontitis. However, all the evaluated implants were placed in patients that had been previously treated for several years before the implants were put in place [7]. Until more evidence becomes available, factors such as time before therapy, the presence of natural treated teeth or immune compromise might render these clinical situations unpredictable.

Reduced alveolar bone height appears as an important consideration when evaluating the prognosis of dental implants. Two different approaches are used to achieve clinical success. In situations of an extremely reduced amount of bone, the surgeon may employ bone augmentation procedures (Figure 3 and 4), which result in higher costs, greater morbidity, and longer treatment times. Another possibility includes the use of short implants, which are defined as devices shorter than 6 to 10 mm [8]. Conflicting information is available about the success of this type of implants, with some authors reporting that short implants are unpredictable in cases of poor bone quality. However, alternative treatments are viable if other favourable factors are considered. A survival rate of approximately 94% for a five-year observation period was observed [8-10]. However, short implants have mechanical disadvantages as a consequence of implant-crown ratios and the amount of osseointegrated area around the implant [11].

In the last decade, there has been an important discussion related to loading implants immediately. Branemark's group postulated that early loading affects the prognosis due to a fibrous capsule that may develop due to micro-movements, thus affecting osseointegration [12]. They introduced the two-phase surgical procedure, in which the implant is submerged under gingival tissue and maintained unloaded during a 3 to 6-month period. However, new approaches have been introduced that include immediate loading based on the primary stability during surgical procedure and surface bioactivity [13].

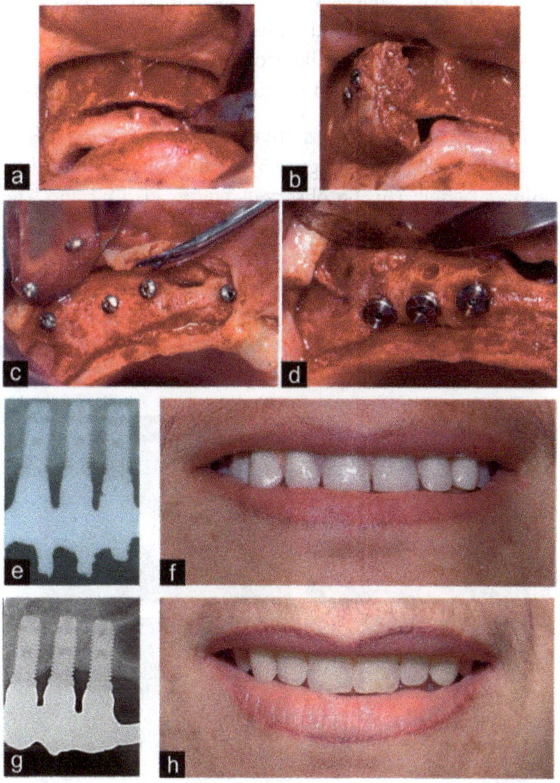

Figure 4. Fourteen-year follow-up of a successful implant-based rehabilitation. (a) Full-thickness flap reflection reveals the extension of missing bone in the buccal aspect of the maxilla. (b) The use of monocortical block graft to reconstruct the horizontal deficiency of the maxilla. (c) After 6 months tissue exhibits good healing. (d) Three implants were placed in this site and submerged for a six-month period. (e,f) Periapical radiography and clinical photography of complete restoration after treatment and (g,h) Periapical radiography and clinical view after 14 years of function.

The long-term success of an implant largely depends on the equilibrium between osseointegration and epithelial/connective tissue attachment. A complete soft tissue seal protects the newly formed bone from bacterial products originated in the oral environment.

Several animal and *in vitro* studies showed similar epithelial and connective structures between the gingiva and the peri-implant mucosa. The outer surface of the peri-implant mucosa is lined by a stratified keratinized oral epithelium that is continuous, with a junctional epithelium attached to the Ti surface by a basal lamina and hemidesmosomes.

The non-keratinized junctional epithelium is only a few cell layers thick in the apical portion and is separated from the alveolar bone by collagen-rich connective tissue. This 3 to 4 mm biological barrier, formed irrespective of the original mucosal thickness, protects the osseointegration zone from factors released by the plaque and the oral cavity [14]. The main differences between the soft tissues around natural teeth and those around implants are the collagen fibres orientation, which run parallel from the implant surface to the crest bone, the low number of fibroblasts, and the reduced vascularization of the scar tissue.

An osseointegrated implant is a good alternative to replace missing teeth, but they are not exempt from failure and complications (Figure 5). Oral implant failures have been classified as: 1) biological failures, which can be observed before loading and are associated with reduced osseointegration. If they take place after loading, they are associated with failing to maintain the achieved osseointegration; 2) mechanical failures, which can be observed as implant or prosthetic structural failures; 3) iatrogenic failures, mainly associated with procedures that affect anatomical structures or the misalignment of implants, which render them impossible to restore; and 4) failures by inadequate patient adaptation that include phonetical, aesthetical or psychological problems [15].

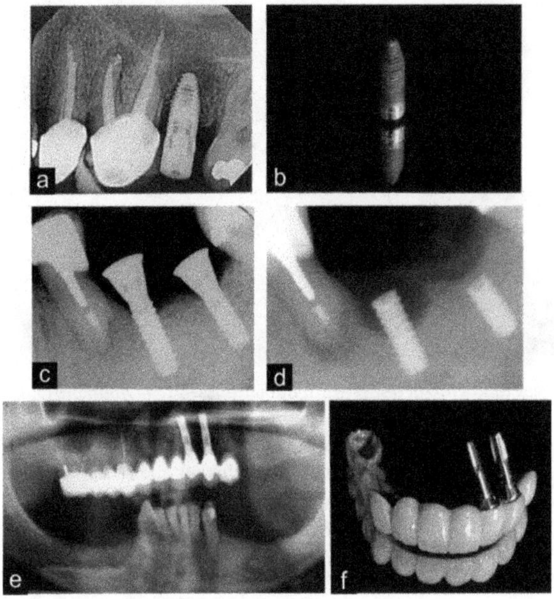

Figure 5. Different failure types in dental implantology. (a,b) Periapical radiography shows a radiolucent lesion on the mesial surface associated with bone resorption and sign of mobility. Photography shows the implant after removal. (c,d) Periapical radiographies show two posterior fractured implants. Fracture was apparently caused by bruxism. Note in the radiography (d) that the more anterior implant exhibits a large radiolucent zone as bone losses and the adjacent tooth shows an apical lesion. (e,f) Failure of full restoration on implants and natural teeth was caused presumably by poor design and support.

Mobility is the most evident sign of implant failure and can be presented as rotational, lateral or horizontal, and axial or vertical mobility [15]. There are different terms in the literature associated with biological implant failure or complications like peri-implant diseases, peri-mucositis and peri-implantitis, where the first two are reversible inflammatory reactions around a functioning implant. Peri-implantitis is a chronic inflammation with loss of the supporting tissues around the implant induced by bacterial colonization, facilitated by the implant/abutment gap and by the chemistry and surface roughness of the restorative components [14, 16].

Bacteria colonize and develop biofilms on the transmucosal abutment of osseointegrated dental implants. Like the gingival crevice around the natural tooth, the peri-implant mucosa covering the alveolar bone is closely adapted to the implant. In partially edentulous subjects, the developing microbiota around implants closely resembles the microflora of natural teeth [14]. In addition to the dark-pigmented, gram-negative anaerobic rods, other bacteria are associated with peri-implant infections (*Bacteroides forsythus, Fusobact. nucleatum, Campylobacter, Peptostreptcc. micros and Prevotella intermedia*) [17], and eventually *Staphylocc. spp, enterics,* and *Candida spp* [18].

Metals, including Ti, may induce non-specific immunomodulation and self-immunity. In immunologic *in vitro* tests, sensitization to Ti was observed [19]. Such problems with Ti and the ever-growing expectations on aesthetics lead to increasing interest in all-ceramic implants.

The surface texture of dental implants affects the rate of osseointegration [20] and biomechanical fixation. Surface roughness may be classified as "macro", "micro" and "nano" sized topologies. The "macro" range, from millimetres to 10 μm, is directly related to implant geometry, with threaded screws and macroporous coatings helping the primary stability of the implants during the early phases of implantation. However, high surface roughness may increase peri-implantitis risk compared with moderate roughness (1-2 μm) within the "micron" range (1-10 μm), maximizing bone/implant interlocking. Surface profiles in the "nano" range play an important role in protein adsorption and osteoblast adhesion and thus, in osseointegration [21]. No reproducible surface roughness is currently clinically available.

Dental implant failure is an active clinical research area. A number of strategies have been studied to modulate cell/material interactions, which play an important role in determining the short- and long-term implant success rate. This chapter will be mainly focused on the basic aspects to study cell/material interactions in dental implants using progenitor cells and *in vitro* biofilm formation approaches, as well as basic information related with micro-engineering technologies to modify dental implant surfaces.

2. Strategies to study cell/material interactions

Mesenchymal Stem Cells (MSCs) are commonly isolated from perinatal tissues (i.e, placenta, umbilical cord and blood from the umbilical cord) and postnatal tissues (bone marrow, trabecular bone, alveolar bone, cartilage, hair follicles, fat, skin and dental pulp) [22]. Bone marrow and pulp derived cells are frequently used in oral biology research to evaluate the biocompatibility of different dental materials. However, the frequency of MSCs in both tissues is not well known (0.001 to 0.01%). Bernardo *et al.* reported that MSCs frequency in bone marrow exhibits an age-related behaviour from 1:10,000 in a new-born to 1:1,000,000 in an 80-year-old subject [23]. Another important aspect is associated with anatomic skeletal site-specific differences [24], where MSCs obtained from calvaria have proven more successful for grafting in craniofacial application than cells obtained from others donor sites.

Different mechanical, chemical, or combined approaches have been used to disrupt the extracellular matrix to isolate MSCs from bone tissue bits or bone marrow. In the mechanical approach, the bone tissue is cut into small pieces using surgical blades, and then either suspended or plated. The main disadvantage of this procedure is shear-stress injury. The enzymatic digestion of the bone chips, with a combination of trypsin and collagenase (chemical-based approach) to obtain a cell suspension, should be avoided because trypsin might damage the cell membrane surface. If these cells are collected for some type of human cell therapy, a complete characterization of these proteases is required by regulatory agencies.

The number of sources for autologous MSCs for dentistry has increased due to the hundreds of mandatory extractions (e.g., third molars and premolars for orthodontic purposes) performed each year, since these tissues are routinely discarded. Many authors have shown that pulp-derived MSCs from deciduous and permanent teeth, as well as from periodontal ligament, might be isolated. These types of MSCs from different niches are heterogeneous and exhibit site-specific features, but in general they are able to produce bone, dentine, cement, and periodontal ligament-like structures [22, 25].

The successful isolation procedure of homogeneous populations is commonly based on morphology at early culture stages, considering that MSCs exhibit a fibroblast-like morphology. In the case of heterogeneous populations, the classification based upon specific markers is more desirable. Although specific and unequivocal markers are not available, an evaluation of non-specific multi-markers allows for a reasonable characterization. Tuan [26] reported that MSCs cells are positive for STRO-1, CD73, CD146, and CD106, and negative for CD11b, CD45, CD34, CD31 and CD117, preferably evaluated by fluorescence-active cell sorting (FACS). Other phenotypic approaches require the evaluation of MSCs capacity for trilineage mesenchymal differentiation (osteoblasts, adipocytes and chondroblasts) under standard *in vitro* differentiating conditions [23, 27].

Dulbecco's Modified Eagle's (D-MEM) and Minimal Essential Medium (α-MEM) are the most used standard cell culture media for *in vitro* bone cell studies. They are supplemented with different percentages of fetal bovine serum (FBS, 10% or 15%), antibiotics, antimycotics, and ascorbic acid. Coelho, *et al.* [28] made a comparative study on the behaviour of human bone marrow (hBM) osteoblastic cells cultured in α-MEM or D-MEM and they found a similar cell proliferation between both media. However, cells cultured in α-MEM exhibited higher ALP levels and earlier formation of mineralized deposits. Serum is essential in promoting or inhibiting cell proliferation and differentiation since it is a complex mixture of proteins, growth factors, ions, lipids and hormones.

Fetal Bovine Serum (FBS) is an inexpensive choice to conduct *in vitro* bone regeneration studies, but Human Serum is another option that can help to obtain more realistic results, and although a direct extrapolation from *in vitro* studies to clinical situations would be desirable, this has not been possible [29]. Recently, Deorosan and Nauman (2011) evaluated the effects of the concentrations of serum and glucose on the metabolic activity of murine MSCs, and concluded that the effects of the serum percentage (e.g. 2%, 5% and 10%) were negligible. However, a high correlation was found between cell viability and glucose concentrations (0.5 to 25 mM) [30].

Antibacterial and antimycotic agents are other types of supplements that can be controversial. Both should be avoided because they may affect cell physiology and mask improper aseptic conditions. However, from a realistic point of view, considering the risk of losing irreplaceable MSCs, the standard use of antibacterials (e.g., Penicilin/Streptomycin, 100 IU/ml and 10mg/ml, respectively; Gentamicin, 10μg/ml) and antimycotics (e.g., Amphotericin B, 0.25μg/ml) is required. Also, tetracyclines are often used to take advantage of their Ca-binding and fluorescence properties under UV light, allowing for the quantification of matrix mineralization. Other tetracyclines, like doxycycline and minocycline, when in low concentrations, may stimulate the proliferation of human osteoblastic cells [31].

The main goal of long-term *in vitro* MSCs cell cultures is their differentiation. To promote osteoblastic differentiation, the most frequently used supplements are ascorbic acid (AA, 50μg/ml), dexamethasone (Dex, 10nM), and β-glycerolphosphate (βGP, 10mM) [32]. When tested *in vitro*, these compounds are able to promote growth and accelerate the differentiation process of osteoprogenitor cells, thus reducing the proliferation period of the developmental sequence in the expression of the osteoblastic phenotype [33]. Coelho and Fernandes studied the proliferation/differentiation of hBMSCs cultured under different conditions of mineralization supplements. Their results show high proliferation in all the tested conditions, but mineralization was only achieved in the presence of βGP and this mineralization was greater in the presence of Dex.[32].

The micro-environment affects the ability of the osteoblast to produce a mineralized matrix, thus compromising the bone repair/regeneration process. Roughness has been considered as a major aspect in the osseointegration of titanium implants. Albrektsson and Wennerberg [34] classified the surface implants as smooth (0.0-0.4 μm), minimally rough (0.5-1.0 μm), moderately rough (1.0-2.0 μm) and rough (>2.0 μm). Theoretically, an increased roughness can be associated with a stronger bone response but also a greater potential for peri-implantitis and a higher risk of ionic leakage [34]. Other materials based on Co-Cr alloys and AISI 316L stainless steel have been used as dental implants. However, their use is controversial because corrosion products from Co-Cr alloys have been demonstrated to affect the cell viability, ALP activity and formation of a mineralized matrix *in vitro* when using osteoblast-like cells from rat, rabbit and human BM-MSCs [35]. These findings have been confirmed with corrosion products from AISI 316L stainless steel (and the isolated Fe, Ni and Cr ions). Deleterious dose- and time-dependent effects on ALP activity and matrix mineralization in rat, rabbit BMSCs and human alveolar bone cell cultures have been shown [33, 36]. On the other hand, the release of ionic species from hydroxyapatite-based materials might be explored to obtain a positive modulation of the osteoblastic cell response. For instance, Si-substituted hydroxyapatite coatings promote osteoblast proliferation and differentiation when compared to hydroxyapatite coatings due, at least in part, to the release of Si ions [37]. The aforementioned studies help to understand the relevance of MSCs and their *in vitro* osteogenic differentiation as tools to assess bone cell response to materials and agents intended for bone repair/regenerations strategies.

2.1. Bacterial adhesion process, biofilm formation and *in vitro* models

Adherence mechanisms of oral bacteria are essential to bacterial colonization of the oral cavity. In their absence, bacteria become part of the salivary flow and are swallowed. As a result, oral bacteria have developed several mechanisms to fulfill this task. The mechanisms are highly specific; the oral cavity is colonized mainly by bacteria that are exclusively found in it. Through retention, these bacteria can form organized, intimate, multispecies communities referred to as dental plaque and biofilms [38, 39]. Microbial adhesion and the accumulation of pathogenic biofilms are considered to play major roles in the pathogenesis of peri-implantitis and implant loss [40]. Therefore, knowledge about the microbiology around dental implants is essential.

After exposure of an osseointegrated implant to the oral cavity, an acquired pellicle is formed on the implant surface through selective adsorption of the environmental macromolecules including glycoproteins (mucins), proline-rich proteins, enzymes like α-amylase, histidine-rich proteins, phosphoproteins like statherin, and other molecules [38, 41]. These are derived mainly from saliva but, in the subgingival region, molecules originate from gingival crevicular fluid [42]. The physicochemical surface properties of a pellicle, including its composition, packing, density, and/or configuration, are largely dependent on the physical and chemical nature of the underlying hard surface.

The adsorption of proteins from an aqueous solution onto a solid surface is the result of various types of interactions that simultaneously occur between all the components, namely the fluid, the solid and the solubilized proteins. The mechanisms involved in pellicle formation include electrostatic, van der Waals, and polarity forces. The polarity of each of these components has great impact on the adsorption process, which is reflected in the hydrophilicity or hydrophobicity of the interacting components [41, 43]. The pellicle plays a decisive role in microbial adhesion, as its constituents may interact with oral micro-organisms, either by direct interaction with them, or indirectly by influencing the thermodynamic conditions for microbial adhesion [44].

After formation of the acquired pellicle, bacterial attachment with initial colonizers followed by cell-to-cell adhesion with secondary colonizers occurs on the implant surface [45]. An initial reversible adhesion involves weak, long-range, non-specific physicochemical interactions between the charge on the microbial cell surface and that of the acquired pellicle. Microorganisms are usually transported passively to the surface by the flow of saliva or gingival crevicular fluid or by active bacterial movement. A few species (e.g *Wolinella, Selenomonas* and *Campylobacter* spp.) found sub-gingivally have flagella and are motile [42]. Alternatively, microorganisms in suspension may also be transported towards each other by microbial (co)aggregates [41].

During the second phase of adhesion, strong, short-range interactions between specific molecules on the bacterial cell surface (adhesins) and complementary molecules (receptors) present in the acquired pellicle can result in irreversible attachment. Oral bacteria generally possess more than one type of adhesin on their cell surface and can participate in multiple interactions both with host molecules and similar receptors on other bacteria (co-adhesion) [46].

Streptococci, the main early colonizers, bind to acidic proline-rich-proteins and other receptors like α-amylase and sialic acid in the acquired pellicle [47]. In addition, *Actinomyces,* which are other primary colonizers, bind to the acquired pellicle and to the streptococci [38, 39, 41]. Consequently, these two groups of primary colonizers are thought to prepare the environment for later colonizers that have more fastidious and slow requirements for growth. Other bacteria, including periodontal pathogens such as *Hamophilus actinomycetemcomitans, Porphyromonas gingivalis, Prevotella intermedia, Treponema denticola, Tannerella forsythensis* or *Fusobacterium* species, bind to *Streptococci* and/or *Actinomyces* [48]. This stage also involves specific inter-bacterial adhesin-receptor interactions and leads to an increase in the diversity of the biofilm [46].

The increase in attached cell numbers leads to biomass augmentation and the synthesis of exopolymers that form a biofilm matrix. This matrix is a common feature of all biofilms, and is more than a chemical scaffold to maintain the shape of the biofilm. It provides a significant contribution to the structural integrity and general tolerance of biofilms to environmental factors (e.g. desiccation) and antimicrobial agents. The close proximity of cells to one another in a biofilm facilitates numerous synergistic and antagonistic interactions between neighboring species. Within the biofilm, oral bacteria do not exist as independent entities but rather as a coordinated, spatially organized, and fully metabolically integrated microbial community, whose properties are more relevant than the sum of the individual composing species [41, 42, 49].

The physico-chemical characteristics of specific material surfaces are known to significantly influence the bacterial adhesion process. Both surface free energy and surface roughness are known to play major roles in this process [41, 50]. High surface roughness values significantly promote bacterial adhesion by reducing the influence of shear forces on initially attaching bacteria, while materials with high surface free energy values are known to increase bacterial adhesion [51]. Furthermore, the bacterial adhesion process is influenced by the chemical composition, surface hydrophobicity, and the zeta potential of the material [52]. An increased zeta potential, which refers to the electrostatic potential generated by the accumulation of ions on the surface, results in decreased bacterial attachment. Generally, hydrophobic microorganisms prefer to attach to hydrophobic substrata, and bacteria with hydrophilic properties prefer hydrophilic materials. Moreover, bacterial adhesion varies between the various bacterial species and strains [50, 53].

Understanding how bacteria relate and act within biofilms is essential for the prevention and proper management of dental and periodontal diseases [54]. In order to increase the knowledge concerning biofilm physiology, the creation of models to study and evaluate this complex consorting under controlled conditions is of great interest. Over the past years, several *in vivo* and *in vitro* biofilm models have been developed with this intent. However, *in vivo* studies in both animals and humans are more restricted due to problems with access and sampling and because of complex ethical issues involved [55, 56].

The currently available *in vitro* models clarify extensively the microbial biofilm physiology, micro-ecology, pathology and behaviour. These models are used to replicate environmental conditions *in vitro* and have served as the major conceptual framework for biofilm research, ranging from static mono-cultures to the development of diverse mixed cultures growing under dynamic conditions. For each particular application, every model has its strengths and weaknesses that can be appropriate for one specific application but not for others [57-59]. Two major biofilm models have been studied in the laboratory, namely biofilms grown without a continuous flow of fresh medium, known as static models, and biofilms grown with a continuous flow of fresh medium, known as flow models [60].

The quantification of biofilms started with simpler methods based on the cultivation of the biofilm in the wall of test tubes or well plates, like the microtiter plate method. With this system, biofilms are grown on the bottom of the walls or in a substrate placed on the wells of a microtiter plate, for a desired period of time. Besides its simplicity, this method has several advantages, such as low cost and the small amount of reagents required and allows to perform a large number of tests simultaneously, remaining among the most frequently used models to assess biofilm formation. By running it under static conditions, the environment in the well will change during the experiment, unless the fluid is regularly replaced. Also, during biofilm formation, bacteria may deposit on the substrate and on the bottom of the well, and not actively attach to the surface [55, 61-63].

Recently, a new model where the substrate can be positioned vertically to assure active attachment of the bacteria to the surface was developed. This simple high-throughout active attachment model consists of a lid with 24 clamps were different substrates can be put. This lid is placed in a common 24-well plate allowing the substrates to be vertically positioned during the period of biofilm formation [59, 63, 64].

Since *in vivo* conditions are almost exclusively dynamic, the use of reactors for the *in vitro* development of biofilms has been largely applied. These systems normally work using continuous flow to provide nutrients to the growing biofilm and can be used for different purposes [57].

A very frequently used model in the dental research field is the constant depth film fermenter (CDFF), a steady-state model. This system consists of a rotating stainless steel disk in which plugs are located. Different materials can be placed and used as substrates for biofilm growth until a maximum thickness has been reached. When reached, the excess biofilm is scraped off and nutrients are distributed into the system. The reactor allows several parameters to be tailored, such as the possibility to choose growth conditions- aerobic or anaerobic- and the option to alter nutrient schemes. This model has been frequently used for dental plaque studies [56, 65, 66].

Another commonly used system is the CDC biofilm reactor developed in the Centers for Disease Control and Prevention by Donlan, *et al.* [67, 68]. The reactor has 24 removable surfaces that allow biofilm formation under moderate to high sheer stresses in batch or continuous flow-conditions. It has been used to evaluate biofilm formation and structure and also to test the effects of antimicrobial agents in the biofilm [67, 68].

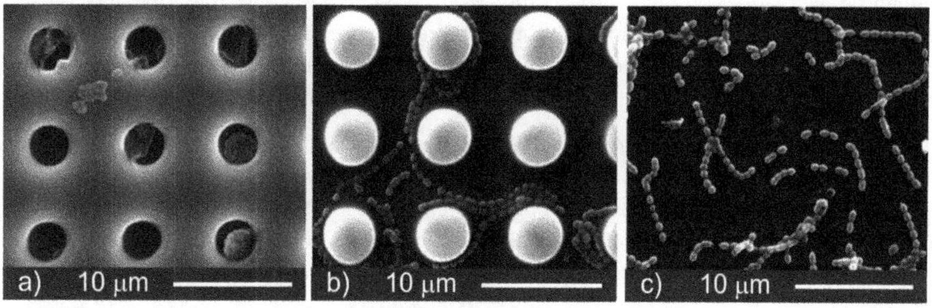

Figure 6. Biofilm formation on a) negative, b) positive relief micropatterned surfaces and c) flat surfaces.

However, the systems described previously lack the possibility of continuously monitoring biofilm growth. Recently, microfluidics systems have been adapted to study biofilm formation. Microfluidics set-ups are normally fabricated through soft-lithography with the size of the channels in the range of 50 to 500 μm and flow rates typically very low, between 0.1–50 μl/min. These devices have the advantage of simulating biological phenomena with physiological flow velocities, low fluid-to-cell volume ratios, and biomimetic micro-/ nano-engineered surfaces (Figure 6). Also, the small size of the chambers allows for real-time microscopic analysis of biofilm formation [55, 69, 70].

3. Micro/nanoscale engineering of cell-material interactions

Understanding the interactions between cells and biomaterial surfaces at the micron-, submicron-, and nano-scale, is crucial for the production of functional biomedical devices (e.g. implants, biosensors, etc.). Biomaterials are known to elicit specific cellular responses (positive or negative) depending upon the surface chemical and/or physical properties. Surface topography (from the micron- down to the nano-scale), for instance, plays a crucial role in controlling important cellular processes such as adhesion, propagation, proliferation, orientation, migration, differentiation, and reactivity to certain hormones, growth factors and drugs, both *in vitro* and *in vivo* [71, 72].

The effects of surface microtopography on cell behaviour have been widely documented. Previous research has shown that microtextured surfaces, independent of surface chemistry, exhibit a strong influence on *in vitro* and *in vivo* cell behaviour [71, 73, 74].

Surface microroughness affects *in vitro* adsorption of albumin and fibronectin [75]. Studies performed on randomly oriented microtopographies showed stronger cell adhesion, altered integrin expression, up-regulation of focal contacts, reorganization of the cytoskeleton, changed proliferation, increased differentiation, and enhanced susceptibility to different hormones and growth factors [76-79]. The implementation of lithographic and dry etching techniques from the silicon microelectronics industry into the cell biology field allowed for further studies on the effects of controlled microtopographies on cell behaviour [73]. For the past two decades, *in vitro* and *in vivo* studies on controlled microtopographies have shown enhanced cell adhesion and proliferation, cell orientation along the direction of the microfeatures, altered migration and motility patterns, up-regulation of certain cytoskeletal and extracellular matrix proteins, reduced immune response, increased mitochondrial activity, augmented differentiation, etc. Such effects are cell and material-dependent [80-85].

The effects of surface nanotopography on cell functions have been studied since the early 1960's [86, 87]. Cells are known to be reactive to objects as small as a few nanometers (~ 5nm) [80]. A number of different topographical patterns at the submicron and nano-scale (both randomly oriented and controlled) have been explored, including columns, dots, pits, pores, meshwork, gratings, nanophase grain, and random surface roughness [88-95].

Some studies have suggested that osseointegration is a function of the initial interactions that occur between the implant surface and blood. Park and collaborators found that platelet adhesion and activation were increased on micro/nanotextured titanium surfaces compared to polished ones [96], presumably due to the increase in actual surface area (leading to increased protein adsorption), and/or topographically-induced cytoskeletal rearrangement, which could have led to downstream intracellular signaling cascades resulting in platelets aggregation and granules release. This enhanced thrombogenic potential is expected to improve endosseous integration, as osteogenic cells reach the implant surface by migrating through the remnants of the initial osteo-conductive/inductive blood thrombus [97, 98]. Additional studies found that surface micro/nanotextures improve osseointegration due to the fact that the initial fibrin clot is mechanically stabilized by the topography [99, 100].

Finally, micro/nanostructured material surfaces have also been shown to provide a greater number of nucleation sites for the precipitation of minerals (e.g. Ca and P) from the blood plasma, which results in the formation of an amorphous apatite layer on the surface of the implant that could potentiate osseointegration [70, 101].

Another way surface micro/nanofeatures could lead to enhanced osseointegration is by directly influencing bone cell responses. Previous studies showed that the initial osteoblast-material interactions (i.e. adhesion, spreading and growth) could play an important role in leading to a long-term positive response at the bone-implant interface [70]. Fewster, *et al.* showed that micron-scale (1-50 μm) pillars and pores on polyethylene terephthalate (PET) and polystyrene (PS) surfaces led to improved osteoblast adhesion [102]. Using a cytodetacher, Wang, *et al.* showed that the adhesion force of osteoblasts to nanostructured titanium surfaces increased (38.5-58.9 nN) in direct proportion to the surface roughness [70]. Webster, *et al.* found increased vitronectin adsorption and enhanced osteoblast cell adhesion on nanophase alumina surfaces [95]. Wan, *et al.* showed that osteoblast adhesion was also enhanced on micro- and nanostructured PLLA and PS surfaces, in comparison to their smooth surface counterparts [103]. In this study, no significant differences in the adhesion were found between micro- and nanostructured surfaces, and the cell proliferation levels were similar on textured surfaces compared to the smooth ones. Other groups showed increased osteoblast cell proliferation on nanostructured ceramic surfaces [104, 105]. This indicates that there may be a synergistic effects between surface topography and chemistry that need to be accounted for in some cases, or perhaps that the specific geometrical parameters (e.g. shape, size) of the nanotexture could also determine the degree of cell responses.

Numerous studies have reported an enhanced osteogenic phenotype in response to surface micro- and nanotextures on polymeric, ceramic, and metallic materials [104-109]. Remarkably, Dalby, *et al.* showed that circular nanostructures can induce osteogenic differentiation in the absence of osteogenic factors in the cell culture medium [110, 111]. More recent reports by Zhao, *et al.* indicated that the combination of micro- and nanostructures on the same surface may result in a cooperative synergy between the micro- and nanotopography that ultimately leads to improved bone cell responses [112].

Micro- and nanotextured surfaces also tend to promote increased adhesion in other cell models (e.g. fibroblasts, smooth muscle cells, and chondrocytes), although the stimulus for topography-mediated increased cell adhesion seems to be more prominent for osteoblast cells [99]. Similarly, topography-mediated changes in cell morphology, gene expression, proliferation, and migration, have been reported for human embryonic stem cells, rat aortic endothelial cells, murine macrophages, epithelial, and glial cells, among others [71, 113].

Although surface nanotopography for the most part has been shown to induce "positive" cell responses (e.g. increased adhesion, proliferation, differentiation), there are other reports that suggest that this phenomenon (i.e. topography-mediated cell responses) may be regulated to some degree by the specific geometrical (and perhaps chemical) properties of the patterns. Curtis, *et al.* studied the effects of different nanofeatures (pillars, pits, randomly distributed Au nanoparticles) on fibroblast and endothelial cell behaviour [114]. The results confirmed increased cell adhesion on the nanogrooves, and decreased adhesion on the nanopillars and nanopits (with adhesion being inversely proportional to the distance between nanofeatures). Cell adhesion on the Au nanoparticles was no different compared to flat surfaces. Dalby, *et al.* studied fibroblast responses to PMMA nanocolumns and found that cell adhesion and spreading were reduced on the nanocolumns compared to smooth surfaces [115, 116]. The cells growing on the nanocolumns exhibited lower actin polymerization, smaller focal adhesions, and increased filopodia formation. Kunzler, *et al.* studied osteoblast responses to silica-based nanotopographies, and found that cell adhesion, spreading, and actin polymerization were reduced on the nanostructured surfaces in comparison to smooth ones [117].

To summarize, there is rather strong evidence suggesting that micro- and nanoscale surface structures have the potential to modulate cell responses, which could be used to better design biomedical devices (e.g. implants, sensors) by turning on/off specific responses depending on the application. However, the lack of systematic and more controlled studies limits the exploitation of this concept, as it is difficult to reach a consensus on a single micro- and/or nanotopography that could lead to optimum cell responses for any given application. Nevertheless, recent advances in the fields of micro- and nanofabrication are enabling the development of studies where different topographical parameters (e.g. feature size, organization, density, geometry) could be evaluated in a more controlled manner, which is expected to ultimately lead to a better understanding of the role of surface topography on cellular responses.

A host of different techniques have been developed to imprint features on the surface of a biomaterial at the submicron and nanoscale: laser irradiation, soft lithography, dip-pen nanolithography, capillary lithography, electron beam (e-beam) lithography, microimprinting, interference lithography, nanoimprint, X-ray lithography, polymer demixing, and colloidal lithography among others [88-90, 92, 93, 118-122].

3.1. Soft Lithography and Sol-Gel Technology

Lithography has been used since ancient years. Initially, photolithography was introduced in the editorial industry to achieve better printing results. However, this technique contributed later to the development of the integrated circuit industry, and it became the main contributor to the information technology. Photolithography is also essential to produce technology for sensors, microsensors, micromechanical systems (MEMS), microanalytical systems, micro-optical systems and integrated circuits [123, 124].

In 2006, Ferrari, et al. wrote "Less than twenty years ago photolithography and medicine were total strangers to one another .... And then, nucleic acid chips, microfluidics and microarrays entered the scene, and rapidly these strangers became indispensable partners in biomedicine" [125].

In basic terms, photolithography helps to create small structures in a massive scale, but it is not always the best option for all applications since it requires expensive technology. Poor results for curved substrates and the fact that it is limited to photosensitive materials are some of its drawbacks. These limitations inspired two important review papers introducing Soft Lithography (Figure 7) [123, 124].

The soft lithography process can be included among other techniques that are basically rapid prototyping processes. Figure 7 shows a 24h flow from the idea to the final prototyping [123]. Establishing borders for soft lithography is a hard task because several techniques, such as microcontact printing (mCP), replica molding, microtransfer molding, micromolding in capillaries (MIMIC) and solvent-assisted micromolding (SAMIM), include the use of stamps or molds as key elements to produce micro- nano-patterns [123].

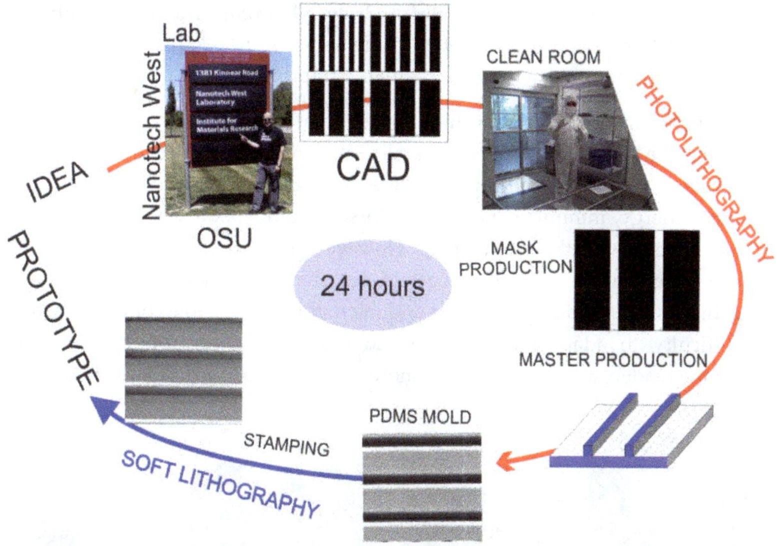

Figure 7. Rapid prototyping using Photolithography and Soft Lithography. Adapted from [123]

Elastomers, such as polydimethylsiloxane (PDMS) have been used in most applications of stamping. Other research groups have also used other elastomers such as polyurethanes, polyimides, and cross-linked Novolac resin (a phenol formaldehyde polymer) [123]. PDMS exhibits advantages such as being chemically inert, good surface reproducibility, limited shrinkage, homogenity, isotropy, transparency, and easily modifiable by plasma. However, its main disadvantages are the swelling in the presence of non-polar solvents, such as toluene and hexane, and forces, like gravity. Adhesion and capillary forces exert stresses on the elastomeric features and cause them to collapse, generating defects in the produced pattern. Therefore, obtaining patterned surfaces with features smaller than 1 μm is difficult [123].

Sol-gel is called *"the wet chemistry"* in processing ceramics, glasses and other materials, but it can be referred to as several chemical reactions occurring during the manufacturing of technological products. Under this term, several techniques of production of materials such as monolithics, powders, fibers, nanospheres, pigments, and coatings have been developed for aerospatial, optical, electronic, refractory, automotive, polymers and sensors, and medical industries [126-130].

First synthesis of silica was described by Ebelmen in 1844, but it found its commercial application in the early 1960's [131]. Since its beginnings in the 1940's, the sol-gel processing has helped to obtain a new generation of materials (ceramic and glasses). Considering the initial precursors, the processing can be divided in aqueous solution of metal salts, metal alkoxide solutions or mixed organic, and inorganic precursors [132]. Frequently, the alkoxides (TEOS/MTES) have been used to produce a hybrid sol. These precursors hydrolyze with the formation of partially hydrolyzed products and they subsequently undergo condensation with formation of an oxide network [133, 134]. The sol stage is used to produce thin films by spin coating, dip coating or imprinting.

In general, these materials present high purity and homogeneity, their particle size distribution may be controlled at the nano-scale level, and they require low temperatures to be prepared. In comparison with high temperature processes, these sol/gel materials and processes save energy, minimize evaporation losses and air pollution, and, in general, do not induce reactions with their containers. However, they are not free of disadvantages. Among these disadvantages, the high cost of raw materials, the large shrinkage during processing, the residual hydroxyls, carbon and microporosity, the long times for processing, and the difficulty to adapt them to produce large pieces are found. For additional information, the readers are invited to read more specialized reviews [134, 135].

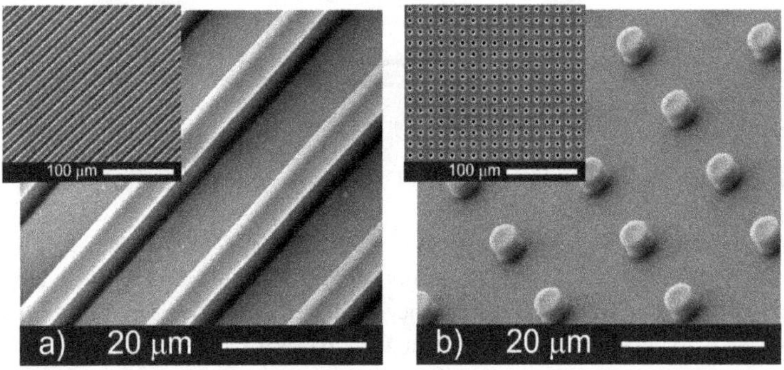

Figure 8. Anisotropic (a) and isotropic (b) silica coatings produced by synergy of soft lithography and sol-gel technology. Inset in both figures shows the PDMS molds.

For optical applications, surface relief features in or on thin films can be used. Specific geometries are used to produce couplers, filters, lenses, beam splitters and mirrors. For biomedical applications, Hench [136] described early that "*A common characteristic of glasses, glass-ceramics and ceramics that bond to living tissues is the development of a bioactive hydroxyapatite layer in vivo at body temperature*". Also, this author proposed the following categories for materials with the potential for biological or medical use: 1) bioactive sol-gel coatings, 2) bioactive sol-gel glasses, 3) doped sol-gel matrices as biological and chemical sensors and 4) sol-gel matrices with entrapped living organisms.

The synergy between soft lithography and sol-gel has been explored before in the production of membranes and waveguides with a feature size of 1 μm [137]. Figure 8 shows anisotropic and isotropic PDMS molds and microstructured silica coatings produced by sol-gel. Specific shapes and dimensions can be selected depending on the application [138].

3.2. Bioactive Micropatterned Surfaces

Ceramic nanoparticles can be added to silica thin films in order to increase bioactivity and contact surface area. Hydroxyapatite (HA) is one of the most widely used synthetic calcium phosphates due to its chemical similarities to the inorganic component of hard tissues.

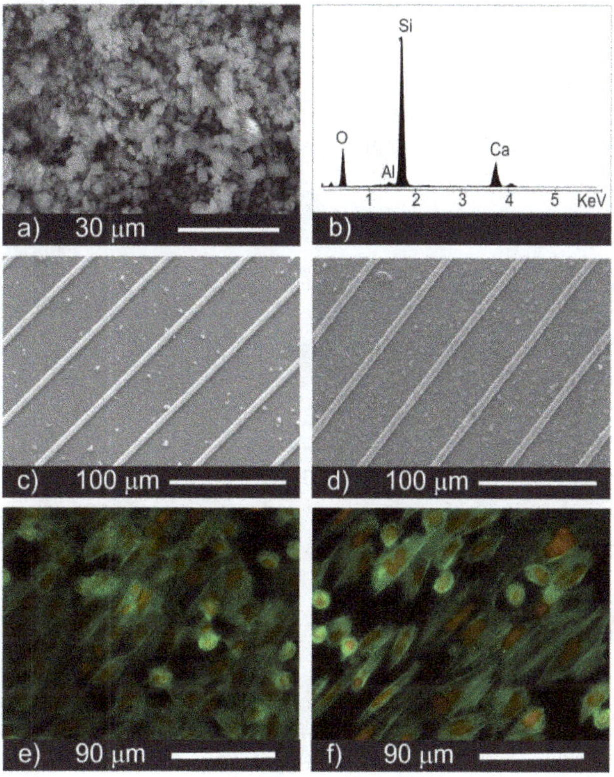

Figure 9. Silica coatings with bioactive Portland cement (PC) particles. SEM images of Portland cement particles (a) Energy Dispersive Spectrum of PC particles (b) micropatterned coating produced by dual (c) and single molding (d) Saos-2 osteoblast cells responses to the produced coatings. Dual (e) and single (f) molding.

HA exhibits exceptional biocompatibility and bioactivity, key features for the formation of a direct and strong interface with bone, and in addition, osteoconductivity, which is the ability to serve as a template for the local formation and growth of new bone [139-141]. Current research is focused on the development of new HA formulations with properties closer to those of living bone, such as nano-sized and monolithic structures [139, 142]. Compared to conventional ceramic formulations, nanophase HA properties such as surface grain size, pore size, wettability, etc., could control protein interactions, thus modulating osteoblast adhesion and long-term functionality [95, 143] . These implant materials are suitable for bone replacement and could be useful for additional functions, such as the release of drugs, growth factors, or other substances [144].

Portland cement based materials (e.g. MTA) have been extensively used in the dentistry field for stimulating the formation of cement and dentin (Figure 9a-b) [145]. The high content of $Ca(OH)_2$ of hydrated Portland cement causes it to have an extremely basic pH (~12.5-13.0). This basic pH has been shown to be advantageous for endodontic applications as it prevents bacterial contamination. Moreover, the $Ca(OH)_2$ released under physiological conditions interacts with phosphates in the medium to produce rapid precipitation of amorphous apatite. However, previous research by Gallego-Perez, et al. showed that the showed that the $Ca(OH)_2$ present in hydrated Portland cement could be extremely cytotoxic [146, 147].

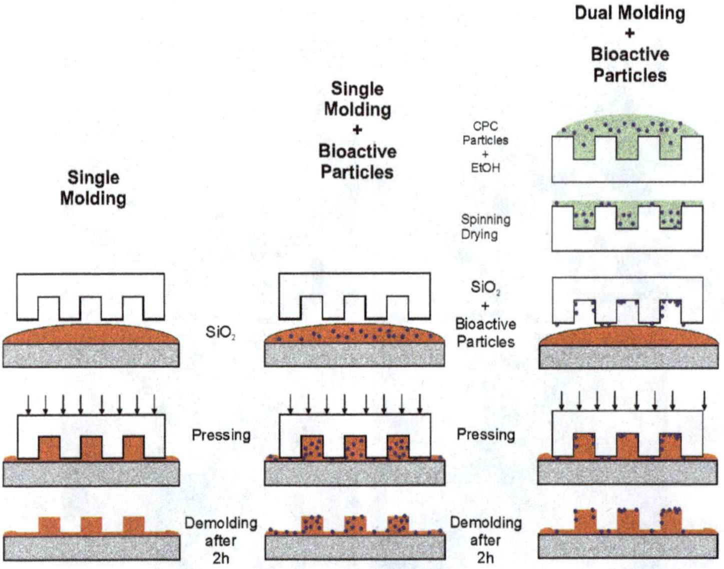

Figure 10. Stamping silica coatings with bioactive particles via single or dual molding technique [148-150].

Such toxic effect effect may not be desirable in certain applications, like implants, as it will prevent adequate cell adhesion and propagation on the surface and cause the formation of a large necrotic zone around the implant after placement, provoking a chronic inflammatory response by the host, which could potentially lead to total rejection of the implant. To avoid this, Gallego-Perez, *et al.* developed a simple strategy to obtain cytocompatible Portland cement based on the carbonation of the paste. The CO_2 introduced during the hydration of Portland cement reacts with the $Ca(OH)_2$ that is being formed to produce calcium carbonate ($CaCO_3$), which decreases the pH of the cement (~7.4), and provides a more compatible environment for cell growth [146, 147, 151]. More recently, a new method was developed (Figure 10) for the production of cytocompatible Portland cement microparticles, which could be incorporated, along with nano-hydroxyapatite particles, in micropatterned bioactive coatings (Figure 9c-f) of interest to dental implantology [148-150].

4. Future considerations

Dental implantology is still an area of active research. A growing number of biomaterials, implant surfaces, and geometries, are currently available in the market. However, independent of the selected implant system, a successful therapy always results from the proper interplay between the implant, surrounding soft/hard tissues, and the oral environment.

Different approaches are constantly being developed to modulate the response of MSCs as precursors of differentiated cells. MSCs cells from craniofacial niches are desirable for the repair or replacement of soft and hard oral tissues. Although a host of surface modification strategies can be implemented, micro-/ nanoengineered surfaces have shown great promise for this application, in part due to their ability to properly control cell and tissue adhesion to the implant surface. A close apposition of gingival tissues helps to prevent apical migration of bacteria, which could be responsible for the resorption of the bone crest and implant failure.

Micro-/nanopatterned surfaces are an interesting model to study the basic phenomena associated with osseointegration and biofilm formation on dental materials. In addition, a number of other micro-/nanoscale technologies also facilitate the development of more complex model systems. As an example, microfluidic devices could help to study biofilm formation on micro-/nanoengineered surfaces under dynamic flow, thus resembling more closely the *in vivo* conditions.

Finally, the synergy between soft lithography and sol-gel chemistry provides several possibilities to develop a new generation of dental implants with micro-/nanopatterned hard surfaces that may lead to improved osseointegration and guided soft/hard tissue regeneration.

Author details

A. Pelaez-Vargas*
INEB - Instituto de Engenharia Biomédica, Universidade do Porto, Porto, Portugal
Universidade do Porto - Faculdade de Engenharia (FEUP), Departamento de Engenharia Metalúrgica e Materiais, Porto, Portugal
Universidad Cooperativa de Colombia, Facultad de Odontología, Medellín, Colombia

A. Carvalho, L. Grenho and F.J Monteiro
INEB - Instituto de Engenharia Biomédica, Universidade do Porto, Porto, Portugal
Universidade do Porto - Faculdade de Engenharia (FEUP), Departamento de Engenharia Metalúrgica e Materiais, Porto, Portugal

D. Gallego-Perez, N. Higuita-Castro and D.J. Hansford
Department of Biomedical Engineering, The Ohio State University and Nanoscale Science and Engineering Center - NSF, Columbus (OH), USA

J.A. Arismendi
Universidad de Antioquia, Facultad de Odontología, Medellín, Colombia

* Corresponding Author

M.H. Fernandes
Universidade do Porto – Faculdade de Medicina Dentária, Porto, Portugal

M.P. Ferraz
INEB - Instituto de Engenharia Biomédica, Universidade do Porto, Porto, Portugal
CEBIMED – Centro de Estudos em Biomedicina, Universidade Fernando Pessoa, Porto, Portugal

Acknowledgements

This work was supported partially by the Portuguese Science and Technology Foundation (Scholarship FCT/SFRH/BD/36220/2007 and Grant No. FCT/PTDC/CTM/100120/2008 "Bonamidi"), CRUP – Acções integradas Luso-Espanholas: E46/09, Acciones integradas Hispano-Portuguesas, MICINN: HP2008-0075. NSF Nanoscale Science and Engineering Center fellowship (NSF, Grant No.EEC-0425626), and an U.S. Air Force Office of Scientific Research MURI (Grant No.F49620-03-1-0421).

5. References

[1] Petersen P E (2003) The World Oral Health Report 2003: continuous improvement of oral health in the 21st century. The approach of the WHO Global Oral Health Programme. Community Dent Oral Epidemiol. 31: 3-24.
[2] Okeson J P (2003), Causes of functional disturbances in the masticatory system in Management of Temporomandibular Disorders and Occlusion, Okeson J P. London: Mosby 149-189 p.
[3] Sennerby L (2008) Dental implants: matters of course and controversies. Periodontol 2000. 47: 9-14.
[4] Albrektsson T, Sennerby L, and Wennerberg A (2008) State of the art of oral implants. Periodontol 2000. 47: 15-26.
[5] Jaffin R A and Berman C L (1991) The excessive loss of Branemark fixtures in type IV bone: a 5-year analysis. J Periodontol. 62: 2-4.
[6] Moy P K, Medina D, Shetty V, and Aghaloo T L (2005) Dental implant failure rates and associated risk factors. Int J Oral Maxillofac Implants. 20: 569-77.
[7] Al-Zahrani M S (2008) Implant therapy in aggressive periodontitis patients: a systematic review and clinical implications. Quintessence Int. 39: 211-5.
[8] das Neves F D, Fones D, Bernardes S R, do Prado C J, and Neto A J (2006) Short implants--an analysis of longitudinal studies. Int J Oral Maxillofac Implants. 21: 86-93.
[9] Arlin M L (2006) Short dental implants as a treatment option: results from an observational study in a single private practice. Int J Oral Maxillofac Implants. 21: 769-76.
[10] Fugazzotto P A (2008) Shorter implants in clinical practice: rationale and treatment results. Int J Oral Maxillofac Implants. 23: 487-96.
[11] Herrmann I, Lekholm U, Holm S, and Kultje C (2005) Evaluation of patient and implant characteristics as potential prognostic factors for oral implant failures. Int J Oral Maxillofac Implants. 20: 220-30.
[12] Branemark P I, Zarb G, and Albrektsson T *Tissue-integrated prostheses. Osseointegration in clinical dentistry*. 1985, Chicago: Quintessence Publishing.

[13] Adell R, Eriksson B, Lekholm U, Branemark P I, and Jemt T (1990) Long-term follow-up study of osseointegrated implants in the treatment of totally edentulous jaws. Int J Oral Maxillofac Implants. 5: 347-59.
[14] Klinge B, Hultin M, and Berglundh T (2005) Peri-implantitis. Dent Clin North Am. 49: 661-76.
[15] Esposito M, Hirsch J-M, Lekholm U, and Thomsen P (1998) Biological factors contributing to failures of osseointegrated oral implants. (I). Success criteria and epidemiology. Eur J Oral Sci. 106: 527-551.
[16] Rimondini L, Cerroni L, Carrassi A, and Torricelli P (2002) Bacterial colonization of zirconia ceramic surfaces: an in vitro and in vivo study. Int J Oral Maxillofac Implants. 17: 793-8.
[17] Tanner A, Maiden M F J, Lee K, Shulman L B, and Weber H P (1997) Dental Implant Infections. Clin Infec Dis. 25: S213-S217.
[18] Leonhardt A, Renvert S, and Dahlen G (1999) Microbial findings at failing implants. Clin Oral Implants Res. 10: 339-45.
[19] Lalor P A, Revell P A, Gray A B, Wright S, Railton G T, and Freeman M A (1991) Sensitivity to titanium. A cause of implant failure? J Bone Joint Surg Br. 73: 25-8.
[20] Boyan B, Lohmann C, Dean D, Sylvia V, Cochran D, and Schwartz Z (2001) Mechanisms involved in osteoblast response to implant surface morphology. Ann Rev Mater Res. 31: 357.
[21] Brett P M, Harle J, Salih V, Mihoc R, Olsen I, Jones F H, and Tonetti M (2004) Roughness response genes in osteoblasts. Bone. 35: 124-33.
[22] Robey P G and Bianco P (2006) The use of adult stem cells in rebuilding the human face. J Am Dent Assoc. 137: 961.
[23] Bernardo M, Locatelli F, and Fibbe W E (2009) Mesenchymal Stromal Cells: A Novel Treatment Modality for Tissue Repair. Ann NY Acad Sci. 1176: 101-117.
[24] Akintoye S O, Lam T, Shi S, Brahim J, Collins M T, and Robey P G (2006) Skeletal site-specific characterization of orofacial and iliac crest human bone marrow stromal cells in same individuals. Bone. 38: 758-68.
[25] Machado E, Fernandes M H, and Gomes P D (2011) Dental stem cells for craniofacial tissue engineering. Oral Surg Oral Med Oral Pathol Oral Radiol Endod. 113: 728-733.
[26] Tuan R (2011) Role of adult stem/progenitor cells in osseointegration and implant loosening. Int J Oral Maxillofac Implants. 26: 5-62.
[27] Dominici M, Le Blanc K, Mueller I, Slaper-Cortenbach I, Marini F C, Krause D S, Deans R J, Keating A, Prockop D J, and Horwitz E M (2006) Minimal criteria for defining multipotent mesenchymal stromal cells. The International Society for Cellular Therapy position statement. Cytotherapy. 8: 315-317.
[28] Coelho M J, Trigo Cabral A, and Fernandes M H (2000) Human bone cell cultures in biocompatibility testing. Part I: osteoblastic differentiation of serially passaged human bone marrow cells cultured in [alpha]-MEM and in DMEM. Biomaterials. 21: 1087-1094.
[29] Aldahmash A, Haack-Sørensen M, Al-Nbaheen M, Harkness L, Abdallah B, and Kassem M (2011) Human Serum is as Efficient as Fetal Bovine Serum in Supporting Proliferation and Differentiation of Human Multipotent Stromal (Mesenchymal) Stem Cells In Vitro and In Vivo. Stem Cell Rev. 7: 860-8.

[30] Deorosan B and Nauman E A (2011) The Role of Glucose, Serum, and Three Dimensional Cell Culture on the Metabolism of Bone Marrow-Derived Mesenchymal Stem Cells. Stem Cells Int. 2011: 429187.

[31] Gomes P S and Fernandes M H (2007) Effect of therapeutic levels of doxycycline and minocycline in the proliferation and differentiation of human bone marrow osteoblastic cells. Arch Oral Biol. 52: 251-259.

[32] Coelho M J and Fernandes M H (2000) Human bone cell cultures in biocompatibility testing. Part II: effect of ascorbic acid,[beta]-glycerophosphate and dexamethasone on osteoblastic differentiation. Biomaterials. 21: 1095-1102.

[33] Amaral M, Costa M A, Lopes M A, Silva R F, Santos J D, and Fernandes M H (2002) Si(3)N(4)-bioglass composites stimulate the proliferation of MG63 osteoblast-like cells and support the osteogenic differentiation of human bone marrow cells. Biomaterials. 23: 4897-906.

[34] Albrektsson T and Wennerberg A (2004) Oral implant surfaces: Part 1 - review focusing on topographic and chemical properties of different surfaces and in vivo responses to them. Int J Prosthodont. 17: 536-43.

[35] Tomas H, Carvalho G S, Fernandes M H, Freire A P, and Abrantes L M (1997) The use of rat, rabbit or human bone marrow derived cells for cytocompatibility evaluation of metallic elements. J Mater Sci Mater Med. 8: 233-8.

[36] Morais S, Sousa J P, Fernandes M H, Carvalho G S, de Bruijn J D, and van Blitterswijk C A (1998) Effects of AISI 316L corrosion products in in vitro bone formation. Biomaterials. 19: 999-1007.

[37] Gomes P S, Botelho C, Lopes M A, Santos J D, and Fernandes M H (2010) Evaluation of human osteoblastic cell response to plasma-sprayed silicon-substituted hydroxyapatite coatings over titanium substrates. J Biomed Mater Res B Appl Biomater. 94: 337-46.

[38] Whittaker C J, Klier C M, and Kolenbrander P E (1996) Mechanisms of adhesion by oral bacteria. Annu Rev Microbiol. 50: 513-52.

[39] Kolenbrander P E, Andersen R N, Blehert D S, Egland P G, Foster J S, and Palmer R J (2002) Communication among oral bacteria. Microbiol Mol Biol Rev. 66: 486-505.

[40] Burgers R, Gerlach T, Hahnel S, Schwarz F, Handel G, and Gosau M (2010) In vivo and in vitro biofilm formation on two different titanium implant surfaces. Clin Oral Implants Res. 21: 156-64.

[41] Teughels W, Van Assche N, Sliepen I, and Quirynen M (2006) Effect of material characteristics and/or surface topography on biofilm development. Clin Oral Implants Res. 17: 68-81.

[42] Marsh P D, Moter A, and Devine D A (2011) Dental plaque biofilms: communities, conflict and control. Periodontol 2000. 55: 16-35.

[43] Hannig C and Hannig M (2009) The oral cavity--a key system to understand substratum-dependent bioadhesion on solid surfaces in man. Clin Oral Investig. 13: 123-39.

[44] Hahnel S, Rosentritt M, Handel G, and Burgers R (2009) Surface characterization of dental ceramics and initial streptococcal adhesion in vitro. Dent Mater. 25: 969-75.

[45] Lee A and Wang H L (2010) Biofilm related to dental implants. Implant Dent. 19: 387-93.

[46] Marsh P D (2004) Dental plaque as a microbial biofilm. Caries Res. 38: 204-11.

[47] Scannapieco F A (1994) Saliva-bacterium interactions in oral microbial ecology. Crit Rev Oral Biol Med. 5: 203-48.
[48] Heuer W, Elter C, Demling A, Neumann A, Suerbaum S, Hannig M, Heidenblut T, Bach F W, and Stiesch-Scholz M (2007) Analysis of early biofilm formation on oral implants in man. J Oral Rehabil. 34: 377-82.
[49] Branda S S, Vik S, Friedman L, and Kolter R (2005) Biofilms: the matrix revisited. Trends Microbiol. 13: 20-6.
[50] An Y H and Friedman R J (1998) Concise review of mechanisms of bacterial adhesion to biomaterial surfaces. J Biomed Mater Res. 43: 338-48.
[51] Quirynen M and Bollen C M (1995) The influence of surface roughness and surface-free energy on supra- and subgingival plaque formation in man. A review of the literature. J Clin Periodontol. 22: 1-14.
[52] Faltermeier A, Burgers R, and Rosentritt M (2008) Bacterial adhesion of Streptococcus mutans to esthetic bracket materials. Am J Orthod Dentofacial Orthop. 133: S99-103.
[53] Buergers R, Rosentritt M, and Handel G (2007) Bacterial adhesion of Streptococcus mutans to provisional fixed prosthodontic material. J Prosthet Dent. 98: 461-9.
[54] Shaddox L M, Alfant B, Tobler J, and Walker C (2010) Perpetuation of subgingival biofilms in an in vitro model. Mol Oral Microbiol. 25: 81-7.
[55] Coenye T and Nelis H J (2010) In vitro and in vivo model systems to study microbial biofilm formation. J Microbiol Methods. 83: 89-105.
[56] Sissons C H (1997) Artificial dental plaque biofilm model systems. Adv Dent Res. 11: 110-26.
[57] McBain A J (2009) Chapter 4: In vitro biofilm models: an overview. Adv Appl Microbiol. 69: 99-132.
[58] Guggenheim B, Guggenheim M, Gmur R, Giertsen E, and Thurnheer T (2004) Application of the Zurich biofilm model to problems of cariology. Caries Res. 38: 212-22.
[59] Exterkate R A, Crielaard W, and Ten Cate J M (2010) Different response to amine fluoride by Streptococcus mutans and polymicrobial biofilms in a novel high-throughput active attachment model. Caries Res. 44: 372-9.
[60] Benoit M R, Conant C G, Ionescu-Zanetti C, Schwartz M, and Matin A (2010) New device for high-throughput viability screening of flow biofilms. Appl Environ Microbiol. 76: 4136-42.
[61] Stepanovic S, Vukovic D, Hola V, Di Bonaventura G, Djukic S, Cirkovic I, and Ruzicka F (2007) Quantification of biofilm in microtiter plates: overview of testing conditions and practical recommendations for assessment of biofilm production by staphylococci. APMIS. 115: 891-9.
[62] Merritt J H, Kadouri D E, and O'Toole G A (2005) Growing and analyzing static biofilms. Curr Protoc Microbiol. Chapter 1: Unit 1B 1.
[63] Silva T C, Pereira A F, Exterkate R A, Bagnato V S, Buzalaf M A, Machado M A, Ten Cate J M, Crielaard W, and Deng D M (2012) Application of an active attachment model as a high-throughput demineralization biofilm model. J Dent. 40: 41-7.
[64] Deng D M, Hoogenkamp M A, Exterkate R A, Jiang L M, van der Sluis L W, Ten Cate J M, and Crielaard W (2009) Influence of Streptococcus mutans on Enterococcus faecalis biofilm formation. J Endod. 35: 1249-52.

[65] ten Cate J M (2006) Biofilms, a new approach to the microbiology of dental plaque. Odontology. 94: 1-9.
[66] Wimpenny J W (1997) The validity of models. Adv Dent Res. 11: 150-9.
[67] Donlan R M, Piede J A, Heyes C D, Sanii L, Murga R, Edmonds P, El-Sayed I, and El-Sayed M A (2004) Model system for growing and quantifying Streptococcus pneumoniae biofilms in situ and in real time. Appl Environ Microbiol. 70: 4980-8.
[68] Goeres D M, Loetterle L R, Hamilton M A, Murga R, Kirby D W, and Donlan R M (2005) Statistical assessment of a laboratory method for growing biofilms. Microbiology. 151: 757-62.
[69] Yawata Y, Toda K, Setoyama E, Fukuda J, Suzuki H, Uchiyama H, and Nomura N (2010) Bacterial growth monitoring in a microfluidic device by confocal reflection microscopy. J Biosci Bioeng. 110: 130-3.
[70] Wang C C, Hsu Y C, Hsieh M C, Yang S P, Su F C, and Lee T M (2008) Effects of nano-surface properties on initial osteoblast adhesion and Ca/P adsorption ability for titanium alloys. Nanotechnology. 19: 335709.
[71] Flemming R G, Murphy C J, Abrams G A, Goodman S L, and Nealey P F (1999) Effects of synthetic micro- and nano-structured surfaces on cell behavior. Biomaterials. 20: 573-88.
[72] Zhao G, Raines A L, Wieland M, Schwartz Z, and Boyan B D (2007) Requirement for both micron- and submicron scale structure for synergistic responses of osteoblasts to substrate surface energy and topography. Biomaterials. 28: 2821-9.
[73] Wilkinson C D W, Riehle M, Wood M, Gallagher J, and Curtis A S G (2002) The use of materials patterned on a nano-and micro-metric scale in cellular engineering. Mater Sci Eng C. 19: 263-269.
[74] Yim E K F and Leong K W (2005) Significance of synthetic nanostructures in dictating cellular response. Nanomedicine. 1: 10-21.
[75] Francois P, Vaudaux P, Taborelli M, Tonetti M, Lew D P, and Descouts P (1997) Influence of surface treatments developed for oral implants on the physical and biological properties of titanium. (II) Adsorption isotherms and biological activity of immobilized fibronectin. Clin Oral Implants Res. 8: 217-25.
[76] Keller J C, Schneider G B, Stanford C M, and Kellogg B (2003) Effects of implant microtopography on osteoblast cell attachment. Implant Dent. 12: 175-81.
[77] Martin J Y, Schwartz Z, Hummert T W, Schraub D M, Simpson J, Lankford J, Jr., Dean D D, Cochran D L, and Boyan B D (1995) Effect of titanium surface roughness on proliferation, differentiation, and protein synthesis of human osteoblast-like cells (MG63). J Biomed Mater Res. 29: 389-401.
[78] Raz P, Lohmann C H, Turner J, Wang L, Poythress N, Blanchard C, Boyan B D, and Schwartz Z (2004) 1alpha,25(OH)2D3 regulation of integrin expression is substrate dependent. J Biomed Mater Res A. 71: 217-25.
[79] Brunette D M (1988) The effects of implant surface topography on the behavior of cells. Int J Oral Maxillofac Implants. 3: 231-46.
[80] Curtis A and Wilkinson C (1997) Topographical control of cells. Biomaterials. 18: 1573-83.
[81] Curtis A S G and Clark P (1990) The effects of topographic and mechanical properties of materials on cell behavior. Crit. Rev. Biocompat. 5: 343-363.

[82] Clark P (1994) Cell behaviour on micropatterned surfaces. Biosens Bioelectron. 9: 657-661.
[83] von Recum A F and van Kooten T G (1995) The influence of micro-topography on cellular response and the implications for silicone implants. J Biomater Sci Polym Ed. 7: 181-98.
[84] Weiss P (1958) Cell contact. Int Rev Cyt. 7: 391-423.
[85] Singhvi R, Stephanopoulos G, and Wang D I (1994) Effects of substratum morphology on cell physiology. Biotechnol Bioeng. 43: 764-71.
[86] Rosenberg M D (1963) Cell guidance by alterations in monomolecular films. Science. 139: 411.
[87] Rosenberg M D (1962) Long-range interactions between cell and substratum. Proc Natl Acad Sci USA. 48: 1342.
[88] Curtis A S G, Gadegaard N, Dalby M J, Riehle M O, Wilkinson C D W, and Aitchison G (2004) Cells react to nanoscale order and symmetry in their surroundings. IEEE Trans Nanobioscience. 3: 61-65.
[89] Kim D H, Kim P, Song I, Cha J M, Lee S H, Kim B, and Suh K Y (2006) Guided three-dimensional growth of functional cardiomyocytes on polyethylene glycol nanostructures. Langmuir. 22: 5419-26.
[90] Wood M A, Wilkinson C D, and Curtis A S (2006) The effects of colloidal nanotopography on initial fibroblast adhesion and morphology. IEEE Trans Nanobioscience. 5: 20-31.
[91] Sapelkin A V, Bayliss S C, Unal B, and Charalambou A (2006) Interaction of B50 rat hippocampal cells with stain-etched porous silicon. Biomaterials. 27: 842-6.
[92] Zhu B, Lu Q, Yin J, Hu J, and Wang Z (2005) Alignment of osteoblast-like cells and cell-produced collagen matrix induced by nanogrooves. Tissue Eng. 11: 825-34.
[93] Zanello L P, Zhao B, Hu H, and Haddon R C (2006) Bone cell proliferation on carbon nanotubes. Nano Lett. 6: 562-7.
[94] Miller D C, Thapa A, Haberstroh K M, and Webster T J (2004) Endothelial and vascular smooth muscle cell function on poly(lactic-co-glycolic acid) with nano-structured surface features. Biomaterials. 25: 53-61.
[95] Webster T J, Schadler L S, Siegel R W, and Bizios R (2001) Mechanisms of enhanced osteoblast adhesion on nanophase alumina involve vitronectin. Tissue Eng. 7: 291-301.
[96] Park J Y, Gemmell C H, and Davies J E (2001) Platelet interactions with titanium: modulation of platelet activity by surface topography. Biomaterials. 22: 2671-2682.
[97] Body S C (1996) Platelet activation and interactions with the microvasculature. J Cardiovasc Pharmacol. 27: 13.
[98] Hong J, Andersson J, Ekdahl K N, Elgue G, AxÃ©n N, Larsson R, and Nilsson B (1999) Titanium is a highly thrombogenic biomaterial: possible implications for osteogenesis. Thromb Haemost. 82: 58-64.
[99] Mendonca G, Mendonca D, Aragao F J L, and Cooper L F (2008) Advancing dental implant surface technology - from micron- to nanotopography. Biomaterials. 29: 3822-3835.
[100] Grew J C, Ricci J L, and Alexander H (2008) Connective-tissue responses to defined biomaterial surfaces. II. Behavior of rat and mouse fibroblasts cultured on microgrooved substrates. J Biomed Mater Res A. 85: 326-335.

[101] Kokubo T and Takadama H (2006) How useful is SBF in predicting in vivo bone bioactivity? Biomaterials. 27: 2907-2915.

[102] Fewster S D, Coombs R R H, Kitson J, and Zhou S (1994) Precise ultrafine surface texturing of implant materials to improve cellular adhesion and biocompatibility. Nanobiology. 3: 201-210.

[103] Wan Y, Wang Y, Liu Z, Qu X, Han B, Bei J, and Wang S (2005) Adhesion and proliferation of OCT-1 osteoblast-like cells on micro-and nano-scale topography structured poly (L-lactide). Biomaterials. 26: 4453-4459.

[104] Popat K C, Leoni L, Grimes C A, and Desai T A (2007) Influence of engineered titania nanotubular surfaces on bone cells. Biomaterials. 28: 3188-97.

[105] Popat K C, Chatvanichkul K I, Barnes G L, Latempa T J, Jr., Grimes C A, and Desai T A (2007) Osteogenic differentiation of marrow stromal cells cultured on nanoporous alumina surfaces. J Biomed Mater Res A. 80: 955-64.

[106] Matsuzaka K, Walboomers F, de Ruijter A, and Jansen J A (2000) Effect of microgrooved poly-l-lactic (PLA) surfaces on proliferation, cytoskeletal organization, and mineralized matrix formation of rat bone marrow cells. Clin Oral Implants Res. 11: 325-33.

[107] Matsuzaka K, Walboomers X F, de Ruijter J E, and Jansen J A (1999) The effect of poly-L-lactic acid with parallel surface micro groove on osteoblast-like cells in vitro. Biomaterials. 20: 1293-301.

[108] Groessner-Schreiber B and Tuan R S (1992) Enhanced extracellular matrix production and mineralization by osteoblasts cultured on titanium surfaces in vitro. J Cell Sci. 101: 209-217.

[109] Abron A, Hopfensperger M, Thompson J, and Cooper L F (2001) Evaluation of a predictive model for implant surface topography effects on early osseointegration in the rat tibia model. J Prosthet Dent. 85: 40-6.

[110] Dalby M J, McCloy D, Robertson M, Agheli H, Sutherland D, Affrossman S, and Oreffo R O (2006) Osteoprogenitor response to semi-ordered and random nanotopographies. Biomaterials. 27: 2980-7.

[111] Dalby M J, Gadegaard N, Tare R, Andar A, Riehle M O, Herzyk P, Wilkinson C D, and Oreffo R O (2007) The control of human mesenchymal cell differentiation using nanoscale symmetry and disorder. Nat Mater. 6: 997-1003.

[112] Zhao L, Mei S, Chu P K, Zhang Y, and Wu Z (2010) The influence of hierarchical hybrid micro/nano-textured titanium surface with titania nanotubes on osteoblast functions. Biomaterials. 31: 5072-5082.

[113] Engel E, Martinez E, Mills C A, Funes M, Planell J A, and Samitier J (2009) Mesenchymal stem cell differentiation on microstructured poly (methyl methacrylate) substrates. Ann Anat. 191: 136-44.

[114] Curtis A and Wilkinson C (2001) Nanotechniques and approaches in biotechnology. Trends Biotechnol. 19: 97-101.

[115] Dalby M J, Riehle M O, Sutherland D S, Agheli H, and Curtis A S G (2004) Changes in fibroblast morphology in response to nano-columns produced by colloidal lithography. Biomaterials. 25: 5415-5422.

[116] Dalby M J, Berry C C, Riehle M O, Sutherland D S, Agheli H, and Curtis A S G (2004) Attempted endocytosis of nano-environment produced by colloidal lithography by human fibroblasts. Exp Cell Res. 295: 387-394.
[117] Kunzler T P, Huwiler C, Drobek T, Voros J, and Spencer N D (2007) Systematic study of osteoblast response to nanotopography by means of nanoparticle-density gradients. Biomaterials. 28: 5000-5006.
[118] Teixeira A I, Nealey P F, and Murphy C J (2004) Responses of human keratocytes to micro- and nanostructured substrates. J Biomed Mater Res A. 71: 369-76.
[119] Lee K B, Park S J, Mirkin C A, Smith J C, and Mrksich M (2002) Protein nanoarrays generated by dip-pen nanolithography. Science. 295: 1702-5.
[120] Karuri N W, Liliensiek S, Teixeira A I, Abrams G, Campbell S, Nealey P F, and Murphy C J (2004) Biological length scale topography enhances cell-substratum adhesion of human corneal epithelial cells. J Cell Sci. 117: 3153-64.
[121] Baac H, Lee J H, Seo J M, Park T H, Chung H, Lee S D, and Kim S J (2004) Submicron-scale topographical control of cell growth using holographic surface relief grating. Mater Sci Eng C. 24: 209-212.
[122] Berry C C, Dalby M J, McCloy D, and Affrossman S (2005) The fibroblast response to tubes exhibiting internal nanotopography. Biomaterials. 26: 4985-92.
[123] Xia Y and Whitesides G M (1998) Soft Lithography. Angew Chem Int Edit. 37: 550-575.
[124] Xia Y and Whitesides G M (1998) Soft lithography. Ann Rev Mater Sci. 28: 153-184.
[125] Ferrari M, Desai T, and Bhatia S (2006), Therapeutic and Micro/Nanotechnology in BioMEMS and Biomedical Technology, Ferrari M. New York: Springer xxi p.
[126] Arkles B (2001) Commercial application of Sol-Gel derived Hybrid materials. MRS bulletin. 5: 402-407.
[127] Chou T, P. and Cao G (2003) Adhesion of Sol-Gel-Derived Organic-Inorganic Hybrid Coatings on Polyester. J Sol-Gel Sci Techn. V27: 31-41.
[128] Nazeri A, Trzaskoma-Paulette P P, and Bauer D (1997) Synthesis and Properties of Cerium and Titanium Oxide Thin Coatings for Corrosion Protection of 304 Stainless Steel. J Sol-Gel Sci Techn. V10: 317-331.
[129] Zhang Q, Whatmore R, and Vickers M E (1999) A Comparison of the Nanostructure of Lead Zirconate, Lead Titanate and Lead Zirconate Titanate Sols. J Sol-Gel Sci Techn. V15: 13-22.
[130] Zhong J and Greenspan D C (2000) Processing and properties of sol-gel bioactive glasses. J Biomed Mater Res. 53: 694-701.
[131] Livage J (1997) Sol-gel processes. Current Opin Solid St Mater Sc. 2: 132-138.
[132] Dimitriev Y, Ivanova Y, and Iordanova R (2008) History of Sol-Gel Science and Technology (Review). J Univ Chem Technol Metal. 43: 181-192.
[133] Guglielmi M (2010), Glass: Sol-gel coatings in Encyclopedia of Materials: Science and Technology, Buschow K H J, Cahn R, Flemings M, Ilschner B, Kramer E, Mahajan S, and Veyssiere P. Amsterdam: Elsevier 3575-3579 p.
[134] Brinker C, Ashley C, Cairncross R, Chen K, Hurd A, Reed S, Samuel J, Schunk PR, RW S, and Scotto C (1996), Sol-gel derived ceramic film - fundamentals and applications in Metallurgical and ceramic protective coatings, Stern K H. London: Chapman & Hall 112-151 p.

[135] Strawbridge I (1992), Glass formation by the sol-gel process in High-performance glasses, Cable M a P, JM. NJ: Chapman and Hall 51-85 p.
[136] Hench L L (1997) Sol-gel materials for bioceramic applications. Curr Opin Solid State Mater Sci. 2: 604-610.
[137] Marzolin C, Smith S P, Prentiss M, and Whitesides G M (1998) Fabrication of Glass Microstructures by Micro-Molding of Sol-Gel Precursors. Adv Mater. 10: 571-574.
[138] Pelaez-Vargas A, Gallego-Perez D, Ferrell N, Fernandes M H, Hansford D, and Monteiro F J (2010) Early spreading and propagation of human bone marrow stem cells on isotropic and anisotropic topographies of silica thin films produced via microstamping. Microsc Microanal. 16: 670-6.
[139] Ferraz M P, Monteiro F J, and Manuel C M (2004) Hydroxyapatite nanoparticles: A review of preparation methodologies. J Appl Biomater Biomech. 2: 74-80.
[140] Best S M, Porter A E, Thian E S, and Huang J (2008) Bioceramics: Past, present and for the future. J Eur Ceram Soc. 28: 1319-1327.
[141] Vallet-Regi M (2001) Ceramics for medical applications. Journal of the Chemical Society-Dalton Transactions. 97-108.
[142] Chevalier J and Gremillard L (2009) Ceramics for medical applications: A picture for the next 20 years. J Eur Ceram Soc. 29: 1245-1255.
[143] Manuel C M, Ferraz M P, and Monteiro F J (2003) Synthesis of hydroxyapatite and tricalcium phosphate nanoparticles preliminary studies. Bioceramics 15. 240-2: 555-558.
[144] Muller-Mai C M, Stupp S I, Voigt C, and Gross U (1995) Nanoapatite and organoapatite implants in bone: histology and ultrastructure of the interface. J Biomed Mater Res. 29: 9-18.
[145] Sarkar N K, Caicedo R, Ritwik P, Moiseyeva R, and Kawashima I (2005) Physicochemical basis of the biologic properties of mineral trioxide aggregate. J Endod. 31: 97-100.
[146] Gallego-Perez D, Higuita-Castro N, Quiroz F G, Posada O M, Lopez L E, Litsky A S, and Hansford D J (2011) Portland cement for bone tissue engineering: Effects of processing and metakaolin blends. J Biomed Mater Res B Appl Biomater. 98B: 308-15.
[147] Higuita-Castro N, Gallego-Perez D, Pelaez-Vargas A, Garcia Quiroz F, Posada O M, Lopez L E, Sarassa C A, Agudelo-Florez P, Monteiro F J, Litsky A S, and Hansford D J (2012) Reinforced Portland cement porous scaffolds for load-bearing bone tissue engineering applications. J Biomed Mater Res B Appl Biomater. 100B: 501-507.
[148] Pelaez-Vargas A, Gallego-Perez D, Higuita-Castro N, Ferrell N, Hansford D, and Monteiro F Silica coatings with Portland cement particles for dental implant surfaces based on sol-gel micromolding. in *Biomedical Engineering Society* 2009. Pittsburg, PA: BMES CD PS 10A-23
[149] Pelaez-Vargas A, Higuita-Castro N, Gallego-Perez D, Hansford D, and Monteiro F (2010) Portland particulated silica coatings – Comparison between two coating techniques. J Dental Research. 89: 4006.
[150] Carvalho A, Pelaez-Vargas A, Gallego-Perez D, Fernandes M, Hansford D, and Monteiro F (2011) Adhesion and proliferation of mesenchymal stem cells on micropatterned thin films modified with nanohydroxyapatite particles. Bone. 48: S106.
[151] Gallego D, Ferrell N, Sun Y, and Hansford D J (2008) Multilayer micromolding of degradable polymer tissue engineering scaffolds. Mater Sci Eng C. 28: 353-358.

Chapter 12

Cell Interaction Analysis by Imaging Flow Cytometry

Cristian Payés, José A. Rodríguez, Sherree Friend and Gustavo Helguera

Additional information is available at the end of the chapter

1. Introduction

Many processes such as cell adhesion, tissue development, cellular communication, inflammation, tumor metastasis, and microbial infection require direct interactions between cells. Some cell-cell interactions are transient, as is the case of the contacts between cells of the immune system, the interactions of white blood cells to malignant cells or to sites of tissue inflammation. These events often entail structural alterations in the point of contact of the cells involved, and may involve the fusion, transfer or exchange of material between the cells; which occur in a scale that is suited for optical microscopy analysis. However, due to its low throughput nature, microscopy often suffers from acquisition bias and limited statistical power. Moreover, because the data is typically analyzed in a qualitative manner, it is difficult to obtain standardized results. Strong scientific conclusions demand objective collection of large amounts of relevant information that can be analyzed in a quantitative, standardized, and statistically robust manner. Flow cytometry overcomes these problems but reduces the rich information available via optical microscopy to a set of intensity measurements. By combining high speed automated image acquisition with quantitative image analysis, Multispectral Imaging Flow Cytometry (MIFC) provides all the elements required for discriminating cells based on intensity and appearance in a standardized and statistical manner. In recent years, the application of this technology for the analysis of cell-cell interaction has multiplied, in particular in the field of immunology, allowing the observation and quantification of events in a way that could not be achieved before. Although studies investigating the interaction of pathogens with host cells have been conducted using MIFC in combination with confocal fluorescence microscopy and immuno cytochemistry microscopy [1], here we will only address the interaction of cells belonging to the same species, therefore host-pathogen cell-cell interactions are beyond the scope of the current manuscript.

2. Combining microscopic imaging and flow cytometry to characterize cell-cell interactions

The characterization of cells and subcellular components using methods that detect fluorescent labels is invaluable to biology. The expansion and adoption of fluorescence microscopy techniques in biology has made routine the detailed characterization of fluorescently labeled structures in single cells. As the modern standard of fluorescence-based microscopic imaging, confocal microscopy is capable of rendering the diffraction-limited three-dimensional representation of a single fluorescently labeled cell; taking up to several minutes to do so. By generating high-resolution images of one or a few cells a minute, fluorescence microscopy is limited in its throughput. In contrast, flow cytometry has revolutionized the identification and measurement of fluorescently labeled cells and subcellular elements via the rapid quantification of bulk fluorescent signal from single cells on a population-wide scale. In fact, modern flow cytometers can routinely measure hundreds of thousands of cells per second [2].

Both fluorescence microscopes and flow cytometers are capable of obtaining multi-spectral measurements of cells. This ability to simultaneously identify multiple fluorescent labels within a sample has facilitated the characterization of many cellular components in multiple cell populations in a single sample [2]. However, biologists have traditionally been forced to chose between the detection of fluorescence signals in cells at high resolution but with low throughput via microscopic imaging, and the rapid quantification of bulk fluorescent signal in a heterogeneous population of cells by flow cytometry [3]. These technologies have thus become largely complementary to each other, supplying information from both ends of the resolution-throughput spectrum.

Given the individual benefits of microscopic imaging and flow cytometry, the ideal measurement should leverage and combine the unique abilities of both technologies. The statistical power of flow cytometry would then be infused with detailed cellular imagery, with each individual point in a dataset representing the detailed image of a single fluorescently labeled cell or group of cells. Until recently, several technical challenges have limited the existence of a robust imaging flow cytometer capable of producing the high resolution fluorescent images expected from a microscope as well as the high throughput measurement associated with flow cytometry. MIFC instruments are now commercially available (see www.amnis.com). [4].

2.1. Elements of multispectral imaging flow cytometry (MIFC)

This chapter describes their use for the characterization of cell-cell interactions: Section 2.1 provides a technical overview of MIFC instruments, section 2.2 provides an overview of the methodology used to characterize cell-cell interactions using MIFC, section 3 describes a specific example of use of MIFC to measure and characterize cell-cell interactions, namely antibody-dependent cell-mediated cytotoxicity, while section 4 covers a broad range of examples in which MIFC has assisted in the characterization of a variety of cell-cell interactions.

2.1.1. Brief history of MIFC

Shortly after the development of flow cytometry, attempts at infusing the technology with microscopic imaging began [3]. The first attempts intended to add morphological information to fluorescence intensity measurements for each object encountered. To do this, slit cyto-fluorometers simultaneously recorded low-resolution measurements of object shapes and bulk fluorescence [5]. Alternatively optical impedance measurements and side-scatter were combined with bulk fluorescence signal to give a rough estimate of particle size and granularity, rapidly and reliably. Strobe lighting combined with fast CCD cameras have allowed for true microscopic imaging of cells in flow [5]. However, these approaches, including scanning, tracking, and strobe illumination techniques, have been limited in their ability to reliably generate high-resolution images of all cells in a flow stream with sufficient fluorescence sensitivity [3]. Recently an imaging system that overcomes most of these barriers has become commercially available and has demonstrated the ability to provide high spatial resolution images, high fluorescence sensitivity, reliable high-throughput acquisition, and the simultaneous analysis of fluorescence signals at various wavelengths [4]. With this system, a new dimension has been effectively added to conventional microscopic analysis, allowing for the execution of previously intractable experiments and the development of completely new forms of analysis.

2.1.2. The MIFC instrument

Multispectral Imaging Flow Cytometry combines the high throughput capability of the flow cytometer with the imaging capability of the microscope. Hydrodyanamically focused cells are trans-illuminated by a brightfield light source and orthogonally by laser light. [4]. Fluorescence emissions, scattered, and transmitted light from the cells are collected using a high numerical aperture objective. The collected light is directed through a spectral decomposition element which places light of different spectral bands into separate positions (channels) on the CCD camera in spatial alignment to each other. [6]. The camera records these images and stores them for subsequent analysis. These systems are capable of measuring over 1000 individual events every second, translating to hundreds of thousands of cells in just a few minutes [4]. Moreover, each cell is comprised of up to 12 images (one for each imaging channel); well over a million images in a five-minute time span [4].

2.1.3. High resolution images in MIFC

When imaging cells moving rapidly within a flow channel, the relative motion of the cells with respect to the detector must be accounted for. A method termed time-delay integration [4] detects the velocity of cells with respect to the detector (a CCD camera) and synchronizes the integration of signal intensity on a per pixel basis with the cell velocity over the length of the camera. This is equivalent to panning the detector alongside passerby cells with a matching speed, in effect neutralizing the relative motion of the cells, which prevents streaking and increases the sensitivity and resolution of the detector [4].

2.1.4. Extended depth of field using MIFC

As with most microscopic imaging systems, proper focus is essential for obtaining high-quality images. This is especially true for cells traveling along a flow channel whose size allows for movement of cells in the direction perpendicular to the flow path, or for cells significantly larger than the depth of focus of the optics used [6]. In such cases, signal from parts of cells that lie outside the focal volume can be blurred or altogether lost. To counteract this effect during standard microscopic analysis of cells, several focal volumes can be collected and assembled or added to generate an effective focal volume that encompasses the entire cell. The imaging flow cytometer employs autofocusing and extended depth of field methods to recover what would normally be out of focus signal, effectively extending the range of a single focal volume from four micrometers to sixteen micrometers at 40X magnification [6].

2.1.5. Multi-parameter image processing using MIFC

While microscopic analysis offers a rich array of parameters that can be interpreted from a single cell, the sampling of enough cells to find statistical significance in these observations may not be tractable. Flow cytometers overcome this statistical limitation by collecting large sample sizes. By quickly sampling thousands of cells within minutes or seconds, MIFC generates a wealth of information in two spatial dimensions for each of its independent imaging channels [4]. The spatial resolution of the images is used to compute hundreds of photometric and morphometric features for every cell. [7]. Thus, MIFC instantly increases the amount of information available for each cell over 1000-fold [3, 4]. In traditional flow cytometry experiments, this information is collapsed to a single point on a scatterplot for each cell detected.

The photometric and morphometric parameters are calculated post acquisition of the images using algorithms that take into account location and intensity of the fluorescence, darkfield and brightfield images. Cell populations can be categorized using multiple parameters. Parameters are available that measure size, shape, texture, signal strength and location. Masks are available that focus the calculations to specific regions of the cells. This means cell morphology can be quantified, allowing cell populations to be categorized using their true structure rather than simple size and granularity measurements typical of flow cytometry analyses. Determinations that are performed by standard flow cytometry such as measurement of intracellular cell calcium or evaluation of tight junctions between cells could also be resolved and analyzed with this technology. Additionally, localization and co-localization of signals from multiple labels can be associated to subcellular compartments such as the nucleus, cell surface, or visible organelles. To identify truly spatially co-existent signals within a subcellular compartment, cells can be measured with or without the use of extended depth of focus technology [6].

2.2. Cell-cell interactions characterized using MIFC

2.2.1. Measuring rare events using MIFC

Many phenomena in biology, including cell-cell interactions, occur as fleeting events or in very low frequencies. The difficulty in characterizing these events is typically overcome by sampling a large number of cells using a high-throughput approach such as flow cytometry. In such a case, the key is to sample as many cells as possible so that the number of rare events observed allows for meaningful information to be extracted from them. A conventional microscope would not serve this purpose well, as it would take an intractable amount of time to find a large enough number of events to make valuable conclusions. However, since MIFC can measure hundreds of thousands of cells from a single population within minutes, even events that only occur at a rate lower than one percent can be investigated [3]. In such a case, when only one out of every one thousand cells forms a cell-cell complex, up to five hundred complexes could be measured within a matter of minutes.

2.2.2. *Detecting a multi-cell complex using MIFC*

The application of extended depth of field to MIFC is particularly important for the accurate detection of cell-cell interactions. The size of most cells lies outside the range of the standard focal volume [6]. Detecting cell-cell interactions when limited by a narrow focal volume would prove to be a debilitating challenge. A cube, whose sides are each 20-30 micrometers in length, depending on the cell type, can represent two or more interconnected cells (Figure 1). In this case, a four micrometer focal volume could likely miss one of the cells altogether and would rarely intercept the cell-cell complex at an appropriate focal plane to resolve the junction between the cells [6]. In contrast, the sixteen-micrometer focal volume offered by the use of the extended depth of field element covers largely the entire volume and should easily resolve the cell-cell junction (Figure 1).

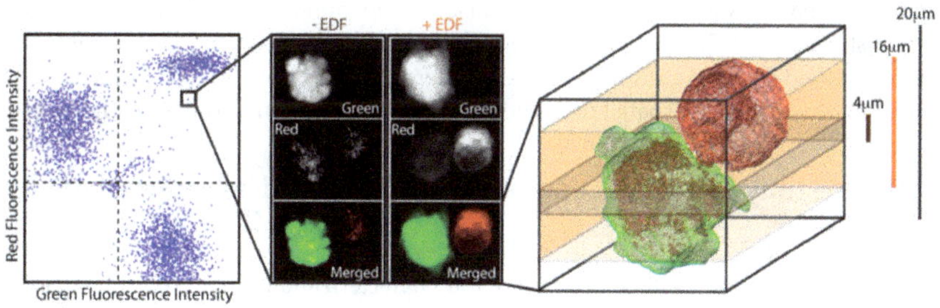

Figure 1. Identification of rare cell-cell interactions using multi-spectral imaging flow cytometry. From two cell populations labelled independently with green and red fluorescent dyes, a scatterplot is shown (left) in which single or dual labelled specimens can be identified. From a single event in the scatterplot, images can be recalled, having been obtained simultaneously for each of the fluorescent channels. Using the extended depth of field technology, the entire complex of one green and one red labelled cell can be imaged as shown in the montage (middle). Without the EDF technology a limited depth of field might not identify the cell-cell interaction or might do so only partially. The range of the interaction volume captured with the aid of the EDF technology is illustrated in the three dimensional cell-cell interaction volume rendered (right).

2.2.3. Multi-color analysis of cell-cell interactions using MIFC

The multi-spectral measurement of bulk cell fluorescence has revolutionized flow cytometry. Populations of cells can be identified as being labeled by unique combinations of fluorescent dyes, implying powerful relationships between the molecules being labeled in single cell populations and the relative quantities in which they are present in these cells [2]. Likewise, MIFC can simultaneously acquire up to twelve independent imaging channels at a time, allowing for a variety of molecular associations to be probed within a single cell population [4]. However, instead of just measuring bulk fluorescence intensity, spatial correlations can be quantitatively determined for each of the fluorescent labels measured (Figure 1). The spatial distribution and co-segregation of each of these independently measured labels can be used to quantitatively paint a true multi-spectral image of a cell population [4]. To achieve quantitative results with single pixel accuracy, corrections are applied to the pixel data for subpixel spatial registration and spectral overlap of the fluorochromes used. Correction values are obtained through routine instrument calibration and the collection of single color compensation controls [4].

The nature of cell-cell interactions necessitates the disambiguation of components from two physically linked cells. Until recently, the true characterization of these interactions required high-resolution fluorescence microscopy or electron micrograph analysis. The challenge lies in spatially separating the signal obtained from each interacting partner. Using spectrally distinct fluorescent labels satisfies this requirement, since emitted signals can be identified from independent cell populations, which when detected in coexistence indicates a physical interaction between the two cell types. Using MIFC, cell-cell interactions can be corroborated by morphometric analysis of the distribution of fluorescence signal and visual analysis of brightfield images [3]. More importantly, both the interaction region and the crossing of signal between interacting cells can also be identified using MIFC by tracking the spatial distribution of fluorophores unique to each interacting partner. Direct interactions can be quantified either as a result of spatial co-segregation of signal or direct signal overlap, and are limited only by the spatial resolution of the optical configuration used (Figure 1).

3. MIFC to study cell interaction in antibody dependent cell-mediated cytotoxicity

The use of MIFC based methodologies for the characterization of cell-cell interactions is enjoying rapid growth. This is true in part because, as described in section 2 of this chapter, MIFC instruments have the unique ability to characterize transient or rare cell-cell interactions in mixed cell populations. In this section we discuss the use of MIFC to quantitatively characterize a particular type of cell-mediated cytotoxicity, termed antibody-dependent cell-mediated cytotoxicity (ADCC), in which immune system cells and their targets on the move exhibit short-lived physical interactions that result in the large-scale loss of target cells.

ADCC is an immune defense mechanism mediated by cells in which effectors cells of the immune system expressing Fc gamma receptors interact with the Fc domain of antibodies bound to antigens present on the surface of the target cell and actively induce their cytotoxicity. This process requires the formation of a direct connection between both cells, also referred as a cytotoxic synapse, in which there is a transient cytoplasmic contact and mutual exchange of membrane lipids between the effector and target cell [8]. The clinical success of many recombinant antibodies used in the therapy of cancer such as the chimeric anti-CD20 IgG1 rituximab and the humanized anti-HER2 IgG1 trastuzumab has been associated in great extent to the mechanism of ADCC [9, 10]. The determination of ADCC is usually performed using methods that quantify the release of a traceable compound from target cells upon their death, or methods that detect the coexistence of different traceable compounds normally found in only effector or target cell populations. Traditionally radio-dosimetry, colorimetry, spectrometry, or flow cytometry have been used to measure these events. However, MIFC has also been used to quantitatively characterize individual ADCC events in mixed populations of immune effector cells and cancer cells, in the presence of a clinically relevant therapeutic antibody [11]. The human monocyte cell line U-937 was used as an effector cell population, and fluorescently labeled with CMTPX red dye, while the human Burkitt's B-cell non-Hodgkin's lymphoma (NHL) cell line Ramos were fluorescently labeled with CFSE green dye. The nuclei of these cells were also fluorescently labeled with the DRAQ5 dye. The two populations of labeled cells were mixed in the presence or the absence of rituximab, to characterize transient effector-target cell interaction events using the ImageStream (multispectral imaging flow cytometer) (Figure 2). MIFC data were processed using the IDEAS software package (Amnis, Seattle WA), which identified individual effector and target cells as well as effector-target cell complexes. The apoptotic index of each cell was also determined through the quantitation of nuclear morphometric parameters (based on the two dimensional distribution of DRAQ5 label within the cell nuclei), including nuclear texture, condensation, and fragmentation (Figure 3). MIFC facilitated the simultaneous analysis of large image datasets representing thousands of events within a measured population. For this analysis a nuclear mask was generated, containing pixels with intensity values in the upper 40% of the intensity range of the DRAQ5 image for $CFSE^+$ (target) or $CMTPX^+$ (effector) cells. Based on this strategy the percentage of apoptotic target cells was determined. These were the apoptotic $CFSE^+$ cells as a fraction of the total number of $CFSE^+$ cells. Since this metric could also be applied to effector cells, the apoptotic status of effector cells within a mixed population could also be quantified. This methodology offers a more complete perspective of cell death within mixed populations, particularly given that current methods ignore the status of the effector cells in the ADCC activity and may overlook some potential toxicity triggered on effector cells. Based on the results presented, MIFC appeared to improve the accuracy with which the biological activity of therapeutic antibody candidates can be evaluated, indicating that the use of traditional methods may provide misleading estimates of the therapeutic benefit of these antibodies.

In addition to providing a quantitative assessment of cell death for each cell in a population, the transfer of cytoplasmic material from effector to target cells was also quantified from images of double-positive ($CMTPX^+$,$CFSE^+$) events. These were identified in multispectral images as the presence of CMTPX dye from the effector cells within the area marked by

CFSE labeled target cells. In the presence of rituximab, a significant increase in the proportion of target cells containing CMTPX signal covering 25-50% of the target cells was detected (Figure 4). In contrast, no cases of signal transfer into effector cells was observed, suggesting that the transfer of cell contents in effector-target pairs is unidirectional, from effector to target cells, as would be expected in cases of ADCC activity.

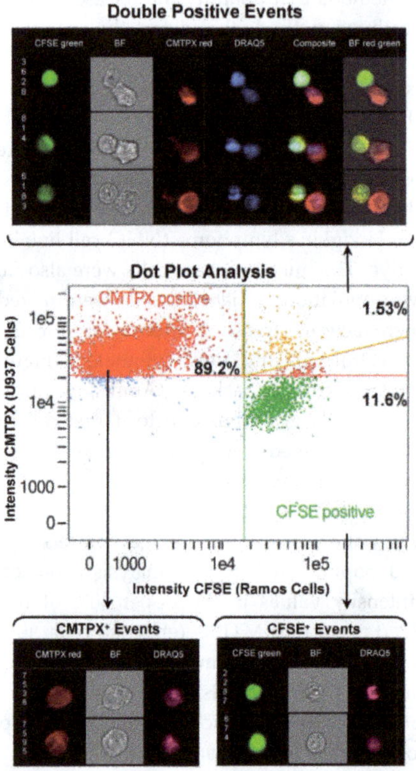

Figure 2. Identification of effector and target cells by imaging flow cytometry. Standard dot plot analysis was used to determine red and green fluorescence intensity in U-937 human monocyte effector cells labeled red with the CMTPX fluorochrome co-incubated for 2 h with Ramos human Burkitt's B-cell NHL target cells labeled green with the CFSE fluorochrome (5:1—E:T ratio) in the presence of 5 μg/ml rituximab. After incubation, cells were fixed and nuclei stained with the DRAQ5 dye. Samples were run in the ImageStream and imagery acquired for 10,000 events. In the middle of the figure we show the dot plot of the treatment with the percentage of events CMTPX+ in the upper left quadrant, CFSE+ in the bottom right quadrant, and CMTPX+/CFSE+ in the top right quadrant. Above the dot plot we show representative imagery of double positive events from the upper right quadrant of the dot plot with CFSE fluorescence (CFSE green), brightfield (BF), CMTPX fluorescence (CMTPX red), DRAQ5 fluorescence (DRAQ5), CFSE, CMTPX, and DRAQ5 fluorescence (composite), and brightfield, CFSE and

CMTPX fluorescence (BF red green). Below left we show representative imagery of single U-937 cells CMTPX positive from the upper left quadrant of the dot plot with CMTPX red, BF, and DRAQ5 pictures. And finally, below right we show representative imagery of single Ramos cells CFSE positive from the bottom right quadrant of the dot plot with CFSE green, BF, and DRAQ5 pictures. Figure is reprinted from Helguera et al. 2011, with permission from Elsevier.

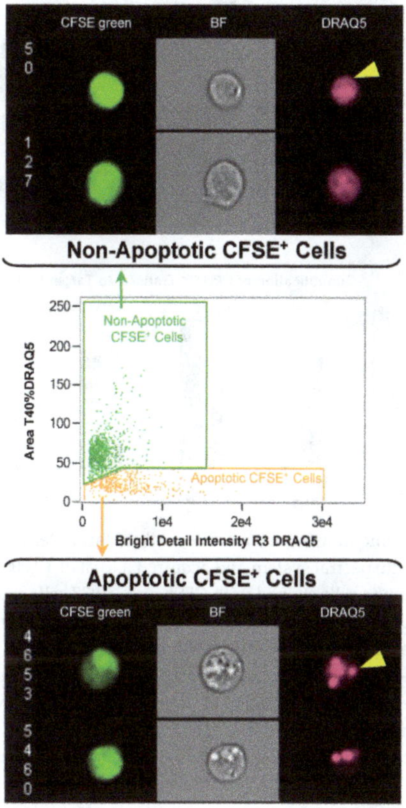

Figure 3. Determination of apoptotic index by nuclear morphology. At the center is a bivariate plot analysis using the IDEAS® software showing Area threshold 40% DRAQ5 intensity versus bright detail intensity R3 DRAQ5 parameters of Ramos cells labeled with CFSE co-incubated for 2 h with U-937 effector cells in the presence of 5 μg/ml rituximab. The non-apoptotic Ramos CFSE⁺ cells are gated in green. Sample CFSE, brightfield and DRAQ5 imagery are shown on top, with the yellow triangle pointing to the homogeneous nucleus stained with DRAQ5 typical of non-apoptotic cells. Apoptotic CFSE⁺ cells are shown in orange, and representative CFSE, brightfield and DRAQ5 images are shown on the bottom, with the yellow triangle pointing to the nucleus fragmented and intensely stained with DRAQ5, typical of apoptotic cells. Note also in BF images the change in morphology, with intense granularity of the cytoplasm. Figure is reprinted from Helguera et al. 2011, with permission from Elsevier.

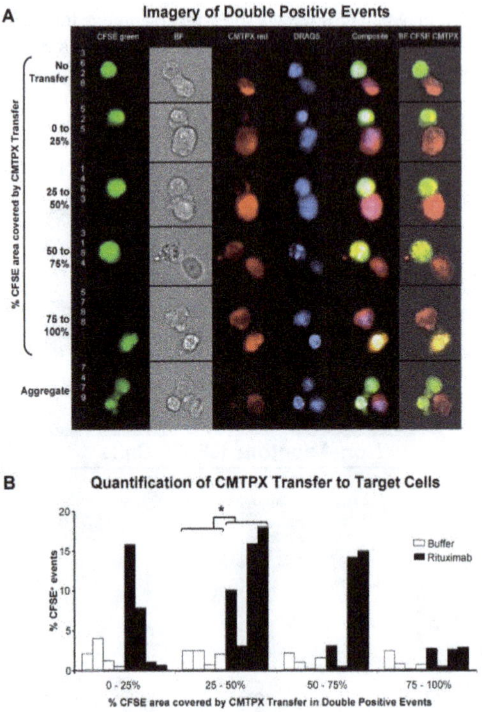

Figure 4. Analysis of cytoplasmic transfer from effector cells to target cells in double positive events. We used an ImageStream multispectral system and acquired a total of 10,000 events to study the double positive events containing Ramos cells stained with CFSE and U-937 cells stained with CMTPX after incubation for 1 h in the absence and presence of rituximab. Panel A shows imagery of double positive events including CFSE fluorescence, brightfield, CMTPX fluorescence, DRAQ5 fluorescence, a composite image of fluorescent stain, and an overlay of brightfield, red and green fluorescence. In the descending rows we show representative imagery of double positive events with increasing percentage of green (CFSE$^+$) masked region covered by red (CMTPX$^+$) mask, evidence of cytoplasmic transfer from effector to target cells. At the bottom we show representative imagery of an aggregate of multiple target cells and an effector cell. Panel B shows the percentage of CFSE$^+$ events with different percentages of CFSE area covered by CMTPX fluorescence comparing buffer treatment (open bars) and rituximab (black bars). Each bar corresponds to different experiments and (*) t-test $p \leq 0.05$. Figure is reprinted from Helguera et al. 2011, with permission from Elsevier.

Similarly, images of double positive events could be analyzed to quantify the contact area between effector and target cells in the context of phagocytic events or antibody-dependent cell mediated phagocytosis (ADCP). The determination of phagocytosis of target cells by effector cells was obtained by measuring the absolute distance between the center of the CMTPX and CFSE signals of cell complexes, which was calculated using the Delta Centroid XY (DC) feature of the IDEAS package. Phagocytic events, in which effector cells surround or engulf target cells, exhibit significantly lower DC values compared to conjugation events.

In events where the DC measurement approaches a 3 μm threshold (smaller than the radius of CMTPX and CFSE images) cell complexes were considered to be in the phagocytosis/internalization range. This was performed for all complexes observed in a mixed population, providing quantitative histogram plots for condition comparisons (Figure 5). In summary, MIFC can be used to simultaneously quantify and visualize the apoptotic index of effector and target cells, together with the transfer of labeled cytoplasmic contents from effector to target cells and the qualitative and quantitative analysis of phagocytic events favored by antibodies, providing a more comprehensive analysis of ADCC activity compared to standard techniques. A related phenomenon observed in the context of cell-cell interaction mediated by antibodies, is the transfer of membrane fragments from a target cell to an acceptor cell, in a process called trogocytosis or "cell nibbling". In this process, the effector cell internalizes the antibody, the captured receptor and membrane fragments from the target cell. This activity was studied for three therapeutic antibodies, the anti-CD20 rituximab, anti-EGFR cetuximab and anti-HER2 trastuzumab using MIFC [12]. MIFC was used to visualize the transfer of portions of the outer membrane of the human B-cell line Raji and its associated protein CD20, into the effector cell line THP-1, in the presence of rituximab. In this case, the membranes of Raji cells were labeled with the PKH26 dye, preincubated in the presence of Alexa 488 labeled rituximab and incubated with monocytic THP-1 cells that were labeled with anti-CD11b and anti-CD14 antibodies coupled to PE-Cy5. Under these conditions, MIFC was performed to visualize and quantify the presence of both PHK26 signal and Alexa 488 labeled rituximab in THP-1 cells alone or complexed with Raji cells, as evidence of receptor and lipid transfer from the target to the effector cell in the process of trogocytosis [12].

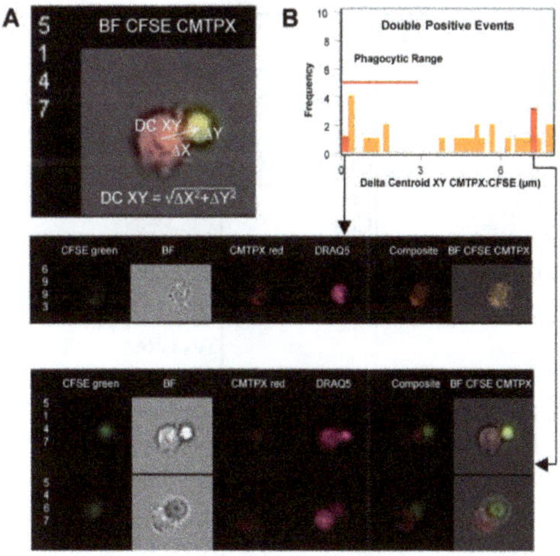

Figure 5. Delta centroid image analysis to identify phagocytosis in double positive events. Panel A shows a composite image of brightfield, CFSE, and CMTPX fluorescence in a double positive event of U-937 effector cells and Ramos target cells incubated for 2 h in the presence of 5 μg/ml rituximab. Superimposed is the vector used to calculate the distance between the centers of the masks of the

effector and the target cell. At the bottom of the image is the equation to calculate the delta centroid XY. Panel B shows a frequency histogram of the delta centroid XY distances of the double positive events. We set a threshold of 3 μm for the delta centroid XY as the phagocytic range, to identify phagocytic events. Below we show imagery of an event inside the phagocytic range in which there is overlapping of green and red signal. At the bottom we show imagery of events in which there is just contact of effector and target cells, resulting in the delta centroid distance equaling the sum of the radii of the green and red fluorescent events (values well beyond the phagocytic range). Figure is reprinted from Helguera et al. 2011, with permission from Elsevier.

4. Other applications of MIFC for the determination of cell-cell interaction

As described in section 3, MIFC offers an in-depth quantitation of previously overlooked aspects of cell-cell interactions that occur as part of ADCC. Likewise, the use of MIFC for the quantification and characterization of cell-cell interactions occurring as a result of a broad range of biological phenomena is expanding our understanding of these events in new and therapeutically meaningful ways. These include cell death arbitrated by other cells, the formation of cell complexes to propagate signals or activate cell populations via the presentation or detection of molecules on cells, the uptake or transfer of cellular materials, the formation of the immunological synapse, and the aggregation of cells to form clusters, plaques, or functional complexes. Cases in which the use of MIFC has led to significant insight into the molecular mechanisms of these phenomena are discussed in this section (Table 1).

Event visualized	Cell types (species)	Identified Marker	Fluorophores (wave length)	Reference
Transfer of Cellular Components	Plasmacytoid dendritic cells (pDC) (human) Monocyte derived dendritic cells (MDDC) (human)	Cell membrane Cytoplasm BDCA-2 CD-123	CFSE (Ex./Em. 492/517nm) PKH67 (Ex./Em. 490/502 nm) PE (Ex./Em. 496/578 nm)	[20]
Transfer of Cellular Components	Raji cells line (from B cells) (human) THP-1 monocytes cells (human)	Cell membrane CD20	PKH26 (Ex./Em. 490/502 nm) Alexa Fluor 488 (Ex./Em. 495/519nm) PE-Cy5 (Ex./Em. 496/785 nm)	[11]
Immunological Synapse	3B11 T hybridoma (mouse) LK35.2 B cells line (mouse)	CD3 LFA-1 F-actin DNA	TexasRed (Ex./Em. 589/615 nm) FITC (Ex./Em. 496/519 nm) Alexa Fluor 568 (Ex./Em. 578/603 nm) Hoechst 33342 (Ex./Em. 350/461 nm)	[15]
Immunological Synapse	T cells (human) Macrophage (human)	Lck CD3 DNA	Alexa Fluor 546 (Ex./Em. 554/570 nm) PE-Alexa Fluor 610 (Ex./Em. 488/628 nm) DRAQ5 (Ex./Em. 631/660 nm)	[14]

Event visualized	Cell types (species)	Identified Marker	Fluorophores (wave length)	Reference
Cell-cell interaction	Mast cells (MCs) (human) Eosinophils (Eos) (human)	Cytoplasm	CFSE (Ex./Em. 492/517nm)	[13]
Cell Aggregate Formation	Regulatory T cells (Tregs) (human) Bone marrow-derived Dendritic cells (BMDCs) (human) Conventional T cells (Tcon) (human)	Cell membrane Cytoplasm CD4 CD11c CD25 IL-2	Alexa Fluor 488 (Ex./Em. 495/519nm) Alexa Fluor 405 (Ex./Em. 401/425nm) Alexa Fluor 633 (Ex./Em. 632/647 nm) Alexa Fluor 647 (Ex./Em. 650/668 nm)	[21]
Cell Aggregate Formation	Leukocytes (human) PBMC (human) of patients with sickle cell disease and healthy individuals	CD45 Glycophorin A (GPA)	FITC (Ex./Em. 496/519 nm) PE (Ex./Em. 496/578 nm)	[23]
Immunological Synapse	$CD4^+$ T cells (mouse) B cells loaded with an antigenic peptide (mouse)	TCRβ LFA-1 DNA	Alexa Fluor 488 (Ex./Em. 495/519nm) PE (Ex./Em. 565/578 nm) 7-AAD (Ex./Em. 546/647 nm)	[16]
ADCC, Apoptosis, Phagocytosis, Transfer of Cellular Components	Ramos cell line, Burkitt's B-cell lymphoma (human) U-937 cell line monocyte (human)	Cytoplasm DNA	CFSE (Ex./Em. 492/517nm) CMTPX (Ex./Em. 577/602nm) DRAQ5 (Ex./Em. 631/660 nm)	[10]
Cell aggregates	Leukocytes (human) with/without CD62 or CD162	CD14 CD36 CD45 DNA	FITC (Ex./Em. 496/519 nm) ECD (Ex./Em. 595/660 nm) PE (Ex./Em. 496/578 nm) DRAQ5 (Ex./Em. 631/660 nm)	[22]
Transfer of Cellular Components	Activated $CD4^+$T cells (human) NK cells (human)	CD3 CD4 CD56 Granzyme K Perforin	Not reported	[10]
Immunological Synapse	1934.4 T cells Antigen presenting cells (APCs) (human)	LFA-1 TCRβ DNA	FITC (Ex./Em. 496/519 nm) PE-TexasRed (Ex./Em. 488/615nm) Hoechst 33342 (Ex./Em. 350/461 nm)	[17]

Table 1. Applications of MIFC in cell-cell interaction

4.1. Analysis of the immunological synapse and antigen presentation

The modulation of cells associated with an immune response often requires a direct cell-cell interaction, in some cases via the formation of an immune synapse. In such cases, MIFC has facilitated the quantification of cross talk between different populations of immune cells [13]. In particular, MIFC has been used to evaluate the interaction between Mast cells (MCs) and eosinophils (Eos), key effector cells of an allergic reaction. MIFC was used to identify direct intercellular MC–Eos communication during an allergic response and whether this interaction exerts functional bidirectional changes on the cells. The evaluation of intercellular MC-Eos conjugation was performed by MIFC using a protocol designed to estimate immunological synapse formation. Human cord blood CB (cord blood) -derived MCs were pre-labeled with CFSE (carboxyfluorescein diacetate succinimidyl ester), and mixed with peripheral blood Eos (PB–Eos). Control MC or Eos monocultures were held under similar conditions. To distinguish homotypic from heterotypic couples, CFSE-intensity was used as the parameter: CFSE-high conjugates were MC-MC pairs, CFSE-low events were interacting MC-Eos, and CFSE-negative couples were Eos-Eos pairs. Each population was individually observed for side-scatter and bright-field parameters, so as to eliminate cell debris and irrelevant events. The study demonstrated that MIFC could identify homotypic MC–MC or Eos–Eos pairs in monocultures, as well as the preferential formation of heterotypic MC–Eos couples in co-cultures. These could be directly visualized in Multispectral images, in which overlapping bright field and CFSE label could be quantified, showing the percentage of conjugation between the cells with bar graph. The authors concluded that the high MC–Eos binding rates detected in vitro suggested that this cell–cell contact may be influential during chronic states [13].

Antigen-presenting cells (APC) activate T lymphocytes via the formation of an immunological synapse. In the process of adaptive immune response, APCs present antigenic peptides via MHC molecules to T lymphocytes. This is required to activate the T-cells. The physical contact between these cells results in the polarization of adhesion and cell signaling molecules to the interface between them. MIFC can visualize and quantify the recruitment of CD3 and Lck signaling molecules during the evolution of an immune synapse [14]. Activated primary human T cells mixed with anti-CD3- coated macrophage targets and labeled with both a nuclear dye (DRAQ5) and T cell specific markers, identifying Lck (Alexa 546) and CD3 (PE Alexa 610). Multispectral images from labeled events were analyzed using two independent algorithmic functions to identify the area of the immunological synapse in T cell/APC conjugates: The Valley function, which identifies the dimmest area between two nuclei, and the Interface function, that identifies the site of contact between two cells and selects only the T cell $CD3^+$ area of the T cell/APC synapse. The Valley function can measure the translocation of proteins derived from both the effector and target cell to the immunological synapse, while the Interface function can selectively locate proteins derived from any of the immune cells. Both analytical approaches were used in combination to detect a time dependent increase in the formation of cell conjugates with mature synapses, revealing that the efficiency of conjugate formation and Lck trafficking to the immunological synapse are calcium dependent. At early times and before adding calcium, fewer than 10% of the conjugates showed both CD3 and Lck concentrated at the immunological synapse. Minutes after calcium was added, approximately 50% of the immunological synapses had matured with both Lck and CD3 accumulating at the interface, although each with its own dynamics [14]. At the T cell/APC interface, the formation of a mature immune synapse involves the accumulation of surface receptors, as well as intracellular signaling and scaffolding molecules defined by clustering of the TCR/CD3 complex at the center of the contact zone (central supramolecular activation cluster, cSMAC), surrounded by a second cluster containing LFA-1 (peripheral SMAC, pSMAC). MIFC implemented through the ImageStream system has been used to study the kinetics of receptor accumulation and immune synapse maturation at the contact zone of 3B11 T cells and LK35.2 B cell couples [15]. In these studies, the interaction between 3B11 T cells and peptide-loaded LK35.2 cells was evaluated. The cell populations were mixed to allow for conjugate formation labeled with molecules detecting CD3 or LFA-1, anti-CD3-biotin plus streptavidin Texas Red and anti-CD18-FITC respectively, F-actin with Phalloidin (Tetramethylrhodamine B isothiocyanate) and Hoechst- 33342 using a nuclear dye. Clustering of CD3 (cSMAC), LFA-1 (pSMAC), and F-actin (pSMAC) were then analyzed using MIFC. The contact zone between 3B11 T and LK35.2 cells was defined as the events with T-cell/B-cell pairs and a Hoechst dye dependent valley mask between the cell pairs. The immune synapse mask was the result of the combination of the valley mask with a T-cell mask that was based on the distribution of CD3 signal. In the absence of presented peptides, most T cells displayed a rather homogeneous surface distribution of CD3 and LFA-1 molecules on their surface as observed by multispectral images of individual events. In contrast, the presence of peptides resulted in the recruitment of both CD3 and LFA-1 signal to the contact zone. A time course study showed that the enrichment of both CD3 and LFA-1 in the contact zone forming the mature synapse, peaked after 30 min of incubation in the presence of the antigenic peptide, decreasing thereafter, but stayed significantly above the level obtained without antigenic peptide. Interestingly, the amount of F-actin in the contact zone, which is crucial for the avidity regulation of integrins, increased over time. The ion chelator EDTA, an inhibitor of both integrin function and TCR/CD3-dependent calcium influx, abolished formation of a mature immune synapse. While the LFA-1 inhibitor, BIRT377 interfered with the accumulation of both LFA-1 and CD3 at the T-cell/B-cell interface and, concomitantly, disturbed mature immune synapse formation [15].

The process of antigen presentation in the context of APC/T cell pairs is compounded by the dual nature of this interaction. T cells are also required to continuously scan DCs presenting a variety of peptides and only be activated in the presence of foreign antigens. To better understand this complex process, MIFC has been used to directly quantify the potency of the immune synapse between T cells and APCs [16]. Recruitment of molecular markers to the intercellular contact zone of APC/T cell pairs was analyzed using MIFC. Because T cell receptor β (TCRβ) and lymphocyte function antigen-1 (LFA-1) are potent activators of immune synapse formation, the authors labeled TCRβ and LFA-1 and measured the enrichment of both markers in the cell-cell interacting domain. The authors isolated CD4 T cells from mice and mixed them with B cells loaded with an antigenic peptide as APCs. The formation of T–B cell conjugates was detected using MIFC, based on the labeling of independent cell populations with anti-TCRβ Alexa Fluor 488, or phycoerythrin-labeled anti- LFA-1 antibody (2D7), and the DNA marker 7-AAD. Complex formation was identified as the co-existence of TCRβ and LFA-1 signals within a multispectral image. No enrichment of TCR and LFA-1 in the intercellular contact zone was observed in the absence of antigenic peptides. While in the presence of peptide, T cells from mice containing normal numbers of cDCs specifically rearranged TCR and LFA-1, resulting in immune synapse generation. TCR and LFA-1 recruitment peaked at 20–30 min after initiation of T cell- APC contact; however, T cells from DC-depleted mice failed to reorganize TCR and LFA-1, resulting in lack of immune synapse maturation even in the presence of peptide [16].

Similarly, in order to study the correlation of the development of binding forces between T-cells and APCs with the maturation of immune synapse, the kinetics of immune synapse formation was studied using MIFC [17]. As a measure of immune synapse formation, the kinetics of accumulation of T-cell receptor (TCR) and LFA-1 at the interphase of conjugates between murine 1934.4 T cells and APCs were determined by MIFC. APCs (L.Au and L.Au.ICAM-1) cells were loaded with MBP-peptide, and mixed with 1934.4 T cells to allow for immune synapse formation. Cells were labeled with CD18-FITC specific for LFA-1, or TCRβ-biotin plus Streptavidin-PE-TXred specific for the TCR, and their nuclei labeled with the Hoechst dye 33342. An extensive redistribution of LFA-1 and TCR on L.Au.ICAM-1/1934.4 conjugates were observed using MIFC in the presence of antigenic peptide. However, in the absence of peptide, LFA-1 and TCR remained evenly distributed over the T cell surface. Likewise, in complexes of peptide-pulsed L.Au cells and 1934.4 T cells, no redistribution of LFA-1 and TCR molecules to the contact zone was observed. Kinetic studies revealed an increase of mature immune synapse at 15 min post conjugate formation, where mature immune synapse formation could be inhibited with anti-LFA-1 antibody, which correlated with the binding forces measured using cell force microscopy by the authors [17]. These studies are a clear demonstration of the combined advantage of using MIFC over standard flow cytometry or fluorescence microscopy, since MIFC is uniquely capable of objectively and rapidly quantifying the complex distribution of molecules in the contact zone between T cells and APCs for T-cell/APC pairs within a mixed population.

4.2. Phagocytosis of cells and cellular fragments

Some types of dendritic cells possess the ability to take up cellular components from other cells, phagocytose apoptotic bodies or entire cells, and to sample surface components of neighboring live cells via scavenger receptors in a mechanism known as cell nibbling or trogocytosis [18, 19], while other DCs are also known for their ability to generate and release exosomes to mediate cellular interactions. MIFC has been used to investigate the internalization of cellular components between interacting plasmacytoid dendritic cells (pDC) derived from blood, with HSV-infected and uninfected monocyte derived dendritic cells (MDDC) [20]. For these studies, BDCA-2-PE labeled pDC prepared from human PBMC were co-cultured with uninfected or HSV-infected MDDC labeled with CFSE (5,6-carboxyfluorescein diacetate succinimidyl ester), a cytoplasmic marker, or with PKH67, a green fluorescent lipophilic dye that labels cell membranes. From multispectral images obtained using MIFC, morphometric and photometric parameters for each cell were determined. From these parameters, pDC were quantitatively shown to preferentially associate with and take up membrane from HSV-infected MDDC cells labeled with PKH67. Moreover, the uptake of cytoplasmic components was directly observed and quantified, showing that pDC preferentially take up cellular material from HSV-infected MDDC cells labeled with CFSE. These studies suggest that both PKH67 and CFSE labeling are suitable for analysis of cellular uptake to evaluate cell-cell interactions using MIFC [20].

4.3. Multi-cell clustering and aggregates

While MIFC technology offers a number of advantages over standard detection method for quantifying the behavior of cell pairs, its ability is not limited to two or a few cells. Regulatory T cells (Tregs) are important effectors of immune tolerance; however the signaling molecules influencing their suppressive activity have yet to be fully characterized. MIFC is being used as an important tool to elucidate Treg activity via the quantification and visualization of cell-cell interactions between Tregs and a cytoplasmic factor involved in T cell activation [21]. In these studies, $CD4^{+}CD25^{+}$Tregs from mice were labeled with anti-CD4 Alexa 488, while bone marrow-derived dendritic cells (BMDCs) were labeled with anti-CD11c Alexa 405. Using this labeling scheme, the formation of BMDC-Tregs conjugates was quantified, the complexes scored according to the number of Tregs (1, 2 or 3) attached to the BMDCs. Analysis of multispectral images showed that the presence of antigenic peptides favored the formation of cell complexes, as did the absence of the cytoplasmic tyrosine phosphatase Src homology region 2 domain-containing phosphatase 1 (SHP-1). The presence of cell complexes with a third cell population of conventional T cells (Tcons) labeled with Alexa 647, and of the cells expressing the cytokine IL-2 was also identified [21].

MIFC technology has also been used to study the cellular composition of cell aggregates, as is the case during the formation of rosettes by the adhesion of platelets and leukocytes [22]. CD36 is frequently used as a marker of monocytes, erythrocytes, or platelets in clinical cytometry, while antibodies binding to CD36 may induce platelet activation and formation of platelet's rosettes on leukocytes. This frequently results in the false identification of expression of platelet markers on white blood cells. MIFC analysis has been used to visualize, quantify, and further confirm platelet rosetting on leukocytes after CD36 activation, and the application of anti-CD162 and anti-CD62 antibodies to prevent the formation of these complexes [22]. Using the "Spot Count" function of the IDEAS software applied to multispectral images acquired using MIFC, a direct relation between the number of adherent platelets and CD36-FITC intensity for neutrophils, monocytes and lymphocytes was discovered. The number of particles on the surface of neutrophils could be directly counted by analyzing morphological features of detected cells. Among the neutrophils, 64% exhibited at least one particle on their surface, ranging from 1 to 12 particles for FITC fluorescent cells in presence of CD36 mAbs. Using the same morphological analysis, inhibition of platelet adhesion by anti-CD162 mAb on cells was quantified. The percentage of neutrophils with adherent platelets decreased from 64 to 3% in the presence of the anti-CD162 mAb, ranging from 1 to 3 particles per FITC-fluorescent cells [22]. These observations suggest that MIFC can be a useful tool to study in depth the formation of cell aggregates, and their constituents.

The analysis of cell aggregate formation is also relevant in disease settings, such as cell clustering of sickle cells. The sickle cell disease is associated with painful vaso-occlusive crises that occur due to abnormal interactions between endothelial cells, red blood cells and leukocytes. MIFC has been used to study the interaction between red blood cells (RBC) and circulating leukocytes from patients with sickle cell disease or healthy subjects, to evaluate the factors that lead to aggregate formation, which enhance the incidence of vaso-occlusive crises [23]. White blood cells labeled with anti-CD45-FITC and erythrocytes with glycophorin A (GPA)-(PE) conjugated antibodies were mixed and analyzed using MIFC and the IDEAS software. Cell populations were identified by gating on cells expressing surface markers and confirmed by visual inspection of multispectral images. This analysis allowed for the quantitative characterization of single cells or aggregates within a mixed population. In all sickle cell disease patients, a high number of events double positive for both CD45-FITC and GPA-PE was observed, but not in healthy subjects [23]. Cell clusters observed in sickle cell disease patients showed that the double-labeled events corresponded to PBMC aggregates, suggesting that this technology can provide helpful information to analyze the unique constituents of cell aggregates in mixed populations and potentially the disease state of cells like the red blood cells of sickle cell disease [24].

5. Conclusion

The advancement of high-throughput imaging flow cytometry provides a new and powerful tool to the arsenal of technologies available to study cell-cell interaction. Due to the combined capabilities of photometric as well as morphometric analysis of cells on thousands of events, MIFC can be uniquely applied to a number of research and clinical applications. MIFC technology allows researchers the opportunity to combine spatial resolution in multicolor images of cells with the statistical analysis of large number of events through combining the features of fluorescence microscopy and modern flow cytometry in one system. Overall, the applications presented in this chapter are evidence of the broad applications and larger potential of imaging flow cytometry in the field of cell-cell interactions. This technology has also been used successfully in the determination of co-localization of cell surface or nuclear proteins, cell signaling, cell cycle, mitosis analysis, cell death, DNA damage, protein-protein interaction with Forster Resonance Energy Transfer (FRET), intracellular calcium, drug uptake, and pathogen internalization among other uses. It is expected that further advancements of this technology and in the tailoring of fluorophores specifically for these applications will provide deeper insights into this continuously evolving field.

Author details

Cristian Payés and Gustavo Helguera*
School of Pharmacy and Biochemistry, University of Buenos Aires, Argentina

* Corresponding Author

José A. Rodríguez
Molecular Biology Institute, University of California, Los Angeles, CA, USA

Sherree Friend
Amnis – Part of EMD Millipore, Seattle, WA, USA

Acknowledgement

G.H. is supported by the grant PICT-PRH 2008-00315 from Agencia Nacional de Promoción de la Ciencia y Tecnología (ANPCyT) FONARSEC, and by the Grant PIP 114-20110100170 from Consejo Nacional de Investigaciones Científicas y Tecnológicas (CONICET), Argentina. J.A.R. is a Howard Hughes Medical Institute Gilliam fellow and is further funded by the UCLA MBI Whitcome Fellowship. S. F. is employed by Amnis - part of EMD Millipore.

6. References

[1] Okagaki, LH, *et al.*, (2010) Cryptococcal cell morphology affects host cell interactions and pathogenicity. PLoS Pathog. vol. 6, p. e1000953.
[2] Elliott, GS, (2009) Moving pictures: imaging flow cytometry for drug development. Comb Chem High Throughput Screen. vol. 12, pp. 849-59.
[3] Basiji, DA, *et al.*, (2007) Cellular image analysis and imaging by flow cytometry. Clin Lab Med. vol. 27, pp. 653-70, viii.
[4] Basiji, DA, (2007) Multispectral Imaging Flow Cytometry. in 4th IEEE International Symposium on Biomedical Imaging: From Nano to Macro. Arlington, VA. USA.
[5] Kubota, F, *et al.*, (1995) Flow cytometer and imaging device used in combination. Cytometry. vol. 21, pp. 129-32.
[6] Ortyn, WE, *et al.*, (2007) Extended depth of field imaging for high speed cell analysis. Cytometry A. vol. 71, pp. 215-31.
[7] McGrath, KE, *et al.*, (2008) Multispectral imaging of hematopoietic cells: where flow meets morphology. J Immunol Methods. vol. 336, pp. 91-7.
[8] Horner, H, *et al.*, (2007) Intimate cell conjugate formation and exchange of membrane lipids precede apoptosis induction in target cells during antibody-dependent, granulocyte-mediated cytotoxicity. J Immunol. vol. 179, pp. 337-45.
[9] Dall'Ozzo, S, *et al.*, (2004) Rituximab-dependent cytotoxicity by natural killer cells: influence of FCGR3A polymorphism on the concentration-effect relationship. Cancer Res. vol. 64, pp. 4664-9.
[10] Gennari, R, *et al.*, (2004) Pilot study of the mechanism of action of preoperative trastuzumab in patients with primary operable breast tumors overexpressing HER2. Clin Cancer Res. vol. 10, pp. 5650-5.
[11] Helguera, G, *et al.*, (2011) Visualization and quantification of cytotoxicity mediated by antibodies using imaging flow cytometry. J Immunol Methods. vol. 368, pp. 54-63.

[12] Beum, PV, *et al.*, (2008) Binding of rituximab, trastuzumab, cetuximab, or mAb T101 to cancer cells promotes trogocytosis mediated by THP-1 cells and monocytes. J Immunol. vol. 181, pp. 8120-32.

[13] Elishmereni, M, *et al.*, (2011) Physical interactions between mast cells and eosinophils: a novel mechanism enhancing eosinophil survival in vitro. Allergy. vol. 66, pp. 376-85.

[14] Ahmed, F, *et al.*, (2009) Numbers matter: quantitative and dynamic analysis of the formation of an immunological synapse using imaging flow cytometry. J Immunol Methods. vol. 347, pp. 79-86.

[15] Hosseini, BH, *et al.*, (2009) Immune synapse formation determines interaction forces between T cells and antigen-presenting cells measured by atomic force microscopy. Proc Natl Acad Sci U S A. vol. 106, pp. 17852-7.

[16] Hochweller, K, *et al.*, (2010) Dendritic cells control T cell tonic signaling required for responsiveness to foreign antigen. Proc Natl Acad Sci U S A. vol. 107, pp. 5931-6.

[17] Hoffmann, S, *et al.*, (2011) Single cell force spectroscopy of T cells recognizing a myelin-derived peptide on antigen presenting cells. Immunol Lett. vol. 136, pp. 13-20.

[18] Harshyne, LA, *et al.*, (2001) Dendritic cells acquire antigens from live cells for cross-presentation to CTL. J Immunol. vol. 166, pp. 3717-23.

[19] Harshyne, LA, *et al.*, (2003) A role for class A scavenger receptor in dendritic cell nibbling from live cells. J Immunol. vol. 170, pp. 2302-9.

[20] Megjugorac, NJ, *et al.*, (2007) Image-based study of interferongenic interactions between plasmacytoid dendritic cells and HSV-infected monocyte-derived dendritic cells. Immunol Invest. vol. 36, pp. 739-61.

[21] Iype, T, *et al.*, (2010) The protein tyrosine phosphatase SHP-1 modulates the suppressive activity of regulatory T cells. J Immunol. vol. 185, pp. 6115-27.

[22] Ouk, C, *et al.*, (2011) Both CD62 and CD162 antibodies prevent formation of CD36-dependent platelets, rosettes, and artefactual pseudoexpression of platelet markers on white blood cells: a study with ImageStream(R). Cytometry A. vol. 79, pp. 477-84.

[23] Chaar, V, *et al.*, (2010) Aggregation of mononuclear and red blood cells through an {alpha}4{beta}1-Lu/basal cell adhesion molecule interaction in sickle cell disease. Haematologica. vol. 95, pp. 1841-8.

[24] Amer, J and Fibach, E, (2005) Chronic oxidative stress reduces the respiratory burst response of neutrophils from beta-thalassaemia patients. Br J Haematol. vol. 129, pp. 435-41.

Permissions

The contributors of this book come from diverse backgrounds, making this book a truly international effort. This book will bring forth new frontiers with its revolutionizing research information and detailed analysis of the nascent developments around the world.

We would like to thank Dr. Sivakumar Joghi Thatha Gowder, for lending his expertise to make the book truly unique. He has played a crucial role in the development of this book. Without his invaluable contribution this book wouldn't have been possible. He has made vital efforts to compile up to date information on the varied aspects of this subject to make this book a valuable addition to the collection of many professionals and students.

This book was conceptualized with the vision of imparting up-to-date information and advanced data in this field. To ensure the same, a matchless editorial board was set up. Every individual on the board went through rigorous rounds of assessment to prove their worth. After which they invested a large part of their time researching and compiling the most relevant data for our readers. Conferences and sessions were held from time to time between the editorial board and the contributing authors to present the data in the most comprehensible form. The editorial team has worked tirelessly to provide valuable and valid information to help people across the globe.

Every chapter published in this book has been scrutinized by our experts. Their significance has been extensively debated. The topics covered herein carry significant findings which will fuel the growth of the discipline. They may even be implemented as practical applications or may be referred to as a beginning point for another development. Chapters in this book were first published by InTech; hereby published with permission under the Creative Commons Attribution License or equivalent.

The editorial board has been involved in producing this book since its inception. They have spent rigorous hours researching and exploring the diverse topics which have resulted in the successful publishing of this book. They have passed on their knowledge of decades through this book. To expedite this challenging task, the publisher supported the team at every step. A small team of assistant editors was also appointed to further simplify the editing procedure and attain best results for the readers.

Our editorial team has been hand-picked from every corner of the world. Their multi-ethnicity adds dynamic inputs to the discussions which result in innovative outcomes. These outcomes are then further discussed with the researchers and contributors who give their valuable feedback and opinion regarding the same. The feedback is then collaborated with the researches and they are edited in a comprehensive manner to aid the understanding of the subject.

Apart from the editorial board, the designing team has also invested a significant amount of their time in understanding the subject and creating the most relevant covers. They scrutinized every image to scout for the most suitable representation of the subject and create an appropriate cover for the book.

The publishing team has been involved in this book since its early stages. They were actively engaged in every process, be it collecting the data, connecting with the contributors or procuring relevant information. The team has been an ardent support to the editorial, designing and production team. Their endless efforts to recruit the best for this project, has resulted in the accomplishment of this book. They are a veteran in the field of academics and their pool of knowledge is as vast as their experience in printing. Their expertise and guidance has proved useful at every step. Their uncompromising quality standards have made this book an exceptional effort. Their encouragement from time to time has been an inspiration for everyone.

The publisher and the editorial board hope that this book will prove to be a valuable piece of knowledge for researchers, students, practitioners and scholars across the globe.

List of Contributors

Yuri B. Shmukler and Denis A. Nikishin
N.K.Koltzov Institute of Developmental Biology, Russian Academy of Sciences, Moscow, Russia

Akio Nishikawa
Shimane University, Japan

Aiala Salvador, Manoli Igartua, José Luis Pedraz and Rosa María Hernández
NanoBioCel Group, Laboratory of Pharmaceutics, University of the Basque Country, School of Pharmacy, Biomedical Research Networking Center in Bioengineering, Biomaterials and Nanomedicine (CIBER-BBN), Vitoria, Spain

Elena Kovalenko, Leonid Kanevskiy, Anna Klinkova, Anastasiya Kuchukova, Maria Streltsova and Alexander Sapozhnikov
Laboratory of Cell Interactions, Shemyakin-Ovchinnikov Institute of Bioorganic Chemistry, Moscow, Russia

William Telford
Experimental Transplantation and Immunology Branch, National Cancer Institute, National Institutes of Health, Bethesda, MD, USA

Rodrigo Pacheco, Francisco Contreras and Carolina Prado
Fundación Ciencia y Vida, Universidad San Sebastián and Universidad Andrés Bello, Santiago, Chile

Fernanda Ledda
Laboratory of Molecular and Cellular Neuroscience, Institute of Cellular Biology and Neuroscience (IBCN)-CONICET, Buenos Aires, Argentina
Laboratory of Molecular and Cellular Neuroscience, Department of Neuroscience, Karolinska Institute, Stockholm, Sweden

Nasra Naeim Ayuob
Department of Anatomy, Faculty of Medicine, King Abdulaziz University, Jeddah, Saudi Arabia Department of Histology and Cytology, Mansoura University, Egypt

Soad Shaker Ali
Department of Anatomy, Faculty of Medicine, King Abdulaziz University, Jeddah, Saudi Arabia
Department of Histology and Cytology, Assuit University, Egypt

P. Pivonka and P.R. Buenzli
Faculty of Engineering, Computing and Mathematics, University of Western Australia, Australia

C.R. Dunstan
Department of Biomedical Engineering, University of Sydney, Australia

J. P. Neves Silva and M. E. C. D. Real Oliveira
Centre of Physics, University of Minho, Campus of Gualtar, Braga, Portugal

A. C. N. Oliveira and A. C. Gomes
Centre of Molecular & Environmental Biology - CBMA, University of Minho, Campus of Gualtar, Braga, Portugal

Hidekazu Kuwayama
Faculty of Life and Environmental Sciences, University of Tsukuba, Japan

A. Pelaez-Vargas
INEB - Instituto de Engenharia Biomédica, Universidade do Porto, Porto, Portugal
Universidade do Porto - Faculdade de Engenharia (FEUP), Departamento de Engenharia Metalúrgica e Materiais, Porto, Portugal
Universidad Cooperativa de Colombia, Facultad de Odontología, Medellín, Colombia

A. Carvalho, L. Grenho and F.J Monteiro
INEB - Instituto de Engenharia Biomédica, Universidade do Porto, Porto, Portugal
Universidade do Porto - Faculdade de Engenharia (FEUP), Departamento de Engenharia Metalúrgica e Materiais, Porto, Portugal

D. Gallego-Perez, N. Higuita-Castro and D.J. Hansford
Department of Biomedical Engineering, The Ohio State University and Nanoscale Science and Engineering Center - NSF, Columbus (OH), USA

J.A. Arismendi
Universidad de Antioquia, Facultad de Odontología, Medellín, Colombia

Cristian Payés and Gustavo Helguera
School of Pharmacy and Biochemistry, University of Buenos Aires, Argentina

José A. Rodríguez
Molecular Biology Institute, University of California, Los Angeles, CA, USA

Sherree Friend
Amnis - Part of EMD Millipore, Seattle, WA, USA

www.ingramcontent.com/pod-product-compliance
Lightning Source LLC
Chambersburg PA
CBHW070728190326
41458CB00004B/1087